AF324604

CLIMATE CHANGE

CLIMATE CHANGE

AN ENCYCLOPEDIA OF SCIENCE, SOCIETY, AND SOLUTIONS

Volume 1
Land and Oceans

Bruce E. Johansen

 ABC-CLIO™

An Imprint of ABC-CLIO, LLC
Santa Barbara, California • Denver, Colorado

Library of Congress Cataloging-in-Publication Data
Names: Johansen, Bruce E. (Bruce Elliott), 1950–
Title: Climate change : an encyclopedia of science, society, and solutions /
    [edited by] Bruce E. Johansen.
Description: Santa Barbara, California : ABC-CLIO, LLC, 2017. |
    Includes bibliographical references and index.
Identifiers: LCCN 2016059526 (print) | LCCN 2016059775 (ebook) |
    ISBN 9781440840852 (hardcopy : alk. paper : set) | ISBN 9781440848698
    (vol. 1) | ISBN 9781440848704 (vol. 2) | ISBN 9781440848711 (vol. 3) |
    ISBN 9781440840869 (ebook)
Subjects: LCSH: Climatic changes—Encyclopedias.
Classification: LCC QC902.92 .C54 2017 (print) | LCC QC902.92 (ebook) |
    DDC 577.2/2—dc23
LC record available at https://lccn.loc.gov/2016059526

ISBN: 978-1-4408-4085-2 (set)
        978-1-4408-4869-8 (vol. 1)
        978-1-4408-4870-4 (vol. 2)
        978-1-4408-4871-1 (vol. 3)
EISBN: 978-1-4408-4086-9

21 20 19 18 17   1 2 3 4 5

This book is also available as an eBook.

ABC-CLIO
An Imprint of ABC-CLIO, LLC

ABC-CLIO, LLC
130 Cremona Drive, P.O. Box 1911
Santa Barbara, California 93116-1911
www.abc-clio.com

This book is printed on acid-free paper ∞

Manufactured in the United States of America

Contents

VOLUME II: WEATHER AND GLOBAL WARMING

Weather

VOLUME III: HUMAN IMPACT AND PRIMARY DOCUMENTS

Agriculture and Food

Human Health

Indigenous Peoples and Cultures

Preface

We have designed a civilization based on science and technology and at the same time have arranged things so that almost no one understands anything at all about science and technology. This is a clear prescription for disaster. We may for a while get away with this mix of ignorance and power but sooner or later it is bound to blow up in our face.

Carl Sagan (Powell 2011, frontispiece)

Nature cannot be fooled.

Richard Feynman (Powell, 2011, frontispiece)

Tim Rinne, editor of the *Nebraska Report*, a publication of Nebraskans for Peace, told me once that the scariest thing about global warming is that we as a species essentially can ruin Earth as we know it by changing nothing. Continuing business as usual will do it, a state of affairs much more difficult to change than resistance to the nuclear arms race (where Tim cut his teeth as an activist). In the days when the bomb was our premier worry (Nebraskans for Peace began as Cat Lovers Against the Bomb), someone would have had to push that proverbial red button. With global warming, all we have to do is press our gas pedals and flip our light switches— keep doing what seems to most of us to be natural, necessary, and convenient.

In 2015, the atmospheric level of carbon dioxide breached 400 parts per million in all areas during all seasons. Levels of methane and nitrous oxides, the other two principal greenhouse gases, also reached record levels by substantial margins. The same year, world temperatures, stoked by El Niño conditions, surged to a new record as well, surpassing 2014's previous highs. The year 2016 soon surpassed all prior records, even after the El Niño had ended. "We're moving into uncharted territory at a frightening speed," said World Meteorological Organization Secretary General Michel Jarraud (Warrick 2015).

The combination of an intense El Niño and rising greenhouse gas levels in the atmosphere was quickly reflected in surface temperatures. The winter of 2015–2016 in the lower 48 United States was the warmest on the 121-year instrumental record, according to the National Oceanic and Atmospheric Administration (NOAA), at nearly 5°F above 20th-century averages. Alaska had its second warmest winter on record, almost 11°F above average. At the end of February, Anchorage had no snow on the ground for the first time on record. Land areas outside the tropics in the Northern Hemisphere were 1.46°C above average, 0.5 degrees above any previous monthly anomaly, a spectacular high. World temperatures hit record highs as well.

During December, 2015, January, 2016, and February 2016 (meteorological winter), temperatures not only set world records but also did so by the largest margins (anomalies) since record keeping began around 1880. February's global temperature was 1.35°C above the 1951–1980 average, exceeding the previous record anomaly set in January, 1.13°C, according to NASA's Goddard Institute for Space Studies, December 2015 was 1.11°C above the same set of averages. As Weather .com noted in February 2016, higher latitudes were the warmest: "Much of Alaska into western and central Canada, as well as eastern Europe, Scandinavia, and much of Russia were at least 4 degrees Celsius (roughly 7 degrees Fahrenheit) above February averages, according to NASA [and the Goddard Institute for Space Studies]."

"The departures are what we would consider astronomical," said NOAA climate scientist Jessica Blunden. "It's on land. It's in the oceans. It's in the upper atmosphere. It's in the lower atmosphere. The Arctic had record low sea ice. Everything everywhere is a record this month, except Antarctica," Blunden said. "It's insane." Georgia Tech climate scientist Kim Cobb added: "When I look at the new February 2016 temperatures, I feel like I'm looking at something out of a sci-fi movie. In a way we are: it's like someone plucked a value off a graph from 2030 and stuck it on a graph of present temperatures. It is a portent of things to come, and it is sobering that such temperature extremes are already on our doorstep" (Borenstein 2016).

As radical rises in worldwide temperatures startled scientists early in 2016, James Hansen and 18 coauthors published a study in the open-access journal *Atmospheric Chemistry and Physics* (from the European Geophysical Union), making a case that several meters in sea level rise could take place within a century, not the several hundred years projected by many scientists. This conclusion is based on a study of paleoclimate during the Eemian interglacial period, 120,000 years ago, a situation analogous to today (except that temperature increases occurred less rapidly then than now).

During the Eemian, Hansen and colleagues assert that excess heat in the deep oceans rapidly eroded ice in Antarctica and Greenland at an accelerating pace. This melting was accelerated by a slowing of the Atlantic Ocean's meridional circulation that usually distributes heat through the world ocean. The slowing of ocean circulation caused stagnation of relatively warm water in some areas in and near the Arctic.

The study's authors point to maps of worldwide warming during the winter of 2015–2016, noting that the only areas with below-average temperatures were over oceans adjacent to Greenland and Antarctica, where rapid melting of ice influences the water's temperature level. "My interpretation is that this is the beginning," Hansen said of these cool patches. "And it's one or two decades sooner than in our model" (Gillis 2016). "I think almost everybody who's really familiar with both paleo and modern is now very concerned that we are approaching, if we have not passed, the points at which we have locked in really big changes for young people and future generations," Hansen said (Mooney 2016).

Limiting global temperature rise to 2°C (3.6°F) over preindustrial levels, as recommended by recent diplomatic efforts such as the 2015 Paris Accords, will not prevent climate-driven changes that will force evacuation of many coastal cities, Hansen and colleagues warned.

Hansen and colleagues (2016) hypothesized that

[M]ass loss from the most vulnerable ice, sufficient to raise sea level several meters, is better approximated as exponential than by a more linear response. Doubling times of 10, 20, or 40 years yield multimeter sea level rise in about 50, 100 or 200 years. These climate feedbacks aid interpretation of events late in the prior interglacial, when sea level rose to +6–9 meters with evidence of extreme storms while Earth was less than 1°C warmer than today. [Climate] modeling, paleoclimate evidence, and ongoing observations together imply that 2°C global warming above the preindustrial level could be dangerous, [including] growing sea level rise, reaching several meters over a time scale of 50–150 years.

"Some of the claims in this paper are indeed extraordinary," said Michael E. Mann, a climate scientist at Pennsylvania State University. "They conflict with the mainstream understanding of climate change to the point where the standard of proof is quite high." However, noting that Hansen and colleagues often have presaged consensus, Mann said, "I think we ignore James Hansen at our peril" (Gillis 2016).

Design of the Work

Climate Change: An Encyclopedia of Science, Society, and Solutions in three volumes is Bruce E. Johansen's fourth encyclopedic work in this area, each volume of which has been expanded and updated. This sequence began with *The Global Warming Desk Reference* (one volume, Greenwood 2001) followed by *Global Warming in the 21st Century* (three volumes, Praeger 2006) and *The Encyclopedia of Global Warming Science and Technology* (two volumes, Greenwood 2009). These works have concentrated on describing changing weather conditions and interpreting global warming science for general readers and students through the undergraduate level. *Climate Change: An Encyclopedia of Science, Society, and Solutions* is organized into topical areas (designated in the table of contents as Key People; Terrain; Flora and Fauna; Oceans; Weather; Global Warming; Agriculture and Food; Human Health; Indigenous Peoples and Cultures; Greenhouse Gases and Energy; Solutions and Actions; Primary Documents; and Bibliography). Detailed entries are listed alphabetically within each area. A detailed table of contents and index will provide aid to readers.

A Broader Context

Increases in levels of greenhouse gases in Earth's atmosphere are part of a broader and intensifying trend in a geological epoch now widely called the *Anthropocene*, in which human activities have been the primary force altering the planet, "sufficiently to produce a stratigraphic signature in sediments and ice that is distinct from that of the Holocene epoch" (Waters et al. 2016). The Anthropocene actually began tens of thousands of years ago with the first use of fire and deforestation, but it intensified with the advent of fossil fuels around the beginning of the 19th century. By

1950, humanity's role in shaping the Earth system was dominant as population and industrialization exploded.

The controlling role of humanity includes the infusion of carbon dioxide, methane, and other greenhouse gases into the atmosphere at levels previously unknown in the natural system after the Pliocene Epoch 2 million to 4 million years ago (with consequent increases in temperatures). Along with this infusion of pollution have come rising levels of several artificial pesticides and herbicides, lead, fly ash and other forms of air pollution, fertilizers, plastics, and radioactivity from nuclear weapons testing. Nuclear radioactivity from past tests will be detectable in ice and sediments for at least 100,000 years. The amount of plastic manufactured each year equals the weight of the entire human race. Most of it decays only slowly and ends up in landfills or the oceans. As human population has increased, the extinction rates of other species have accelerated. "Unlike with prior subdivisions of geological time, the potential utility of a formal Anthropocene reaches well beyond the geological community," commented Waters and colleagues in *Science* (2016). "It also expresses the extent to which humanity is driving rapid and widespread changes to the Earth system that will variously persist and potentially intensify into the future."

Humanity's role has shaped climate so fundamentally that some research suggests the fossil-fuel age has played a major role in short-circuiting the onset of a new glacial cycle, and that any new glaciation may not occur for at least 100,000 years, outside any period that scientists can reliably model (Crucifix 2016; Ganopolski et al. 2016, 200).

Writing in *Nature Climate Change* in 2016, several well-known climate scientists asserted that our decisions on energy use during the next century will lock in changes in Earth's environment for millennia to come. "The next few decades offer a brief window of opportunity to minimize large-scale and potentially catastrophic climate change that will extend longer than the entire history of human civilization thus far," wrote the 22 climate researchers led by Peter Clark from Oregon State University (Mooney 2016).

According to these scientists,

> Most of the policy debate surrounding the actions needed to mitigate and adapt to anthropogenic climate change has been framed by observations of the past 150 years as well as climate and sea-level projections for the twenty-first century. The focus on this 250-year window, however, obscures some of the most profound problems associated with climate change. Here, we argue that the twentieth and twenty-first centuries, a period during which the overwhelming majority of human-caused carbon emissions are likely to occur, need to be placed into a long-term context that includes the past 20 millennia, when the last Ice Age ended and human civilization developed, and the next ten millennia, over which time the projected impacts of anthropogenic climate change will grow and persist. *This long-term perspective illustrates that policy decisions made in the next few years to decades will have profound impacts on global climate, ecosystems and human societies—not just for this century, but for the next ten millennia and beyond* (emphasis added). (Clark et al. 2016)

The power of the status quo is global warming's greatest ally. After the Obama administration decided to permit Shell Oil's drilling for new offshore oil in the Arctic, climate activist Bill McKibben remarked that it "shows why we may never win the fight against climate change. Even in this most extreme circumstance, no one seems able to stand up to the power of the fossil fuel industry" (McKibben 2015). "This is not climate denial of the Republican sort, where people simply pretend the science isn't real," McKibben said. "This is climate denial of the *status quo* sort, where people accept the science, and indeed make long speeches about the immorality of passing on a ruined world to our children. They just deny the meaning of the science, which is that we must keep carbon in the ground" (McKibben 2015).

This was the same White House that said in May 2015 that climate change would act as "an accelerant of instability around the world" by prompting water scarcity and food shortages that could escalate tensions globally and lead to overpopulation. The administration also said that rising temperatures would "change the nature of U.S. military missions," increasing the demand for resources in the Arctic and other coastal regions that would be affected by rising sea levels, and resulting in humanitarian crises that are larger and more frequent (Davis 2015).

Although politicians and parties shuck and jive over climate change, accelerating warming of the planet has a way of getting into everything. It is, of course, a worldwide problem that provokes changes in temperature—and more. A casual observer may be surprised at how a change in proportion of a trace gas (carbon dioxide, in this case) can have such wide-ranging influences. Raise the level of carbon dioxide in the oceans, for example, and acidity rises, imperiling any creature that lives in a calcium shell. This includes tiny phytoplankton, the basis of the maritime food chain.

Raise the amount of carbon dioxide, methane, and other heat-retaining gases, and temperatures eventually rise, melting ice and raising sea levels. Human beings have an affinity for the oceans, and a large proportion of us live on or near coastlines. Glance at a map of the world and large cities that will be in peril as sea levels rise a few feet—Shanghai, Kolkata, London, New York City, Miami, and many others—are readily apparent. Increase the proportion of greenhouse gases in the atmosphere and change its circulation patterns, expanding convection patterns that meteorologists call "Hadley cells," and see declines in rainfall over some areas with expanding deserts as a result. Harvests fail, and people go hungry. We are adapted to the climate, as are all of Earth's flora and fauna. When the climate changes, everything changes.

According to the United Nations' chief climatic body, the Intergovernmental Panel on Climate Change (IPCC), "The gathering risks of climate change are so profound they could stall or even reverse generations of progress against poverty and hunger if greenhouse gas emissions continue at a runaway pace" (Gillis 2014). Change the proportion of greenhouse gases in the air and "threaten society with food shortages, refugee crises, the flooding of major cities and entire island nations, the mass extinction of plants and animals, and a climate so drastically altered it might become dangerous for people to work or play outside during the hottest times of the year" (Gillis 2014).

A change in the balance of trace gases in the atmosphere creates demands to fundamentally change the way energy is obtained and used by everyone on Earth—the ways in which we transport ourselves, heat and cool our homes, and manufacture nearly everything we use in daily life. Such a change creates and compounds debate and dissension as scientists' warnings become more ominous, as the IPCC warns that burning even a fraction of proven fossil fuel reserves will damage the ecosystem beyond repair.

At the same time, levels of greenhouse gases continue to rise even as fossil fuel companies find new ways to acquire (and profit from) oil, natural gas, and coal, "and they are spending some $600 billion a year to find more. Utilities and oil companies are still building coal-fired power plants and refineries, and governments are spending another $600 billion directly subsidizing the consumption of fossil fuels" (Gillis 2014).

Even as the scientists cast the situation in their starkest terms, diplomats have been able to do little except agree to allow each individual nation to set its own goals—and even that will not *begin* until at least 2020. In the meantime, the Keeling Curve—the proportion of carbon dioxide in the atmosphere—continues to accelerate. "If they choose not to talk about the carbon budget, they're choosing not to address the problem of climate change," said Myles R. Allen, a scientist at Oxford University in Great Britain who helped write the IPCC's report in 2014. "They might as well not bother to turn up for these meetings" (Gillis 2014). In the U.S. Congress, climate change has not been seriously discussed in more than a decade. The Republican Party has quit considering science seriously, even as the principles of thermal inertia tell scientists that today's climate is a result of carbon dioxide emissions 50 years ago when emissions were much lower than today.

Change the proportion of trace gases in the atmosphere and famine spreads, riots ensue, and some governments fall. In agricultural regions, wrote Gillis (2014), "climate change had already become a small drag on overall global production, and could become a far larger one if emissions continue unchecked. . . . [In] recent years the world's food system has shown signs of instability, with sudden price increases leading to riots and, in a few cases, the collapse of governments." Even as politicians argue over whether humankind is responsible for rising temperatures and wild weather, a change in the proportion of trace gases in the air warms forests in western North America enough to speed the reproductive cycle of pine bark beetles that kill hundreds of thousands of acres and cast a pall of ghostly gray from Alaska to Colorado.

A tropical storm named Patricia exploded into a category 5 hurricane with sustained winds of 200 miles an hour within 36 hours, the strongest in the recorded history of the Western Hemisphere, as it smashed into the west coast of Mexico. News reports indicated astonishment at the speed of intensification by weather forecasters whose models missed most of it, with an observation that the storm had incubated in water warmed to 86°F during the third week of October 2015.

At the same time that politicians flush with oil-company cash deny that global warming exists, wind and solar power grow as well. In a scattershot way, many people have come to realize that building a sustainable future is not a luxury. The creation of such a future requires public awareness of the problem rooted in an understanding of science. *Climate Change: An Encyclopedia of Science, Society, and Solutions* is offered to address a fundamental disconnect between concepts of global warming developed in scientific journals and much of the popular debate in the popular realm. This disconnect often vexes scientists, who realize that while they can propose, it is the politicians, businesspeople, and other nonscientists who will ultimately decide how much and how soon the system by which we acquire and use energy (which drives greenhouse gas production) will change.

In scientific journals, the subject is studied with reference to the way in which the Earth system operates and invokes such concepts as thermal inertia, feedback loops, and various aspects of oceanic and atmospheric circulation in the context of paleoclimate (Earth's climatic history). Understanding such things is necessary at the popular level because the atmosphere will react to today's atmospheric greenhouse gas emissions a half-century from now, which demands that popular opinion anticipate the future and not merely react to present conditions. Even our terminology may seem archaic in the future. The term "climate change" is now in vogue, although, having written on this topic for some 20 years, I find it to be a highly imprecise phrase because climate changes all the time for many different reasons (or combinations of "forcings," the scientific term for provocations). We are talking here about an overload of greenhouse gases in the atmosphere caused in large part by human use of fossil fuels. This leads to heating, which can influence the course of climate in many ways. Some important effects of rising carbon dioxide levels (such as acidification of the oceans) would take place even in the absence of rising temperatures.

The scope of this work is at once global and local, involving all of Earth's more than 7 billion people as well as each individual on a personal level. In 100 years, students of history may remark at the nature of the fears that stalled responses to climate change early in the 21st century. Deniers of global warming kept change at bay, it may be remarked, by appealing to most people's fear of change that might erode their comfort and employment security, all of which were wedded psychologically to the massive burning of fossil fuels. A necessary change in our energy base may have been stalled, they may conclude, beyond the point where climate change forced attention, comprehension, and action. In retrospect, the nature of political debate over climate change in our time will seem remarkably, even pathetically, short-sighted and even comical—until we realize how such antics compromised the life, liberty, and happiness of future generations.

Today we are witnessing an energy-system paradigm shift. In one of many examples, in 2016 Iowa's electrical generation surpassed half wind. At the same time, technological change, as always, generates fear of unemployment. Paradoxically, such changes always generate economic activity, and a change in our basic energy paradigm in the 21st century will not cause the ruination of our economic base, as some deniers of climate change believe, any more than the coming of the railroads in the 19th century ruined an economy in which the horse once was the major land-based vehicle of transportation. The advent of mass automobile ownership early in the 20th century propelled economic growth, as did the transformation of information gathering with computers recently. The same developments also put out of work blacksmiths, keepers of hand-drawn accounting ledgers, and anyone who repaired manual typesetters.

Before the end of this century, the urgency of global warming will become manifest to everyone. Solutions to our fossil fuel-dilemma—solar, wind, hydrogen, and others—will evolve during this century. Within our century, necessity will *compel* invention. Other technologies may develop that have not yet even broached the realm of present-day science fiction, any more than digitized computers had in the days of the Wright Brothers more than 100 ago. We will take this journey because the changing climate, along with our own innate curiosity and creativity, will compel a changing energy paradigm.

As we took this set to press in early 2017, climate weirdness continued to make headlines. Ice melted at the North Pole in November 2016. A few weeks later in December, Arctic air was displaced (courtesy of the Arctic Oscillation) over the middle latitudes of North America. One location in North Dakota reported a temperature of –31°F and a wind chill of 52°F. On December 22, ice was melting around a weather buoy 90 miles from the North Pole as a reporter quipped in the *Washington Post*: "Santa may need water skis instead of a sleigh this year" (Samenow 2016). Christmas Day brought Nebraska and Kansas something really unusual: strong, spring-style thunderstorms. Nebraska recorded its first Christmas tornadoes, three EF-1s, about 200 miles southwest of Omaha. The same week, a location in Alaska notched a very old-fashioned –59°F. After five consecutive years of withering drought, California in January 2017, was swamped by damaging, record deluges.

In the meantime, Donald J. Trump declared a "mandate" in a United States presidential election that he actually lost by nearly 3 million popular votes. When he called global warming a Chinese hoax, climate scientists working for the federal government backed up their data on private flash drives lest appointees of the new regime destroy their public data. Who might have guessed 20 years ago that our imaginary glittering bridge to the third millennium would have an off-ramp to the dark ages of scientific denial? Carbon dioxide, if it had a sense of humor, would be enjoying a monstrous laugh over the games that people play. It does not laugh, of course, and it is not having a debate with Mr. Trump. It merely holds heat.

Since the advent of the Atomic Age 70 years ago, *The Bulletin of the Atomic Scientists* has used a "doomsday clock" to gauge humankind's risk of nuclear war. The clock now also factors climate change into its recipe for apocalypse. In early 2017, *The Bulletin* placed the clock's reading at two and one-half minutes to midnight. This is the closest the clock has been to midnight since 1953 when both the Soviet Union and the Unites States tested their first thermonuclear weapons within six months of each other. The clock keepers said that Trump's loose lips on nuclear weapons and ignorance of climate science provoked them to raise the chances of human apocalypse. Trump, however, cannot repeal the laws of atmospheric physics.

Bruce E. Johansen
Omaha, Nebraska
May 2017

Further Reading

Borenstein, Seth. "Beyond Record Hot, February [2016] was 'Astronomical' and 'Strange.'" Associated Press. March 17, 2016. http://phys.org/news/2016-03-hot-february-astronomical-strange.html.

Clark, Peter U., et al. "Consequences of Twenty-First-Century Policy for Multi-Millennial Climate and Sea-Level Change." *Nature Climate Change*, February 2016. http://www.nature.com/nclimate/journal/v6/n4/abs/nclimate2923.html.

Crucifix, Michael. "Earth's Narrow Escape from a Big Freeze." *Nature* 529 (January 14, 2016): 162–163.

Davis, Julie Hirschfeld. "Obama Recasts Climate Change as Peril with Far-Reaching Effects." *The New York Times*, May 20, 2015. http://www.nytimes.com/2015/05/21/us/obama-recasts-climate-change-as-a-more-far-reaching-peril.html.

Ganopolski, A., R. Winkelmann, and H. J. Schellnhuber. "Critical Insolation: CO_2 Relation for Diagnosing Past and Future Glacial Inception." *Nature* 529 (January 14, 2016): 200–203.

Gillis, Justin. "Scientists Warn of Perilous Climate Shift Within Decades, Not Centuries." *The New York Times*, March 22, 2016. http://www.nytimes.com/2016/03/23/science/global-warming-sea-level-carbon-dioxide-emissions.html.

Gillis, Justin. "U.N. Panel Issues Its Starkest Warning Yet on Global Warming." *The New York Times*, November 2, 2014. http://www.nytimes.com/2014/11/03/world/europe/global-warming-un-intergovernmental-panel-on-climate-change.html.

Hansen, J., et al. "Ice Melt, Sea Level Rise, and Superstorms: Evidence from Paleoclimate Data, Climate Modeling, and Modern Observations That 2°C Global Warming Could Be Dangerous." *Atmospheric Chemistry and Physics* 16(3) (March 22, 2016), 3761–3812. doi:10.5194/acp-16-3761-2016.

McKibben, Bill. "Obama's Catastrophic Climate-Change Denial." *The New York Times*, May 13, 2015. http://www.nytimes.com/2015/05/13/opinion/obamas-catastrophic-climate-change-denial.html?hp&action=click&pgtype=Homepage&module=c-column-top-span-region®ion=c-column-top-span-region&WT.nav=c-column-top-span-region&_r=0.

Mooney, Chris. "We Had All Better Hope These Scientists Are Wrong about the Planet's Future." *Washington Post*, March 22, 2016. https://www.washingtonpost.com/news/energy-environment/wp/2016/03/22/we-had-all-better-hope-these-scientists-are-wrong-about-the-planets-future/?wpmm=1&wpisrc=nl_evening.

Mooney, Chris. "What the Earth Will Be Like in 10,000 Years, According to Scientists." *Washington Post*, February 10, 2016. https://www.washingtonpost.com/news/energy-environment/wp/2016/02/08/what-the-earth-will-be-like-in-10000-years-according-to-scientists/.

Powell, James Lawrence. *The Inquisition of Climate Science*. New York: Columbia University Press, 2011.

Samenow, Jason. "Weather Buoy Near North Pole Hits Melting Point." *Washington Post*, December 22, 2016. https://www.washingtonpost.com/news/capital-weather-gang/wp/2016/12/22/weather-buoy-near-north-pole-hits-melting-point/?utm_term=.41638c867f45&wpisrc=nl_evening&wpmm=1.

Warrick, Joby. "Greenhouse Gases Hit New Milestone, Fueling Worries about Climate Change." *Washington Post*, November 9, 2015. https://www.washingtonpost.com/national/health-science/greenhouse-gases-hit-new-milestone-fueling-worries-about-climate-change/2015/11/08/1d7c7ffc-8654-11e5-be39-0034bb576eee_story.html.

Waters, Colin N., et al. "The Anthropocene is Functionally and Stratigraphically Distinct from the Holocene." *Science* 351 (January 8, 2016). http://science.sciencemag.org/content/351/6269/aad2622. doi: 10.1126/science.aad2622.

Weather.com. "February 2016 Was the Most Abnormally Warm Month Ever Recorded, Topping January 2016, NASA Says." March 14, 2016. https://weather.com/news/climate/news/record-warmest-february-global-2016.

Introduction

The fossil fuel age dawned just as the United States became Earth's most powerful economy, built across an expanding territory with surging immigration (mainly, but not entirely, from Europe). The exploitation of coal and then oil and natural gas between the mid-19th and early 20th centuries introduced machine labor that was the equivalent of 1 billion horses (or 3 billion human slaves). Not coincidently, perhaps, human slavery became economically as well as politically obsolete. To understand just how much human labor was transferred to fossil-fueled machines between 1800 and 1970, consider that the number of human hours of labor going into an acre of wheat declined from 56 to 2.9. For an acre of cotton, the same figure declined from 185 to 24. The food-producing economy has become as mechanized as the manufacture of anything else: seven calories of energy (mainly fossil fuels) by 2014 was required to produce one calorie of food (Johnson 2014, 14, 19, 39). This revolution in energy generation increased the production of heat-retaining greenhouse gases in Earth's atmosphere.

As part of Earth's natural cycle, the greenhouse effect (which scientists call "infrared forcing") is critical to life on Earth. Without it, the planet's average temperature would be minus 2°F. It is the added warming provoked by human combustion of fossil fuels that causes a problem. Like chocolate, a little is a good thing; too much is toxic to the system. Fossil fuels provide us comfort and convenience, and altering their use in a fundamental way presents the challenge of the century—and, most probably, for several centuries to come. Unless we wean ourselves from fossil fuels and do so quickly, the *real* problems will begin after the middle of the 21st century. Sir John Houghton, one of the world's leading experts on global warming, told *The Independent* (London): "We are getting almost to the point of irreversible meltdown, and will pass it soon if we are not careful" (Lean 2004, 8).

Climatic Bills Coming Due

The due bills for our use of fossil fuels are now being served. By 2015, scientists had figured that "burning the currently attainable fossil fuel resources is sufficient to eliminate the [Antarctic] ice sheet" (Winkelmann et al. 2015). This study is directed at Antarctica only, but all other ice would melt at the same time. How much time may be required to produce an ice-free planet? No one really knows. At current rates of increase, the actual burning of fossil fuel reserves may take place within a thousand years. Complete melting of the ice, factoring in delays of thermal inertia, may require several thousand years—but the momentum of this inertia would be irreversible.

Ken Caldeira, a researcher at Stanford University's Carnegie Institute of Science and one of the study's four coauthors, told Chelsea Harvey of the *Washington Post*, "The legacy of what we're doing over the next decades and the next centuries is really going to have a dramatic influence on this planet for many tens of thousands of years" (Harvey 2015).

As the world's overload of human-generated carbon dioxide continues to rise, its geography of generation has been changing. The relative shares of carbon dioxide emissions illustrate the rise of China and India in a short time as well as the ongoing role of the United States and Europe since the beginning of the Industrial Age. This is important because once CO_2 is emitted, it remains in the atmosphere for several centuries. In 2014, according to figures compiled by the Carbon Dioxide Information Analysis Center, Oak Ridge National Laboratory and, updated by BP, China had a 25 percent share of world CO_2 fossil fuel emissions, the United States had 15 percent, and Europe (plus a small part of Eurasia) had 13 percent. The cumulative share (1751–2014) are Europe and a small part of Eurasia, 29 percent; United States, 20 percent; China, 10 percent; and India, 3 percent.

Climatic Consequences

Scientists are projecting that ice may melt even faster than first thought given existing emissions. The scientists were surprised when their model anticipated that half the melting of Antarctica's ice (and, by implication, the world's) could occur within 1,000 years, causing the seas to rise on average a foot per decade for centuries, a pace that "would almost certainly throw human society into chaos, forcing a rapid retreat from the world's coastal cities." "To be blunt: If we burn it all, we melt it all," said Winkelmann, a researcher at the Potsdam Institute for Climate Impact Research in Germany (Gillis 2015, September 12). Many coastal cities would drown, perhaps including London, Berlin, Paris, Shanghai, Sydney, Rome, Tokyo Miami, New York City, Boston, New Orleans, Houston, Amsterdam, Stockholm, Venice, Buenos Aires, Beijing, and Washington, D.C. "This is humanity as a geologic force," said Caldeira. "We [human beings] are not a subtle influence on the climate system—we are really hitting it with a hammer" (Gillis 2015, September 12).

The consequent melting of ice would be provoked by an average global temperature rise of some 20°F—more at higher latitudes and inland areas, less in the tropics and along shorelines. It would be an increase, wrote Justin Gillis in *The New York Times*, "that would likely render vast stretches of the Earth too hot and humid for human habitation, cause food production to collapse, and drive much of the plant and animal life of the planet to extinction" (Gillis 2015, September 12).

By 2014, scientists were reporting, "The effects of human-induced climate change are being felt in every corner of the United States, with water growing scarcer in dry regions, torrential rains increasing in wet regions, heat waves becoming more common and more severe, wildfires growing worse, and forests dying under assault from heat-loving insects" (Gillis 2014, May 6).

The National Climate Assessment put forth by the U.S. Global Change Research Program has said, "Summers are longer and hotter, and extended periods of unusual heat last longer than any living American has ever experienced. . . . Winters are generally shorter and warmer. Rain comes in heavier downpours. People are seeing changes in the length and severity of seasonal allergies, the plant varieties that thrive in their gardens, and the kinds of birds they see in any particular month in their neighborhoods." "Yes, climate change is already here," said Richard B. Alley, a climate scientist at Pennsylvania State University who was not involved in writing the report but reviewed a late draft. "But the costs so far are still on the low side compared to what will be coming under business as usual by late in this century" (Gillis 2014, May 6).

The year 2016 was the hottest year in recorded history by far, the third straight year of record temperatures. In fact, the only areas with colder-than-average temperatures in 2015 were parts of the oceans off Greenland and Antarctica where rapidly melting ice has been cascading into the oceans, chilling the air above. In geophysical terms, the margin of the new record was astonishing—0.23°F (0.23°C) (according to NASA) and 0.29°F (0.16°C) (as measured by NOAA). New global highs or lows are often measured in few hundredths of a degree. A strong El Niño was an important factor, but so was long-term planetary warming caused by human emissions of greenhouse gases. "The whole system is warming up, relentlessly," said Gerald A. Meehl, a scientist at the National Center for Atmospheric Research in Boulder, Colorado (Gillis 2016).

The year 2016 extended 2015's hot streak, with the greatest departure from average of any month on record. Extremes were greatest in the Arctic, with some areas more than 6°C (13°F) above the 1951–1980 mean. The Arctic as a whole was 4.0°C (7.2°F) above the same period.

Global warming is not merely a matter of rising temperatures. Warming temperatures also change the behavior of Earth's hydrological cycle, increasing the severity of storms as well as the frequency and intensity of droughts and deluges. A warmer atmosphere can hold more moisture, increasing the explosive nature of precipitation because warming also increases evaporation. The result: drought and deluge may increase at the same time in different areas—or even alternate in the same place. Theory, as well as an increasing number of daily weather reports, strongly indicate that changes in precipitation patterns may vary wildly across time and space. The hydrological cycle seemed to be changing more rapidly than temperatures. Such changes will be highly uneven, episodic, and often nasty.

One of the National Climate Assessment's most dramatic findings concerned the rising frequency of torrential rains. Scientists have expected this effect for decades because more water is evaporating from a warming ocean surface, and the warmer atmosphere can hold the excess vapor, which then falls as rain or snow. But even the leading experts have been surprised by the magnitude of the effect. Justin Gillis wrote in *The New York Times* on May 6, 2014, that the National Climate Assessment found that "the eastern half of the [United States] is receiving more precipitation in general. And over the past half-century, the proportion of precipitation that is falling in very heavy rain events has jumped by 71 percent in the Northeast, by 37 percent in the Midwest, and by 27 percent in the South."

Such changes are occurring worldwide. As floods devastated Pakistan in the summer of 2010, for example, an extremely rare deluge inundated the town of Leh, in Ladakh, India, which sits in what is usually one of the driest deserts on the planet.

The village sits in what is usually a high-altitude desert protected from most precipitation by surrounding mountains. The average rainfall there in August is 15 millimeters, a fraction of an inch. In the early morning of August 6, 2010, however, a half-hour deluge swept much of the village away, killing 150 people and leaving several hundred missing. The storm was so intense—but so isolated—that it missed a weather station in the valley and went unmeasured.

In one instance, large parts of Nashville, Tennessee, were inundated by almost 20 inches of rain in 2013, and parts of Colorado received a year's worth of average rainfall in one week. Parts of the Florida panhandle were devastated by as much as two feet of rain in 24 hours. At the same time, much of the Navajo reservation in New Mexico and Arizona, which is usually dry, was afflicted by surging sand dunes as precipitation decreased from scanty to nearly none—except for a rare but violent downpour. Parts of South Carolina received two feet of rain in three days during early October 2015. Two weeks later, 20 inches fell in some areas of southeastern Texas.

Geophysics Outpaces Diplomacy

Global warming is a deceptively backhanded crisis in which thermal inertia delivers results a half-century or more after our burning of fossil fuels provokes them. Our political and diplomatic debates react *after* we see results. Political inertia plus thermal inertia thus presents the human race and the planet we superintend with a challenge to fashion a new energy future *before* raw necessity—the hot wind in our faces—compels action. Global warming is dangerous because it is a sneaky, slow-motion emergency that demands we acknowledge a reality centuries in the future with a system of individual, legal, and diplomatic reaction that reacts in the past tense.

In September 2008, two scientists with the Scripps Institution of Oceanography at the University of California–San Diego published research in the *Proceedings of the National Academy of Sciences* that indicated that if greenhouse gas emissions had stopped completely by 2005, the world's average temperature still would increase by 2.4°C (4.3°F) by the end of the 21st century. Richard Moss, vice president and managing director for climate change at the World Wildlife Fund, said the new carbon dioxide figures and research show that "we're already locked into more warming than we thought" (Eilperin 2008; Ramanathan and Feng 2008). These calculations are now more than a decade old. As far as world diplomatic change is concerned, these have been lost years.

In addition to thermal inertia, a second important principle that influences climate change involves feedbacks, including albedo (light reflectivity). Melting Arctic ice exposes open water in summer when the sun shines at the top of the world. Dark ocean water absorbs more heat than lighter ice and snow, causing even more heating and more melting. In the meantime, around the Arctic Circle, permafrost on land melts, adding still more carbon dioxide and methane, accelerating the natural process that feeds on itself. To these natural processes add the trigger of increasing human emissions and the problem is magnified.

Climate change is cumulative. Many of the feedbacks that provoke rising temperatures tend to accelerate over time, compounding one another. For example, rising emissions from human sources provoke melting permafrost, which adds even more carbon dioxide and methane to the atmosphere. Melting ice makes surfaces darker, which absorb more heat. In the meantime, "in low-lying island nations, like the Marshall Islands, . . . rising seas are soaking coastal soil, killing crops and contaminating fresh water supplies. 'The groundwater that supports our food crops is becoming inundated with salt,' said Tony deBrum, foreign minister of the Marshall Islands. 'The green is becoming brown'" (Davenport 2014).

By 2014 and 2015, reports of the U.N. Intergovernmental Panel on Climate Change (IPCC) were raising the ante for action because, as Justin Gillis wrote in *The New York Times* (2014, August 26), "Runaway growth in the emission of greenhouse gases is swamping all political efforts to deal with the problem, raising the risk of 'severe, pervasive and irreversible impacts' over the coming decades, according to a draft of a major new United Nations report." The toll already includes significant reductions in grain production, intensifying heat waves, and increasing instances of flooding rains in some areas and drought in others.

As temperatures rise and rampaging weather becomes a staple of daily headlines, diplomats and climate scientists gather every year in an attempt to arrange a worldwide pact to stall the rise in greenhouse-gas emissions amid a rising chorus of warnings that the results will be too little, too late. By 2015, the warming of the climate and its effects on daily weather was racing ahead of global diplomacy's efforts to cope with it. A renewable energy system was developing as wind and solar spread across the planet (Germany, the world's fourth-largest economy, derived one-third of its electricity from renewable sources by 2016), but too slowly to contain the rise in temperatures.

Here is the basic question: can humanity change its energy paradigm quickly enough to avoid intrinsic harm to the environment? Even as James Hansen contends that the 2°C target is too high, other scientists have said it will never be achieved given the worldwide momentum of greenhouse-gas emissions. "The goal is fundamentally unachievable," wrote David Victor (a professor at the University of California–San Diego) and Charles Kennel (of the Scripps Institute of Oceanography) in *Nature* (Kolbert 2015, 30). Meeting the goal would require global emissions of greenhouse gases to fall essentially to zero during the second half of the 21st century.

One study issued in 2015 said that even if all diplomatic proposals to limit greenhouse-gas emissions issued in 2015 were enacted, global warming by the end of the 21st century would be limited to 6.3°F compared to the 8.1°F if emissions continued at present levels ("Climate Scoreboard" 2015). Global emissions would be declining within a decade or two if all nations met their pledges—but only slowly and not nearly enough to prevent thermal inertia from raising temperatures, melting ice worldwide, raising sea levels, and devastating flora and fauna. The pledges countries made before the global climate summit in Paris at the end of 2015 were "a big step forward, but not sufficient—not even close," said John D. Sterman, a professor of management at the Massachusetts Institute of Technology (Gillis and Sengupta 2015).

The United States and China reached a first-time agreement in 2014, including a joint announcement that the United States will cut its emissions as much as 28 percent by 2025 and China's emissions will peak by 2030. In India, the world's third-largest source of greenhouse gases in 2014, the government does not expect that emissions will peak until at least 2040. Action to reduce emissions is all "on speculation"—in the future—as the levels of carbon dioxide and methane continue to rise. And as long as these levels rise, humankind is losing its battle against warming.

The Political System Lags Behind

Nearly all of the actions in the United States have been undertaken by President Barack Obama with neither advice nor consent from a Republican-controlled Congress, where climate-change denial (and refusal to believe the basic geophysical facts of the matter) has become a political litmus test. Obama's actions include an Environmental Protection Agency order in June 2014 that forces major emissions cuts from coal-fired power plants.

Henry M. Paulson, former secretary of the U.S. treasury (and a Republican), compared the climate crisis to the financial collapse of 2008:

> We are building up excesses (debt in 2008, greenhouse gas emissions that are trapping heat now). Our government policies are flawed (incentivizing us to borrow too much to finance homes then, and encouraging the overuse of carbon-based fuels now). Our experts (financial experts then, climate scientists now) try to understand what they see and to model possible futures. And the outsize risks have the potential to be tremendously damaging (to a globalized economy then, and the global climate now).
>
> Back then, we narrowly avoided an economic catastrophe at the last minute by rescuing a collapsing financial system through government action. But climate change is a more intractable problem. The carbon dioxide we're sending into the atmosphere remains there for centuries, heating up the planet. That means the decisions we're making today—to continue along a path that's almost entirely carbon-dependent—are locking us in for long-term consequences that we will not be able change but only adapt to, at enormous cost. To protect New York City from rising seas and storm surges is expected to cost at least $20 billion initially, and eventually far more. And that's just one coastal city. (Paulson 2014)

Paulson was echoing an analysis by the Prince of Wales, who also argued that the greatest financial collapse since the Great Depression of the 1930s will be dwarfed by ecological problems, including climate change. Accepting an honorary degree from the London Business School, Prince Charles warned: "This [the financial collapse] we can address relatively easily."

> But there is another systemic risk which is, to my mind, much more serious in the long term: the threat of increasing and accumulating environmental

collapse and with it the devastating consequences for us, as a species, but also for the myriad others which shape this planet with us and on which we depend for our survival. . . . Our myopic determination to ignore the facts and to continue with business as usual is, I fear, creating the risk of a crash which will be far more dramatic, and far harder to recover from, than anything we have experienced over the past few years. (Prince of Wales 2011)

With Michael R. Bloomberg (former New York City mayor and investment firm owner) and Tom Steyer (retired hedge fund manager), Paulson compiled an economic analysis of climate change costs titled *Risky Business* in 2014. They favor a tax on carbon dioxide emissions and a phase out of subsidies for fossil fuels.

"The biggest lag is in the political system," said geoscientist Michael Oppenheimer of Princeton University. The seriousness of the threat has been discussed for the past two decades, he said, and another 20 years may pass before a worldwide diplomatic program that is up to the challenge is in place. In the meantime, the window of time before feedbacks take control narrows. "We can't really afford to do a 'wait and learn' policy," Oppenheimer said. "The most important question is: when do we commit to [contain global warming to] 2 [°C]. Really, there isn't a lot of headroom left. We better get cracking." The current pace, said Roger Pielke Jr. of the University of Colorado at Boulder, "isn't going to do it" (Kerr 2007).

Reaching Tipping Points

Scientists now generally agree that warming will not stop at the 3.6°F (2°C) target that is widely considered crucial for preventing fundamental, long-term damage.

According to a large body of scientific research, that is the tipping point at which the world will be locked into a near-term future of drought, food and water shortages, melting ice sheets, shrinking glaciers, rising sea levels and widespread flooding—events that could harm the world's population and economy. . . . The objective now, negotiators say, is to stave off atmospheric temperature increases of 4 to 10 [°C]; at that point, they say, the planet could become increasingly uninhabitable. (Davenport 2014)

The idea of variable standards that accommodate each country's domestic politics is not producing the requisite worldwide reductions. Even limited goals take years to negotiate—as greenhouse-gas emissions continue to rise on a global scale. A report issued in November 2014 by the U.N. Environment Program concluded that global emissions must peak within the next 10 years and then drop to half of current levels by 2050 to avoid the 3.6°F increase.

Scientific reports raise the specter that feedbacks may already have guaranteed major erosion of ice sheets and glaciers in Greenland and western Antarctica that probably will raise coastal sea levels around the globe by several feet before 2100 (and more after that), inundating several major urban areas. According to Richard A. Kerr, writing in *Science*, "Ice losses from Greenland and West Antarctica have been accelerating, showing that some ice sheets are disconcertingly sensitive to warming" (Kerr 2012, November 30a,1138). Changes once estimated to take place in centuries are now being forecast to take place within the lives of our children and grandchildren. A decade ago, ice-free summers in the Arctic Ocean were anticipated by the end of the 21st century, but after several years of accelerated ice melt, two decades from now begins to look more likely. Greenland, which contains most of the Northern Hemisphere's ice, has been melting more quickly. Even the island's northeastern quadrant, which had been stable, has been losing ice—10 billon tons a year between April 2003 and April 2012 (Khan et al. 2014; Koch 2014).

NASA satellites have detected a widening rift in Antarctica's Pine Island glacier. This glacier acts as a plug—like a cork in a bottle—between the ocean and the West Antarctic Ice Sheet. Should this ice sheet melt into the sea, the world ocean would rise 16 to 18 feet. This "cork" is now breaking up. Greenland's ice is also melting at an accelerating rate. By 2014, more scientists were regarding the dissolution of the West Antarctic ice sheet as highly likely, probably beginning in the 22nd century. Two papers published during May 2014 in *Science* and *Geophysical Research Letters* described upwelling of warm water that is expected to provoke the ice sheet's slow-motion collapse (Joughin et al. 2014; Rignot et al. 2014). This collapse was forecast in 1978 by glaciologist John H. Mercer (who died in 1987), who was an outlier at the time. In the decades since, his ideas have gathered credibility. By 2014, scientists were setting a tough timetable for the ice sheet's dissolution—200 to 900 years.

A decrease in the melt rate back to earlier levels would be "too little, too late to stabilize the ice sheet," said Ian Joughin, a glaciologist at the University of Washington and lead author of the paper in *Science*. "There's no stabilization mechanism" (Gillis and Chang 2014). "The basic problem is that much of the West Antarctic ice sheet sits below sea level in a kind of bowl-shaped depression in the Earth," explained Justin Gillis and Kevin Chang in *The New York Times*. "As Dr. Mercer outlined in 1978, once the part of the ice sheet sitting on the rim of the bowl melts and the ice retreats into deeper water, it becomes unstable and highly vulnerable to further melting" (Gillis and Chang 2014).

Forgoing Fossil Fuel Reserves

Along with future ice-melt rates, scientists also have been anticipating that three-quarters of known fossil fuel reserves must remain in the ground if warming is to be kept at a tolerable level—something that would require companies that exploit these reserves to eventually write down their asset values to zero. In the meantime, the same companies have been developing new sources of oil and natural gas such as tar sands and hydraulic fracturing ("fracking"). A report from the International Energy Agency said in 2012 that "world oil production could increase through 2035 and meet the world's growing demand for energy as oil" (Kerr 2012b, November 30). "The scarcity of fossil fuels will not solve our problem. We are in the middle of a fossil fuel renaissance," said said Ottmar Edenhofer, a German economist and co-chair of the IPCC in 2014. "We cannot afford to lose another decade," said Edenhofer. "If we lose another decade, it becomes extremely costly to achieve climate stabilization" (Gillis 2014, April 13).

World emissions of greenhouse gases were rising steadily until recent years when they jumped: "From 1970 to 2000, global emissions of greenhouse gases grew at 1.3 percent a year. But from 2000 to 2010, that rate jumped to 2.2 percent a year . . . and the pace seems to be accelerating further in this decade" (Gillis 2014, August 26). Consumption of coal, the dirtiest fossil fuel in terms of emissions per unit of energy delivered, increased in China until 2014—and then declined for the first time. Coal consumption also appears to have peaked in the United States, and by 2010 its use began to decline. Use of coal continued to rise in India, however.

This Cake Is Already Being Baked

The proportion of carbon dioxide in the atmosphere continues to rise worldwide, however, a trend that has not changed since the beginning of the Industrial Revolution. The level reached 400 parts per million (ppm) in 2015, as high as it was in the Pliocene 2 million to 4 million years ago when sea levels were 100 feet higher and temperatures 4 to 6°F warmer. This is a key figure and one that indicates how much change has yet to be experienced because of thermal inertia within the next few centuries. This cake is already being baked. In terms of geologic time, the change is coming about remarkably quickly. Carbon dioxide is a trace gas, a tiny proportion of the atmosphere; at 400 ppm, it comprises only one-tenth of one-third of 1 percent of the air. It is, however, a remarkably efficient retainer of heat and is now increasing at a rate more quickly than at any time in the geophysical record going back hundreds of millions of years.

"The changes we're seeing are really drastic," Oksana Tarasova, chief of the World Meteorological Organization's Global Atmospheric Watch, told the *Washington Post*. The carbon dioxide level rose just short of 3 parts per million in 2013, much more than the 1 ppm to 2 ppm annual average of the past few decades. "We are seeing the growth rate rising exponentially," Tarasova said. Carbon dioxide is now 40 percent higher than the usual cyclical peak before the advent of the Industrial Age. Methane, the second most prevalent greenhouse gas, is 150 percent higher (Warrick 2014). "A continued rapid growth of emissions in coming decades could conceivably lead to a global warming exceeding 8 degrees Fahrenheit," Justin Gillis reported in *The New York Times* (2014, August 26). "The warming would be higher over land areas, and higher still at the poles. Warming that substantial would almost certainly have catastrophic effects, including a mass extinction of plants and animals, huge shortfalls in food production, extreme coastal flooding and many other problems" (Gillis, August 26, 2014).

Even with such warnings, no worldwide system exists to require and enforce reductions in greenhouse gas emissions for all countries by set dates—even despite growing evidence of the problem. "This spring [2014], a report by 13 federal agencies [the National Climate Assessment] concluded that climate change would harm the American economy by increasing food prices, insurance rates, and financial volatility. In China, the central government has sought to quell citizen protests related to coal pollution" (Davenport 2014).

Rising Sea Levels and Ocean Acidity

In 2014, the IPCC projected that by 2100 rising sea levels and storm surges probably will swamp some of Asia's largest cities, among them Mumbai, Bangkok, Kolkata, Dhaka, Shanghai, Ho Chi Minh City, and Yangon. In Europe, London will be at risk, and New York City, New Orleans, Miami, and others will succumb to seas that rise in storm surges. The IPCC also projects that a rise in atmospheric carbon dioxide level to 430 ppm to 480 ppm will raise acidity levels in the world's oceans to a level by 2100 that will imperil nearly everything that grows in a shell, from phytoplankton (the basis of the maritime food web) to corals, many of which are also in trouble because of rising water temperatures. Carbon dioxide is being injected into the oceans far faster than nature can neutralize it. Seawater is usually slightly alkaline at a pH around 8.2. The pH of the oceans has fallen 0.1 during industrial times. The scale is logarithmic, so a 0.1 change means a 30-percent increase in the concentration of hydrogen ions. Under a business-as-usual scenario, the pH may fall by roughly 0.5 by 2100.

By the early years of the 21st century, carbon dioxide levels were rising in the oceans more rapidly than any time since the age of the dinosaurs, according to work published by Ken Caldeira and Michael E. Wickett (2003): "We find that oceanic absorption of CO_2 from fossil fuels may result in larger pH changes over the next several centuries than any inferred in the geological record of . . . 300 million years, with the possible exception of those resulting from rare, extreme events such as bolide impacts or catastrophic methane hydrate degassing." A "bolide" is a large extraterrestrial body (usually at least a half mile in diameter) that impacts the Earth at a speed roughly equal to that of a bullet in flight. A "bolide" might be a large asteroid.

Can Human Technology and Diplomacy Catch Up?

Can human technology and diplomacy catch up? Even as emissions of greenhouse gases continue to rise worldwide, costs of wind and solar power have been falling so quickly that they are competitive with fossil fuels in many markets. Even as greenhouse gas concentrations race upward, an energy revolution is underway, using alternative sources such as solar, wind, and geothermal. William Moomaw, a lead author of a chapter in the IPCC's 2007 assessment on energy options and a professor of international environmental policy at Tufts University, said,

> Here in the early years of the 21st century, we're looking for an energy revolution that's as comprehensive as the one that occurred at the beginning of the 20th century when we went from gaslight and horse-drawn carriages to light bulbs and automobiles. In 1905, only 3 percent of homes had electricity. Right now, 3 percent is about the same range as the amount of renewable energy we have today. None of us can predict the future any more than we could in 1905, but that suggests to me it may not be impossible to make that kind of revolution again. (Revkin 2007)

In Omaha, for example, when I sat down to write my last global warming encyclopedia, the electricity in my home was two-thirds coal generated and the rest nuclear.

By the time this work reaches print, our power will be one-third generated by the enduring and nonpolluting prairie wind. States around us are doing even better; Iowa will be half wind powered.

A large proportion of the diplomatic conflict relates to a feud between rich and poor nations over how much the affluent should pay to help mitigate the problem: "Many rich countries argue that $100 billion a year is an unrealistic demand; it would essentially require them to double their budgets for foreign aid, at a time of economic distress at home. That argument has fed a rising sense of outrage among the leaders of poor countries, who feel their people are paying the price for decades of profligate Western consumption" (Gillis 2014, March 30).

Despite the U.S. Republican Party's willful ignorance, recognition of global warming's impacts is spreading among establishment circles in such precincts as the International Monetary Fund and the World Bank. Insurance companies, whose profits depend on gauging future risks, have been sensitive to such perils as prospective damage by rising seas. "Unless we take action on climate change, future generations will be roasted, toasted, fried, and grilled," Christine Lagarde, managing director of the International Monetary Fund told a World Economic Forum audience in Davos-Klosters, Switzerland, in January 2013 ("World Economic" 2013).

Possible damage from climate change has become a major subject of discussion in several world economic forums. The Climate Vulnerable Forum estimated that by 2013 climate change was contributing to 5 million deaths per year and costing the world economy more than $1.2 trillion U.S. (approximately 1.6 percent of annual global gross domestic product, or GDP). The same group projects that the proportion will double by 2030 to 3.2 percent of GDP. Less-developed countries will be losing more than 10 percent of GDP to air pollution and climate change by that time, according to the same set of estimates ("World Economic" 2013).

"Greening the economy is the only way to accommodate nine billion people by 2050," said Thomas Kerr, director of Climate Change Initiatives at the World Economic Forum. "There are many successful cases where governments have strategically targeted their public funds to mobilize significant sums of private investment for green infrastructure. It's now time to scale up these proven solutions" ("World Economic" 2013).

World Bank President Jim Yong Kim laid out the organization's action plan to combat climate change, something it was pursing with $7 billion worth of climate investment funds by 2013, including spreading solar energy and promoting green expansion for cities, among other initiatives:

The world's top priority must be to get finance flowing and get prices right on all aspects of energy costs to support low-carbon growth. Achieving a predictable price on carbon that accurately reflects real environmental costs is key to delivering emission reductions at scale. Correct energy pricing can also provide incentives for investments in energy efficiency and cleaner energy technologies. A second immediate step is to end harmful fuel subsidies globally, which could lead to a 5 percent fall in emissions by 2020. Countries spend more than $500 billion annually in fossil fuel subsidies and an additional

$500 billion in other subsidies, often related to agriculture and water, that are, ultimately, environmentally harmful. That trillion dollars could be put to better use for the jobs of the future, social safety nets or vaccines. A third focus is on cities. The largest 100 cities that contribute 67 percent of energy-related emissions are both the center of innovation for green growth and the most vulnerable to climate change. We have seen great leadership, for example, in New York and Rio de Janeiro on low-carbon growth and tackling practices that fuel climate change. (Kim 2013)

Concern about the cascading effects of climate change has long exceeded the bounds of the disparaged hippie tree hugger. By 2014, the U.S. Department of Defense had issued several reports detailing the national security implications of climate change. As early as 2008, the National Intelligence Council issued a secret "national intelligence estimate" on climate change. Climate change deniers not only ignore this analysis but also that by the National Research Council (the United States' top scientific research group), which said in a study commissioned by the Central Intelligence Agency (CIA), "Climate change is accelerating, and it will place unparalleled strains on American military and intelligence agencies in coming years by causing ever more disruptive events around the globe" (Broder 2012). Republicans in Congress objected when the CIA created a climate-change center. In 2012, Hurricane Sandy provided a foretaste of what can be expected more often in the near future, said the report's lead author, John D. Steinbruner, director of the Center for International and Security Studies at the University of Maryland. "This is the sort of thing we were talking about," said Steinbruner, a veteran national security expert.

You can debate the specific contribution of global warming to that storm. But we're saying climate extremes are going to be more frequent, and this was an example of what they could mean. We're also saying it could get a whole lot worse than that . . . states will fail, large populations subjected to famine, flood, or disease will migrate across international borders, and national and international agencies will not have the resources to cope. (Broder 2012)

Bill McKibben, leader of the world-wide climate organization 350.org, points out that Donald Trump's election as president of the United States comes at a very crucial time for the world in its struggle to avert a global climate catastrophe. As of this writing, he had withdrawn the United States from the Paris accords and was dismantling Barack Obama's environmental legacy.

"So let's be entirely clear about what those actions would represent," McKibben wrote in the *Washington Post*. "The biggest, most against-the-odds and most irrevocable bet any president has ever made about anything."

It's the biggest because of the stakes. This year [2016] has been the hottest year recorded in modern history, smashing the record set in 2015, which smashed the record set in 2014. The extra heat has begun to steadily raise sea levels, to the point where some coastal U.S. cities already flood at high tide even in calm weather. Global sea-ice levels are at record lows, and the oceans are 30 percent more acidic. And that's just so far. Virtually every scientific forecast says that without swift action in the next few years to cut carbon emissions, this crisis will grow to be catastrophic, with implications for everything from agriculture to national security that dwarf our other problems. (McKibben, 2016)

The drumbeat of disaster beats more often these days. The Amazon River valley suffers 200-year droughts in five years, and pine forests across western North America fall to waves of pine beetles whose reproductive cycles accelerate with warmer temperatures. The evidence accumulates relentlessly. In one week late in 2013, reports indicated that mass loss from Greenland's ice sheet had quadrupled in 20 years and that warmer-than-average ocean waters had stoked supertyphoon Haiyan's intensity before it became the strongest tropical cyclone to hit land in recorded history. Regulators also voted to cancel the 2014 shrimping season in the Gulf of Maine because of overharvesting and warming waters. The value of the harvest had shrunk from more than $10 million in 2011 to $1.2 million during a shortened harvest in 2013. This drumbeat of news is not unusual (Straneo and Heimbach 2013, 36; Normile 2013; "Maine: Shrimp" 2013).

At nearly the same time (2011), the U.S. House of Representatives voted 240 to 184 (largely along party lines) to defeat a resolution that simply stated, "Climate change is occurring, is caused largely by human activities, and poses significant risks for public health and welfare" (McKibben 2011). As the U.S. House of Representatives invented its own geophysical reality, Russia's grain harvest failed in a record heat wave, and Australia's Queensland lost its grain harvest to drought one year and deluge the next. The Missouri River ran at record flood levels one year, followed by searing heat and drought the next.

The Speed of Change

During more than two decades of study, writing, and commentary on global warming, I have been amazed at how quickly the endgame has chased us. Twenty years ago, the problem seemed comfortably in the future tense. Yes, the Arctic ice cap might melt, but not until the end of the 21st century. Having lost nearly a quarter of its mass in between 1979 and 2015, projections of the ice cap's life in summer now range in the low two figures. In 1995, 450 parts per million carbon dioxide in the atmosphere seemed a tolerable turning point before Earth's ecosystems sustained major damage. The latest scientific assessments have lowered that limit to 350 as the figure passed 400 ppm and continued to rise rapidly.

I am trying here to interpret the science and technology of the subject to students and other nonspecialist readers. I have chosen this emphasis because there exists a great need for public understanding of the scientific basis for what has become an enduring environmental issue and a highly acrimonious public policy debate. Nonscientists need to be conversant with the basics of an issue that is so vital to our future. The scientific world has become more united over the years in its insistence that politicians, diplomats, and drivers of public opinion generally need to deal with this issue. Every day of inaction will make the lives of future generations more intolerable. The geophysical facts of the matter will hold all of us to account.

Every day, month, and year of business-as-usual fossil fuel consumption increases the probability of a climatic emergency within this century. A study published in February 2008 in *Geophysical Research Letters* by H. Damon Matthews (of the Department of Geography, Planning and Environment at Concordia University, Montreal) and Ken Caldeira (Carnegie Institution senior scientist) asserted that dangerous consequences (droughts, deluges, and coastal floods, among others) may be avoided only if humankind stops adding fossil fuel emissions to the atmosphere *completely* by 2050.

> To hold climate constant at a given global temperature requires near-zero future carbon emissions. Our results suggest that future anthropogenic emissions would need to be eliminated in order to stabilize global-mean temperatures. As a consequence, any future anthropogenic emissions will commit the climate system to warming that is essentially irreversible on a centennial timescale. (Matthews and Caldeira 2008)

Across Alaska, northern Canada, and Siberia, scientists have already found telltale signs that permafrost is melting more quickly than expected only a few years ago. As permafrost melts, additional carbon dioxide and methane convert from their solid forms as stored in the ground to gas in the atmosphere that retains more heat. Once again, human contributions of greenhouse gases are provoking a natural process—like the trigger of a gun. This process compounds itself and accelerates over time.

Global warming can be tricky, not only because of its delayed effects but also because it is not simplistically linear. Weather is variable, and many other forcings, or influences, come into play. The association between rising carbon dioxide levels and temperature is not seamless—a mistaken assumption made by deniers of the geophysical facts who seem to jump on every cold wave as evidence of a new ice age or at least as proof that global warming's back has been broken. Other deniers take such variations as proof that carbon dioxide has nearly nothing to do with warming because its rise and temperatures are not year by year exact matches.

Worldwide temperatures hit an exceptional peak in 1998 with an especially strong El Niño, for example, and then spent the next 10 years backing and filling as frequent cooler La Niña conditions set in. In 2014, 2015, and 2016, however, an El Niño even stronger than that of 1997–1998 provoked global temperature rises of extreme proportions. "Too many think global warming means monotonic relentless warming everywhere year after year," said Kevin Trenberth, a climate scientist at the National Center for Atmospheric Research in Boulder, Colorado. "It does not happen that way" (Revkin 2008). "We're learning that internal climate variability is important and can mask the effects of human-induced global change," said Noel Keenlyside of the Leibniz Institute of Marine Sciences in Kiel, Germany, and lead author of a paper on the subject in *Nature*. "In the end this gives more confidence in the long-term projections" (Keenlyside et al. 2008; Revkin 2008).

As Kofi Annan, secretary general of the United Nations until 2007, said during his last speech in that position:

The scientific consensus, already clear and incontrovertible, is moving toward the more alarmed end of the spectrum. Many scientists long known for their caution are now saying that warming has reached dire levels, generating feedback loops that will take us perilously close to a point of no return. A similar shift may be taking place among economists, with some formerly circumspect analysts saying it would cost far less to cut emissions now than to adapt to the consequences later. Insurers, meanwhile, have been paying out more and more each year to compensate for extreme weather events. And growing numbers of corporate and industry leaders have been voicing concern about climate change as a business risk. The few skeptics who continue to try to sow doubt should be seen for what they are: out of step, out of arguments and just about out of time. (Annan 2006, November 8 and 15)

These words are now more than a decade old now, and they are more salient than ever.

Earth's rising temperature is gradually raising sea level through both thermal expansion of the oceans and melting of glaciers and ice sheets. Scientists are particularly concerned by the melting of the Greenland and West Antarctica ice sheets, which has accelerated sharply in recent years. If these ice sheets, a mile thick in some places, were to melt entirely, sea levels would rise by several meters. Even a one-meter (39-inch) sea level rise would inundate vast areas of low-lying coastal land, including many of the rice-growing river deltas and floodplains of India, Thailand, Viet Nam, Indonesia, and China (Brown 2006).

A one-meter rise in sea level could inundate half of Bangladesh's rice-growing land. Forty million Bangladeshis would be forced to migrate. As Lester Brown has written, "Several hundred cities, including some of the world's largest, would be at least partly inundated by a one-meter, rise in sea level, including London, Alexandria, (Egypt), and Bangkok. More than a third of Shanghai, a city of 15 million people, would be underwater. A one-meter rise combined with a 50-year storm surge would leave large portions of Lower Manhattan and the National Mall in Washington, D.C. flooded with seawater" (Brown 2006).

James E. Hansen, long-time director of NASA's Goddard Institute for Space Studies, and the first person to discuss global warming in a scientific context (Hansen et al. 1981) has said, "The bottom line is this: business-as-usual, if it continues for even another decade, will be disastrous for the planet. We can have a stable climate, clean air, and an unpolluted ocean. And clean energies yield good jobs. It is up to the public to make sure that we get onto a path that stabilizes climate and allows all the creatures of Creation to continue to thrive on this planet" (Johansen 2007). The effects of climate change are not theoretical, and they are not speculative problems that can be handed off to future generations. Economic activity around the world, as well as the lives of animals and plants, are being affected today by rising temperatures. This is not merely a matter of a few degrees on the thermometer but of alterations in an environment that sustains all of us.

Greenhouse gases have no morals, loyalty, or party affiliation. Carbon dioxide is not having a debate with us. It merely retains heat. Thomas Friedman, a columnist for *The New York Times*, visited Greenland and remarked,

My trip with Denmark's minister of climate and energy, Connie Hedegaard, to see the effects of climate change on Greenland's ice sheet leaves me with a very strong opinion: Our kids are going to be so angry with us one day. We've charged their future on our Visa cards. We've added so many greenhouse gases to the atmosphere, for our generation's growth, that our kids are likely going to spend a good part of their adulthood, maybe all of it, just dealing with the climate implications of our profligacy. (Friedman 2008)

When Our Children Are Grandparents

Thus, in 50 years—when our children are grandparents—the planetary emergency that we are just beginning to taste will be a dominant problem in everyone's life unless we act now. Within a decade or two, thermal inertia will take off on its own, portending a hot and miserable future for coming generations. Thermal inertia explains why so many scientists find the problem so urgent now.

Eventually, in two or three centuries, given "business as usual," large-scale melting of the world's mountain glaciers, along with a large part of the Greenland and parts of the West Antarctic ice sheets could add as much as 25 meters to sea levels worldwide. Shall we lay bets on the date that first-floor toilets will back up from flooding at the White House, which is only some 50 feet above the tidal basin in Washington, D.C.? The British government is already discussing how much time will pass before its seat of government in London may have to abandon a city swallowed by rising waters.

According to Hansen,

Global warming differs from previous pollution problems in two fundamental ways. With water pollution or common air pollution (smog), the problems occur immediately when the pollutants are emitted. If we decide there is a problem and stop emitting them, the problem goes away. However, global warming is caused by greenhouse gases that have a lifetime of hundreds of years. So we can't wait until we have a full-blown problem and then say "Oh, we better stop emitting these." It's too late then. The gases will hang around for centuries.

The second major difference with the global warming problem is that the climate system responds slowly to the gases that we add to the air. Because of the great thermal inertia of the ocean, only about half of the eventual warming due to gases already in the air has been realized. The Earth has warmed 1.5 degrees Fahrenheit so far, but there is another one degree already in the pipeline. Moreover, there are surely more gases in the pipeline, because of power plants that we have in place and vehicles that we are not going to take off the road.

One and one-half degrees! Who cares about that? Even with another degree or two in the pipeline, who cares about a few degrees? Well, we had better all care about it, because we have already brought the planet close to some tipping points. If we pass those tipping points there will be dramatic consequences. We will leave an impoverished planet for our children, we will have been lousy stewards of creation, we will have destroyed creation for future generations. (Johansen 2007)

Many climate scientists believe that the middle of the 21st century will witness dramatic acceleration of global warming partly because natural carbon sinks will be exhausted. At about the same time, various feedback loops are also expected to accelerate increases in atmospheric greenhouse gas levels and consequently worldwide temperatures. These include several natural processes that add greenhouse gases to the atmosphere, such as melting permafrost in the Arctic, as well as an increasingly dark Arctic Ocean that will absorb more heat as the ice cap melts.

In each of these cases, human-provoked warming caused by an overload of greenhouse gases in the atmosphere is expected to aggravate existing problems—sort of like a bank account drawing an environmentally dangerous form of compound interest. Evidence is accumulating that these processes already have begun. According to many people who are familiar with the paleoclimatic record, the danger is this: Once this journey has begun in earnest, any return trip may become a matter of many centuries as well as considerable pain and suffering for humans, plants, and other animals.

Frozen bogs in Siberia contain an estimated 70 billion tons of methane. If the bogs become drier as they warm, according to one observer, the "methane will oxidize and the emissions will be primarily CO_2. But if the bogs stay wet, as they have been recently, the methane will escape directly into the atmosphere . . . with twenty times the heat-trapping power of carbon dioxide" (Romm 2007). About 600 million tons of methane are emitted each year from human and natural sources, so release of even a small fraction of the 70 billion tons of methane in the Siberian bogs will accelerate warming dramatically.

Another example of feedback loops that will reinforce warming temperatures (and probably drought) occurs in the tropics. A rain forest or jungle creates its own weather, as the foliage recycles moisture. In our time, however, human beings seeking food and shelter compete with nature. Roughly 20 percent of the Amazon rain forest has been destroyed, and approximately another 20 percent has been damaged by logging, which allows drying sunlight to reach the forest floor. Some models suggest that after 50 percent of the Amazon's rain forests have been destroyed. Increased heat and drought will reinforce each other in a feedback loop to further reduce local rainfall, accelerating drought and local temperatures and increasing release of carbon currently locked in Amazon soils and vegetation in a vicious cycle (Romm 2007, 72).

Feedback loops are dangerous because they compound themselves. Ongoing warming liberates carbon dioxide and methane from Arctic and tropical soils and, ultimately, with enough atmospheric warming, even from the oceans. These feedbacks become self-reinforcing and feed on themselves. The atmosphere is only one of many places where the Earth system stores carbon dioxide. Roughly one-third of Earth's carbon is stored in far northern latitudes (mainly in tundra and boreal forests), so the speed with which warming of the ecosystem releases this carbon dioxide to the atmosphere is vitally important to forecasts of global warming's speed and effects.

The amount of carbon stored in Arctic ecosystems equals two-thirds of the amount currently found in the atmosphere. The Arctic has been the most rapidly warming region of Earth. The possibility of a runaway greenhouse effect has provoked a lively scientific debate over the "tipping point" at which such an effect, having been triggered by human consumption of fossil fuels, will develop a life of its own, increasing the level of greenhouse gases in the atmosphere (and consequent warming) beyond any possibility of human mediation. At that point, we will find ourselves riding a runaway train that will continue to accelerate as long as the levels of greenhouse gases in the atmosphere continue to rise—*and* for several decades after they stabilize or fall because of thermal inertia.

No one yet knows with any certainty how long temperatures and sea levels will continue to rise even after greenhouse gas levels stabilize. One scientific observer believes that the "rise in mean global SAT [surface average temperature] and the world ocean level (due to thermal expansion) may . . . continue for several centuries after the stabilization of the carbon dioxide concentration [in the atmosphere], due to the gigantic thermal inertia of the oceans." The same observer also asserts that "The response of ice sheets to earlier climate changes may continue for several centuries after the climate stabilizes" (Kondratyev et al. 2004).

The Future's Chamber of Horrors

Should our fossil fuel industries continue their current course—an accelerating, unchecked development of all available reserves, including conventional, tar sands, fracked shale, and so forth—and if the atmosphere's proportion of greenhouse gases continues rise to levels unknown for many millions of years, we can anticipate an uninhabitable world.

The ultimate climatic nightmare becomes possible after the oceans warm enough to convert methane deposits that are now solid in the oceans to atmospheric gas, further accelerating warming because of melting permafrost, melting ice sheets, and desiccating rain forests. Geophysical evidence suggests that Earth has suffered bouts of severe warming in the distant past from natural causes that were intensified by release of the planet's stores of greenhouse gases that usually reside in solid form, beginning with peat and permafrost on land and followed by methane hydrates in the oceans.

In cold water, methane clathrates form crystal structures that are somewhat similar to water ice. Warming temperatures could destabilize the clathrates and release some of their stored methane. Roughly 10 trillion tons of methane is trapped under pressure in crystal structures in permafrost or on the edges of the oceans' continental shelves, constituting "Earth's largest fossil-fuel reservoir," according to Gerald Dickens, a geologist at James Cook University in Townsville, Australia (Pearce

1998). The greenhouse potential of all the methane stored in clathrates on the continental shelves and in permafrost worldwide is roughly equal to that of all the world's coal reserves.

The "methane burp" will not be tomorrow's news, but climate scientists pay attention to such things because rising greenhouse gas emissions could be taking us down a similar path eventually, even if a matter of many centuries. James Hansen believes that even a 2°C worldwide temperature increase could "whipsaw" Earth into a climate regime via feedback loops that would constitute "a different planet." At that point, increases in greenhouse gases would engage a self-perpetuating feedback loop that cannot be stopped. The final destination of this climatic journey, as well, as the time of arrival, is not known.

Ice Free, Human Free, Eventually?

Most scientific projections of global warming's effects on our atmosphere, our climate, and our lives—the latest report of the IPCC is an example—do not extend past the end of this century. The projections do not stop there because global warming will end. It is not a matter of humanity having had enough and then setting a thermostat and reversing things. In fact, thermal inertia and various other feedbacks virtually guarantee a rapid amplification of warming into the next century and beyond if fossil fuels continue to be burned at anything like today's rate. Consensus science does not have the tools to forecast such things, however.

James Hansen is still intruding on the consensus. Lately, he has been asking what could happen to climate (as well as Earth's flora and fauna) if the human race burns every single ounce of available fossil fuels—that is, all of the coal, oil, tar sands, oil shale, fracked oil, and natural gas? The fossil fuels industry in the last few decades has vastly improved its technology and has increased its access to new reserves.

What will be the eventual outcome of "Drill, baby, drill!"? A recent examination by Hansen concludes that Earth eventually could be ice free and all but bereft of living things, including human beings. A few acres of extremely high-priced real estate may remain on the highest mountains and Antarctica. Perhaps, as in the film *Elysium*, a fraction of the "one percent" may escape to space stations. In the real world, however, they would not have to worry about invasion by Earth-bound rabble. The scalding surface would be too hot for that (Hansen 2013).

Hansen's consideration of this question takes him to the climate of Venus, which was the subject of his doctoral dissertation. It was the greenhouse warming of Venus, in fact, that led him during the 1970s to devote his scientific life to study global warming on Earth. Millions of years ago, Venus may have been more like Earth—a little warmer perhaps owing to its position closer to the sun, but not too warm for liquid oceans. The geophysics of Venus differs from Earth's in one important way, however: it has no plate tectonics. Plates on Earth allow subsurface pressures to be discharged piecemeal and expressed in volcanic eruptions and earthquakes that can be deadly close at hand but help maintain a rough equilibrium for the entire Earth.

On Venus, subsurface pressures are said to have built up to enormous levels, then exploded in a single, spectacular blast that provoked a runaway greenhouse effect. Temperatures rose, and the oceans boiled away. As a result, today's Venus has a surface atmosphere hot enough to melt lead. Most of Venus' carbon is now in the atmosphere, not in its crust (as on Earth), observes Hansen (Hansen 2013, April 15).

So that's Venus—no place for a picnic. What about Earth? Here, with technological ingenuity and a desire for profit, comfort, and convenience, human fossil fuel corporations are supplying us with products manufactured from carbon extracted from the crust, combusted as energy, and released to the atmosphere as carbon dioxide and methane. We are, in other words, mimicking natural processes on Venus.

At some point feedbacks may take off on their own and accelerate these changes, perhaps hyperthermally. Exactly when that is we do not know, although the IPCC now says roughly 2040 if we continue on our current emissions path. By that time, the climate deniers will have been discredited, but it will be too late. We will be on our way to a world where the air in most places will eventually become simply too hot to sustain human (or most other) life. "It is not an exaggeration to suggest," wrote Hansen in an e-mail post September 26, 2013, "based on best available scientific evidence that burning all fossil fuels could result in the planet being not only ice-free but human-free."

The geophysical facts do not cease to matter because some politicians do not recognize them. This set of volumes is meant to increase understanding of science and possible solutions, with a realization that humankind's survival instincts will run congruent with a sustainable future.

Bruce E. Johansen
Omaha, Nebraska
May 2016

Further Reading

Annan, Kofi. "As Climate Changes, Can We?" *Washington Post*, November 8, 2006, A27. http://www.washingtonpost.com/wp-dyn/content/article/2006/11/07/AR2006110701229_pf.html.

Annan, Kofi. "Global Warming an All-Encompassing Threat." Address to United Nations Conference on Climate Change, Nairobi, Kenya. Environment News Service, November 15, 2006. http://www.ens-newswire.com/ens/nov2006/2006-11-15-insann.asp (no longer available).

Broder, John M. "Climate Change Report Outlines Perils for U.S. Military." *The New York Times*, November 9, 2012. http://www.nytimes.com/2012/11/10/science/earth/climate-change-report-outlines-perils-for-us-military.html.

Brown, Lester R. "The Earth Is Shrinking." Environment News Service, November 20, 2006. http://www.ens-newswire.com (no longer available).

Caldeira, Ken, and Michael E. Wickett. "Oceanography: Anthropogenic Carbon and Ocean pH." *Nature* 425 (September 25, 2003): 365.

"Climate Scoreboard." Climate Interactive. 2015. https://www.climateinteractive.org/tools/scoreboard/.

Davenport, Carol. "Optimism Faces Grave Realities at Climate Talks." *The New York Times*, December 1, 2014. http://www.nytimes.com/2014/12/01/world/climate-talks.html.

Eilperin, Juliet. "Carbon Is Building Up in Atmosphere Faster Than Predicted." *Washington Post*, September 26, 2008, A2. http://www.washingtonpost.com/wp-dyn/content/article/2008/09/25/AR2008092503989_pf.html.

Friedman, Thomas. "Learning to Speak Climate." *The New York Times*, August 6, 2008. http://www.nytimes.com/2008/08/06/opinion/06friedman.html.

Gillis, Justin. "Panel's Warning on Climate Risk: Worst Is Yet to Come." *The New York Times*, March 30, 2014. http://www.nytimes.com/2014/03/31/science/earth/panels-warning-on-climate-risk-worst-is-yet-to-come.html.

Gillis, Justin. "Climate Efforts Falling Short, U.N. Panel Says." *The New York Times*, April 13, 2014, A1.

Gillis, Justin. "U.S. Climate Has Already Changed, Study Finds, Citing Heat and Floods." *The New York Times*, May 6, 2014. http://www.nytimes.com/2014/05/07/science/earth/climate-change-report.html?hp.

Gillis, Justin. "Greenhouse Gas Emissions Are Growing, and Growing More Dangerous, Draft of U.N. Report Says." *The New York Times*, August 26, 2014. http://www.nytimes.com/2014/08/27/science/earth/greenhouse-gas-emissions-are-growing-and-growing-more-dangerous-draft-of-un-report-says.html.

Gillis, Justin. "Study Predicts Antarctica Ice Melt If All Fossil Fuels Are Burned." *The New York Times*, September 12, 2015. http://www.nytimes.com/2015/09/12/science/climate-study-predicts-huge-sea-level-rise-if-all-fossil-fuels-are-burned.html.

Gillis, Justin. "2015 Was Hottest Year in Recorded History, Scientists Say." *The New York Times*, January 20, 2016. http://www.nytimes.com/2016/01/21/science/earth/2015-hottest-year-global-warming.html.

Gillis, Justin, and Kenneth Chang. "Scientists Warn of Rising Oceans from Polar Melt." *The New York Times*, May 12, 2014. http://www.nytimes.com/2014/05/13/science/earth/collapse-of-parts-of-west-antarctica-ice-sheet-has-begun-scientists-say.html.

Gillis, Justin, and Somini Sengupta. "Limited Progress Seen Even as More Nations Step Up on Climate." *The New York Times*, September 28, 2015. http://www.nytimes.com/2015/09/28/world/limited-progress-seen-even-as-more-nations-step-up-on-climate.html.

Hansen, James E. "Making Things Clearer: Exaggeration, Jumping the Gun, and The Venus Syndrome." April 15 2013. http://www.columbia.edu/~jeh1/mailings/2013/20130415_Exaggerations.pdf.

Hansen, James E. "An Old Story, but Useful Lessons." September 26, 2013. http://www.columbia.edu/~jeh1/mailings/2013/20130926_PTRSpaperDiscussion.pdf.

Hansen, James E., et al. "Climate Impact of Increasing Atmospheric Carbon Dioxide," *Science* 213 (1981), 957–956.

Harvey, Chelsea. "Scientists Confirm There's Enough Fossil Fuel on Earth to Entirely Melt Antarctica." *Washington Post*, September 11, 2015. http://www.washingtonpost.com/news/energy-environment/wp/2015/09/11/scientists-confirm-theres-enough-fossil-fuel-on-earth-to-melt-all-of-antarctica/?wpmm=1&wpisrc=nl_headlines.

Johansen, Bruce E. "Global Warming, 'Thermal Inertia,' and Tomorrow's News." *Nebraska Report*, October 2007, 7.

Johnson, Bob. *Carbon Nation: Fossil Fuels in the Making of American Culture.* Lawrence: University Press of Kansas, 2014.

Joughin, Ian, Benjamin E. Smith, and Brooke Medley. "Marine Ice Sheet Collapse Potentially Under Way for the Thwaites Glacier Basin, West Antarctica." *Science* 344 (May 16, 2014): 683.

Keenlyside, N.S., et al. "Advancing Decadal-Scale Climate Prediction in the North Atlantic Sector." *Nature* 453 (May 1, 2008): 84–88.

Kerr, Richard A. "How Urgent Is Climate Change?" *Science* 318 (November 23, 2007): 1231.

Kerr, Richard A. "The Many Dangers of Greenhouse Acid." *Science* 323 (January 23, 2009): 459.

Kerr, Richard A. "Experts Agree Global Warming Is Melting the World Rapidly." *Science* 338 (November 30, 2012a): 1138.

Kerr, Richard A. "An Oil Gusher in the Offing, but Will It Be Enough? *Science* 338 (November 30, 2012b): 1139.

Khan, Shfaqat Abbas, et al. "Sustained Mass Loss of the Northeast Greenland Ice Sheet Triggered by Regional Warming," *Nature Climate Change,* March 2014. doi: 10.1038/nclimate2161.

Kim, Jim Yong. "Make Climate Change a Priority." *Washington Post,* January 24, 2013. http://www.washingtonpost.com/opinions/make-climate-change-a-priority/2013/01/24/6c5c2b66-65b1-11e2-9e1b-07db1d2ccd5b_print.html.

Koch, Wendy. "Greenland's Ice Decline Accelerates." *USA Today*, March 17, 2014, 4A.

Kolbert, Elizabeth. "The Weight of the World: Can Christina Figures Persuade Humanity to Save Itself?" *The New Yorker*, August 24, 2015, 24–30.

Kondratyev, Kirill, Vladimir F. Krapivin, and Costas A. Varotsos. *Global Carbon Cycle and Climate Change.* Berlin: Springer/Praxis, 2004.

Lean, Geoffrey. "Global Warming Will Redraw Map of the World." *The Independent* (London), November 7, 2004, 8.

"Maine: Shrimp Season Is Called Off." *The New York Times*, December 4, 2013, A21.

Matthews, H. Damon, and Ken Caldeira. "Stabilizing Climate Requires Near-Zero Emissions." *Geophysical Research Letters* 35 (February 27, 2008): L04705. doi: 10.1029/2007 GL032388.

McKibben, Bill. "Donald Trump is Betting Against All Odds on Climate Change." *Washington Post*, November 26, 2016. https://www.washingtonpost.com/opinions/donald-trump-is-betting-against-all-odds-on-climate-change/2016/11/17/8f7bf4ee-acf4-11e6-977a-1030f822fc35_story.html.

Normile, Dennis. "Clues to Supertyphoon's Ferocity Found in the Western Pacific." *Science* 342 (November 29, 2013): 1027.

Paulson, Henry M. "The Coming Climate Crash: Lessons for Climate Change in the 2008 Recession." *The New York Times*, June 21, 2014. http://www.nytimes.com/2014/06/22/opinion/sunday/lessons-for-climate-change-in-the-2008-recession.html.

Pearce, Fred. "Nature Plants Doomsday Devices." *The Guardian* (London), November 25, 1998. http://go2.guardian.co.uk/science/912000568-disast.html.

"Prince of Wales Warns Climate Crash Could Dwarf Financial Crisis." Environment News Service, May 24, 2011. http://www.ens-newswire.com/ens/may2011/2011-05-24-01.html.

Ramanathan, V., and Y. Feng. "On Avoiding Dangerous Anthropogenic Interference with the Climate System: Formidable Challenges Ahead." *Proceedings of the National Academy of Sciences* 105(38) (September 23, 2008): 14245–14250.

Revkin, Andrew C. "Climate Panel Reaches Consensus on the Need to Reduce Harmful Emissions." *The New York Times*, May 4, 2007. http://www.nytimes.com/2007/05/04/science/04climate.html.

Revkin, Andrew C. "In a New Climate Model, Short-Term Cooling in a Warmer World." *The New York Times*, May 1, 2008. http://www.nytimes.com/2008/05/01/science/earth/01climate.html (no longer available).

Rignot, E., et al. "Rapid Grounding Line Retreat of Pine Island, Thwaites, Smith, and Kohler Glaciers, West Antarctica from 1992 to 2011." *Geophysical Research Letters*, May 2014. doi: 10.1002/2014GL060140.

Romm, Joseph. *Hell and High Water: Global Warming—the Solution and the Politics—and What We Should Do.* New York: William Morrow, 2007.

Straneo, Fiammetta, and Patrick Heimbach. "North Atlantic Warming and the Retreat of Greenland's Outlet Glaciers." *Nature* 504 (December 5, 2013): 36–43.

Warrick, Joby. "CO_2 Levels in Atmosphere Rising at Dramatically Faster Rate, U.N. Report Warns." *Washington Post*, September 9, 2014. http://www.washingtonpost.com/national /health-science/co2-levels-in-atmosphere-rising-at-dramatically-faster-rate-un-report -warns/2014/09/08/3e2277d2-378d-11e4-bdfb-de4104544a37_story.html.

Winkelmann, Ricarda, et al. "Combustion of Available Fossil Fuel Resources Sufficient to Eliminate the Antarctic Ice Sheet." *Science Advances* 1(8) (September 11, 2015). e1500589. http://advances.sciencemag.org/content/1/8/e1500589. doi: 10.1126/sciadv.1500589.

"World Economic Forum Feels Climate Change Pressure." Environment News Service, January 25, 2013. http://ens-newswire.com/2013/01/25/world-economic-forum-feels -climate-change-pressure/.

KEY PEOPLE

ARRHENIUS, SAVANTE AUGUST (1859–1927)

In 1896, Savante August Arrhenius, a Swedish chemist, published a paper in *The London, Edinburgh, and Dublin Philosophical Magazine and Journal of Science* titled "On the Influence of Carbonic Acid in the Air upon the Temperature of the Ground" (Arrhenius 1896). In his paper, Arrhenius theorized that a rise in the atmospheric level of carbon dioxide could raise the temperature of the air. He was not the only person thinking along these lines at the time; Swedish geologist Arvid Hogbom had delivered a lecture on the same idea three years earlier, which Arrhenius incorporated into his article.

Arrhenius was a well-known scientist in his own time, not for his theories describing the greenhouse effect but for his work in electrical conductivity, for which he was awarded a Nobel Prize in 1903. Later in his life, Arrhenius directed the Nobel Institute in Stockholm. His work in global-warming theory was not much discussed during his own life. Arrhenius, however, used the available measurements of absorption and transmission by water vapor and carbon dioxide to develop the first quantitative mathematical model of Earth's greenhouse effect and obtained results of acceptable accuracy by today's standards for equilibrium climate sensitivity to carbon dioxide changes.

Arrhenius developed his theory through the use of equations, by which he calculated that a doubling of carbon dioxide in the atmosphere would raise air temperatures approximately 10°F. Arrhenius thought 3,000 years would have to pass before human-generated carbon dioxide levels would double, a miscalculation that shares something with Benjamin Franklin's belief, late in the 18th century, that European–American expansion across North America would take a thousand years.

Arrhenius applauded the possibility of global warming, telling audiences that a warmer world "would allow all our descendants, even if they only be those of a distant future, to live under a warmer sky and in a less harsh environment than we were granted" (Christianson 1999, 115). In 1908 in his book *Worlds in the Making*, Arrhenius wrote: "By the influence of the increasing percentage of carbonic acid in the atmosphere, we may hope to enjoy ages with more equable and better climates, especially as regards the colder regions of the Earth, ages when the Earth will bring forth much more abundant crops than at present for the benefit of rapidly propagating mankind" (Christianson 1999, 115).

Further Reading

Arrhenius, Svante. "On the Influence of Carbonic Acid in the Air upon the Temperature of the Ground." *The London, Edinburgh, and Dublin Philosophical Magazine and Journal of Science*, 5th ser. (April 1896), 237–276.
Christianson, Gale E. *Greenhouse: The 200-Year Story of Global Warming.* New York: Walker & Co., 1999.
Oppenheimer, Michael, and Robert H. Boyle. *Dead Heat: The Race against the Greenhouse Effect.* New York: Basic Books, 1990.

CALDEIRA, KEN

From his post as senior scientist in the Department of Global Ecology at the Carnegie Institution for Science at Stanford University since 2005, Ken Caldeira has become a master of straight talk and inconvenient truth, whether as a pioneer of ocean acidification or warning that humankind must reduce greenhouse gas emissions radically or face sulfur injection into the stratosphere to cool the global fever. Few scientists can match his talent for communicating with the public with quick wit and pithy quotes to describe the truly exceptional nature of the problem. Caldeira, who invented the phrase "ocean acidification," also invented another phrase, describing the sulfur-as-savior solution as "solar radiation management" (Funk 2014, 266).

Serious consideration is not the same as adopting sulfur as a solution, however. Caldeira also has said: "If we keep emitting greenhouse gases with the intent of offsetting global warming with ever-increasing loadings of particles in the stratosphere, we will be heading to a planet with extremely high greenhouse gases and a thick stratospheric haze that we would need to maintain more or less indefinitely. This leads to a dystopian world out of a science-fiction story" (Goodell 2010, 20). The fact that solar radiation management would add acid to the oceans has not escaped Caldeira. Even so, he helped pitch Nathan Myhrvold's vision of a giant hose spewing sulfur into the stratosphere as a "senior inventor" with Intellectual Ventures, a Seattle-based invention and patent company (Klein 2014, 264).

A tone of desperation is palpable in climate-change science when well-known people seriously propose that filling the stratosphere with sulfur dioxide may be the only way to stop runaway greenhouse warming. Do we really want to pump the stratosphere full of sulfur to shroud the surface from warmth and then live in a perpetually acidic mist? When Dutch atmospheric chemist Paul J. Crutzen advanced the sulfur shield idea in 2006, he cited a "grossly disappointing international political response" to increasing evidence of global warming (Kerr 2006, 401). Caldeira said that countries need to "undertake studies on what we might do" in a climate crisis, given the current trajectory of carbon concentrations in the atmosphere. "Nobody likes the idea of engineering Earth's climate. . . . Unfortunately, at some point, our other options may be even more unpleasant," said Caldeira (Eilperin 2010).

Education and Early Research

Caldeira earned his BA from Rutgers College, followed by an MS (1988) and PhD (1991), both in atmospheric sciences at New York University. Before and during his studies during the 1980s, Caldeira worked at developing computer software in New York's financial district. Caldeira did postdoctoral research in the Department of Geosciences at Pennsylvania State University as well as at the Energy and Environment Directorate at the Lawrence Livermore National Laboratory, where he then worked on staff from the early 1990s until he joined the Department of Global Ecology at the Carnegie Institution for Science at Stanford University in 2005. While working at the institution, he won that lab's highest award, the Edward Teller Fellowship, in 2004.

Caldeira also served on a U.S. National Academy of Sciences panel, Geoengineering Climate: Technical Evaluation and Discussion of Impacts. He has been a contributing author to the Intergovernmental Panel on Climate Change (IPCC) AR5 report, *Climate Change 2013: The Physical Science Basis*. In 2010, Caldeira was elected as a fellow of the American Geophysical Union. He was a coauthor of the 2010 U.S. National Academy's *America's Climate Choices* report. Caldeira also served as coordinating lead author of the "oceans" chapter in the IPCC report *Carbon Capture and Storage* in 2005.

Caldeira has arranged a seminar series for Bill Gates, cofounder of Microsoft Corporation, on climate and energy issues, and Gates has funded part of Caldeira's work. Caldeira also serves as a professor in Stanford University's Department of Environmental Earth System Sciences, where he teaches and advises students. In addition to his role as scientist and public intellectual, Caldeira, according to his institute's Web page, "has a wide-spectrum approach to analyzing the world's climate systems. He studies the global carbon cycle; marine biogeochemistry and chemical oceanography, including ocean acidification and the atmosphere–ocean carbon cycle; land-cover and climate change; the long-term evolution of climate and geochemical cycles; climate intervention proposals; and energy technology" ("Ken Caldeira" no date).

Work on Ocean Acidification

Caldeira makes a strong case that the oceans are facing a degree of acidification worse than the paleoclimatic record as far back as 300 million years. By the early years of the 21st century, carbon dioxide levels were rising in the oceans more rapidly than any time since the age of the dinosaurs, according to work published by Caldeira and Michael E. Wickett. According to them, "We find that oceanic absorption of CO_2 from fossil fuels may result in larger pH changes over the next several centuries than any inferred in the geological record of the past 300 million years, with the possible exception of those resulting from rare, extreme events such as bolide impacts or catastrophic methane hydrate degassing" (Caldeira and Wickett 2003, 365). A bolide is a large extraterrestrial body—typically a bright asteroid or meteor—usually at least a half mile in diameter, sometimes much larger, that impacts Earth at a speed roughly equal to that of a bullet in flight. "Methane hydrate degassing" involves the rapid conversion of solid methane deposits on ocean floors to gaseous form in the atmosphere by warming temperatures.

Caldeira and colleagues also have presented evidence that the oceans are reaching their limits as "sinks," or absorbers, of carbon dioxide and methane. Caldeira and Philip B. Duffy asserted in *Science* that "uptake," or removal, of human-induced carbon dioxide by the oceans is less than many earlier investigators had assumed and that carbon uptake will diminish further as temperatures warm. In addition, Caldeira and Duffy contend that absorption of carbon into the oceans of the Southern Hemisphere (the focus of their study) has been diminishing since fossil fuel effluvia became a factor in the composition of the atmosphere around 1880. "Ventilation of the deep Southern Ocean was much more vigorous in the period from about 1350 to 1880 than in the recent past," they wrote (Caldeira and Duffy 2000, 620).

Furthermore, according to Caldeira, the effects of acidity in the oceans also will continue long after fossil fuel burning peaks on land. Caldeira modeled ocean acidification for fossil fuel burning that peaks in the year 2100 and found that (given the assumptions of this study) the oceans will continue to become more acidic for centuries after that. At the surface, acidity will peak at about 2750. A kilometer deep in the ocean, acidification will rise for a thousand years. "People should know that the consequences of what we're doing in the next decade will last for thousands of years," said Caldeira (Ruttimann 2006, 979–980).

Caldeira points out directly and forcefully that acidification threatens the phytoplankton that form the basis of the oceanic food chain, as well as the future of coral reefs. He has made a case that polar waters overburdened by carbon-provoked acidification will no longer be able to sustain viable populations of pteropods (plankton). Once acidity reaches levels that dissolve calcium shells in the tropics, something he calls a doomsday scenario for coral reefs that have not already killed by rising water temperatures. "If you look at the business-as-usual scenario for emissions and its impact with respect to aragonite in surface waters, by the end of the century there is no place left with the kind of chemistry where corals grow today," he has told interviewers (Henderson 2006, 31).

Caldeira anticipates that coral reefs may survive only in walled-off enclosures where acidity has been controlled by humankind. "Our emissions are huge compared with natural fluxes," said Caldeira. "If you could stop emissions and wait 10,000 years, natural processes would probably take care of most of it" (Holland 2001, 111). Emissions, however, are not being curtailed.

By the end of the 21st century, according to Caldeira, surface acidity around Antarctica will be roughly double preindustrial levels (a 0.2 decrease in pH), threatening the ability of aquatic life forms to maintain their shells (Kolbert 2006, 70). Such a level would put about two-thirds of cold-water corals in corrosive waters (Kintsch and Stokstad 2008, 1029).

A "Hero Scientist"

Caldeira was named as a "hero scientist " in 2008 by *New Scientist* magazine and as one of 100 "agents of change" in *Rolling Stone* magazine in 2009. His work was described extensively in *The New Yorker* (Spector 2012; Kolbert 2006, 66–75).

Caldeira argues that a zero-carbon-dioxide emissions policy is the only solution in the long term. In 2005, he said,

If you're talking about mugging little old ladies, you don't say, "What's our target for the rate of mugging little old ladies?" You say, "Mugging little old ladies is bad, and we're going to try to eliminate it." You recognize you might not be a hundred percent successful, but your goal is to eliminate the mugging of little old ladies. And I think we need to eventually come around to looking at carbon dioxide emissions the same way. (Kolbert 2006)

In 2014, Caldeira said, "It is time to stop building things with tailpipes and smokestacks. It is time to stop using the sky as a waste dump for our carbon dioxide pollution" (Shields 2014).

Further Reading

Caldeira, Ken, and Philip B. Duffy. "The Role of the Southern Ocean in the Uptake and Storage of Anthropogenic Carbon Dioxide." *Science* 287 (January 28, 2000): 620–622.

Caldeira, Ken, and Michael E. Wickett. "Oceanography: Anthropogenic Carbon and Ocean pH." *Nature* 425 (September 25, 2003): 365.

Eilperin, Juliet. "Geo-Engineering Sparks International Ban, First-Ever Congressional Report." *Washington Post,* October 29, 2010.

Funk, McKenzie. *Windfall: The Booming Business of Global Warming.* New York: Penguin Press, 2014.

Goodell, Jeff. *How to Cool the Planet. Geoengineering and the Audacious Quest to Fix Earth's Climate.* Boston: Houghton-Mifflin Harcourt, 2011.

Henderson, Casper. "The Other CO_2 Problem." *New Scientist,* August 5, 2006, 28–33.

Holland, Jennifer S. "The Acid Threat: As CO_2 Rises, Shelled Animals May Perish." *National Geographic,* November 2001, 110–111.

"Ken Caldeira." Carnegie Institution for Science, Department of Global Ecology, Stanford University. No date. http://globalecology.stanford.edu/labs/caldeiralab/Caldeira_bio.html. Accessed February 25, 2015.

Kerr, Richard A. "Pollute the Planet for Climate's Sake?" *Science* 314 (October 20, 2006): 401–403.

Kintisch, Eli, and Erik Stokstad. "Ocean CO_2 Studies Look beyond Coral." *Science* 319 (February 22, 2008): 1029.

Klein, Naomi. *This Changes Everything: Capitalism and the Climate.* New York: Simon & Schuster, 2014.

Kolbert, Elizabeth. "The Darkening Sea: What Carbon Emissions Are Doing to the Oceans." *The New Yorker,* November 20, 2006, 66–75. http://www.newyorker.com/magazine/2006/11/20/the-darkening-sea.

Ruttimann, Jacqueline. "Oceanography: Sick Seas." *Nature* 442 (August 31, 2006): 978–980.

Shields, Craig. "Maybe It Is Time to Stop Building Things with Tailpipes and Smokestacks." 2Green Energy, October 12, 2014. http://2greenenergy.com/2014/10/12/tailpipes-and-smokestacks/.

Spector, Michael. "The Climate Fixers: Is There a Technological Solution to Global Warming?" *The New Yorker,* May 14, 2012. http://www.newyorker.com/magazine/2012/05/14/the-climate-fixers.

CRUTZEN, PAUL JOZEF (DECEMBER 3, 1933–)

Dutch atmospheric chemist Paul J. Crutzen has played a leading role in describing humanity's growing impact on the atmosphere from his post in the Department of Atmospheric Chemistry at the Max Planck Institute for Chemistry in Mainz, Germany. In 1995, He shared a Nobel Prize in chemistry with Americans M. Molina and F. S. Rowland for pioneering work alerting the world to the dangers of stratospheric ozone depletion. Later, he warned that drastic steps may be necessary to geoengineer the stratosphere if greenhouse gas emissions are not reduced soon and by large amounts.

Crutzen has developed a worldwide reputation in atmospheric chemistry during his work at the Scripps Institution of Oceanography, University of California–San Diego; Seoul University, South Korea; the Georgia Institute of Technology; and Stockholm University in Sweden. He is a public intellectual who speaks and signs petitions frequently in defense of scientific issues.

Crutzen also has coined the word *Anthropocene* to describe the geologic epoch in which human beings are the primary force shaping Earth's future. The term has superseded *Holocene*, the epoch in which most geology textbooks said we lived until *anthropocene* became widely accepted after 2010. The advent of the Anthropocene can be dated to James Watt's invention of the steam engine in 1784 and the ensuing rise in atmospheric greenhouse gases that followed its increasing use, which was initially fueled by coal and by oil in the 19th century (Zalasiewicz et al. 2010, 2228).

Crutzen has emphasized the danger inherent in taking chances with Earth's climate system without understanding its chemistry. The history of atmospheric chemistry during the last few decades, he said, has been one of surprises. "There may be more of these things around the corner," he said (McFarling 2001). Beginning in 1982, Crutzen also explored the concept of nuclear winter, pointing out that such a war would shroud Earth in sooty smoke for several years, promoting intense cooling at the surface, devastating food production, and ensuring mass starvation (Crutzen and Birks 1982).

Early Life and Work

Crutzen was born in Amsterdam, the Netherlands, the son of Anna Gurk and Jozef Crutzen (of mixed German and Polish ancestry) on December 3, 1933. He entered elementary school in the Netherlands in 1940 as the Germans were invading during World War II. Crutzen described his boyhood during the war:

> Our school class had to move between different premises in Amsterdam after the German army had confiscated our original school building. The last months of the war, between the fall of 1944 and Liberation Day on May, 5, 1945, were particularly horrible. During the cold "hongerwinter" (winter of famine) of 1944–1945, there was a severe lack of food and heating fuels. Also water for drinking, cooking, and washing was available only in limited

quantities for a few hours per day, causing poor hygienic conditions. Many died of hunger and disease, including several of my schoolmates. Some relief came at the beginning of 1945 when the Swedish Red Cross dropped food supplies on parachutes from airplanes. To welcome them we waved our red, white, and blue Dutch flags in the streets. (Crutzen 1996)

Crutzen enrolled in Hogere Burgerschool (Higher Citizens School) during 1946 and learned English, German, and French while also developing his knowledge of natural sciences. He graduated in 1951. Crutzen then enrolled in a civil engineering program at Middle Technical School, interrupting his education to take part in 21 months of military training as required of all young Dutch men. In February 1958, he married Terttu Soininen, with whom he had a daughter, Liona, nine months later. A second daughter, Sylvia, was born to Soininen in 1964.

Crutzen took up studies at the Meteorology Institute of Stockholm University and the associated International Meteorological Institute, which housed many well-known researchers and was headed at the time by Bert Bolin, later director of the Intergovernmental Panel on Climate Change (IPCC). By 1963, he had earned a *filosofie kandidat* (master's of science) degree, combining mathematical statistics with meteorology. Work followed there on his *filosofie licentiat* thesis (comparable to a PhD dissertation in the United States). Later, he directed research at the National Center of Atmospheric Research in Boulder, Colorado (1977–1980); after 1980, he was at the Max Planck Institute for Chemistry in Mainz, Germany.

During the middle 1960s, while he was developing a numerical model of the oxygen allotrope distribution in the stratosphere, mesosphere, and lower thermosphere, Crutzen's interest was drawn to the photochemistry of atmospheric ozone, which shaped the studies that led to his Nobel Prize. "I picked stratospheric ozone as my subject, without the slightest anticipation of what lay ahead," Crutzen told the Nobel Committee in Stockholm as he, Molina, and Rowland received their Nobel Prize in 1995 (Crutzen 1996). In 1971, he began to publish work that implicated the increasing use of fertilizers as a factor in nitrous oxide emissions that were reaching the stratosphere and depleting the protective ozone layer, thereby exposing life on the surface to dangerous ultraviolet radiation.

Crutzen has asserted that problems with the stratospheric ozone layer could have been much worse if chemists had developed substances based on bromine, which is 100 times as dangerous for ozone when compared with chlorine atom to atom. "This brings up the nightmarish thought that if the chemical industry had developed organochlorine compounds instead of the [chlorofluorocarbons]—or, alternatively, if chlorine chemistry had behaved more like that of bromine—then without any preparedness, we would have faced a catastrophic ozone hole everywhere and in all seasons during the 1970s, probably before atmospheric chemists had developed the necessary knowledge to identify the problem . . ." (Crutzen 2001, 10). Given the fact that no one seemed overly worried about this problem before 1974, wrote Crutzen, "We have been extremely lucky." This shows, he wrote, "That we should always be on our guard for the potential consequences of the release of new products into the environment . . . for many years to come" (Crutzen 2001, 10).

The "Sulfur Solution"

Crutzen prepared a paper discussing the sulfur solution to *Climatic Change* that was widely circulated before its publication (2006, 211–219). Although he acknowledged that bombing the atmosphere with sulfur should be a strategy of last resort, Crutzen argued that humankind was not mounting an adequate response to global warming. He said that 5.3 million tons of sulfur per year delivered by airplane (or otherwise)— at an annual cost of $50 billion—would compensate for the greenhouse gas warming expected during the 21st century. The idea is not new (Russian scientist Mikhail Budyko had proposed it in the 1970s), but this was the first time that a Nobel Prize–winning scientist with environmental credentials had made such a serious proposal.

When Crutzen advanced the sulfur shield idea in 2006, he cited a "grossly disappointing international political response" to increasing evidence of global warming (Kerr 2006, 401). Caldeira said that countries need to "undertake studies on what we might do" in a climate crisis, given the current trajectory of carbon concentrations in the atmosphere. "Nobody likes the idea of engineering Earth's climate. . . . Unfortunately, at some point, our other options may be even more unpleasant," he said (Eilperin 2010).

In a special August 2006 issue of the scientific journal *Climatic Change*, Crutzen discussed the theoretical basis of the idea, possible methodologies, and advantages and disadvantages. Several other authors also discussed the history of such proposals, the practical and ethical considerations of various approaches, and how best to evaluate them. The authors make it clear that geoengineering climate is a less desirable potential solution to warming than controlling greenhouse emissions and that geoengineering be a better choice only if warming causes sufficiently harmful impacts.

In a draft of his paper, Crutzen estimated the annual cost of his sulfur proposal at up to $50 billion, or about 5 percent of the world's annual military spending. "Climatic engineering, such as presented here, is the only option available to rapidly reduce temperature rises" if international efforts fail to curb greenhouse gases, Crutzen wrote. "So far," he added, "there is little reason to be optimistic" (Broad 2006). Supporters of this idea contend that any increase in sulfur at Earth's surface would be small compared with the tons already being emitted from the smokestacks of coal-fueled power plants (Broad 2006).

Further Reading

Broad, William J. "How to Cool a Planet (Maybe)." *New York Times*, June 2, 2006. http://www.nytimes.com/2006/06/27/science/earth/27cool.html.

Crutzen, Paul J. "Biographical." In Tore Frängsmyr, ed., *The Nobel Prizes 1995*. Stockholm: Nobel Foundation, 1996. nobelprize.org/nobel_prizes/chemistry/laureates/1995/crutzen-bio.html.

Crutzen, Paul J. "The Antarctic Ozone Hole, a Human-Caused Chemical Instability in the Stratosphere: What Should We Learn from It?" Pp. 1–11 in Lennart O. Bengtsson and Claus U. Hammer, eds., *Geosphere-Biosphere Interactions and Climate*. Cambridge, UK: Cambridge University Press, 2001.

Crutzen, Paul. "Albedo Enhancement by Stratospheric Sulfur Injections: A Contribution to Resolving Policy Dilemma?" *Climatic Change* 77 (2006): 211–220.

Crutzen, Paul J., and John W. Birks. "The Atmosphere after a Nuclear War: Twilight at Noon." *Ambio* 11(2–3) (1982): 114–125.

Eilperin, Juliet. "Geo-Engineering Sparks International Ban, First-Ever Congressional Report." *Washington Post,* October 29, 2010. http://www.washingtonpost.com/wp-dyn /content/article/2010/10/29/AR2010102906365_pf.html.

Kerr, Richard A. "Pollute the Planet for Climate's Sake?" *Science* 314 (October 20, 2006): 401–403.

McFarling, Usha Lee. "Fear Growing over a Sharp Climate Shift." *Los Angeles Times,* July 13, 2001, A1.

Zalasiewicz, Jan, et al. "The New World of the Anthropocene." *Environmental Science & Technology,* 2010, 44(7): 2228–2231.

GORE, ALBERT (MARCH 31, 1948–)

Al Gore, U.S. vice president under Bill Clinton, narrowly missed being elected president in the hotly contested 2000 election. Before and since the election he has been a major political figure in the global warming debate through his public advocacy of the issue and participation in diplomacy. Gore also became a leading figure worldwide as an interpreter of global warming science to the public. Gore's leading role in publicizing global warming was recognized in 2007 with a Nobel Prize for Peace, which he shared with the Intergovernmental Panel on Climate Change (IPCC). In its formal citation, the Nobel committee called Gore "probably the single individual who has done most to create greater worldwide understanding of the measures that need to be adopted" (Gibbs and Lyall 2007, A13).

Accepting the Nobel Prize for Peace on December 12, 2007, seven years to the day after the U.S. Supreme Court called a halt to a Florida vote recount involving "hanging chads" and sealed George W. Bush's victory as president (despite Gore winning the popular vote), Gore said: "I read my own political obituary in a judgment that seemed to me harsh and mistaken—if not premature. But that unwelcome verdict also brought a precious if painful gift: an opportunity to search for fresh new ways to serve my purpose" (Kolbert 2007, 43).

As Author and Activist

Gore wrote *Earth in the Balance* (1992) and was narrator of a well-known documentary film, *An Inconvenient Truth* (2006), that won an Oscar. Writing in *The New Yorker*, film critic David Denby said of *An Inconvenient Truth*, "The science is detailed, deep-layered, vivid, and terrifying. . . . If even half of what Gore says is true, this may the most galvanizing documentary you will see in your lifetime" (Denby 2006). Directed by Davis Guggenheim, *An Inconvenient Truth* was released in May 2006 and had gross receipts of more than $46 million, making it one of the top-grossing documentaries ever made. A companion book by Gore quickly became a best seller, reaching number 1 on *The New York Times* best-seller list (Broad 2007). The film was credited with raising the salience of global warming as a major political issue in the United States.

An Inconvenient Truth was made after Laurie David, a prominent Hollywood environmentalist, saw Gore give a short version of his presentation at an event held just before the premiere of the climate disaster movie *The Day After Tomorrow*. David said she was stunned by the power of Gore's talk and helped organize presentations in New York and Los Angeles for people involved in the news media, environmental groups, business, and entertainment. By the time she held the Los Angeles event, "I realized we had to make a movie out of it," she said. "What's the guy going to do? There are not physically enough hours in the day to travel to every town and city to show this thing" (Revkin 2006, May 22). David helped recruit a team of filmmakers and investors and persuaded Gore to be followed by a film crew.

In late September 2006, Gore endorsed a popular movement in the United States to seek an immediate freeze in greenhouse gases. He made the endorsement during a speech at the New York University Law School.

> Merely engaging in high-minded debates about theoretical future reductions while continuing to steadily increase emissions represents a self-delusional and reckless approach. In some ways, that approach is worse than doing nothing at all, because it lulls the gullible into thinking that something is actually being done, when in fact it is not. (Revkin 2006, September 19)

During 2008, Gore also played a leading role in shaping and financing a $300 million publicity campaign against global warming that ranked among the most expensive public advocacy campaigns in United States history.

Scientists Debate Gore's Work

Some scientists have said Gore's film is by turns inaccurate and alarmist. "I don't want to pick on Al Gore," Don J. Easterbrook, an emeritus professor of geology at Western Washington University, told hundreds of experts at the annual meeting of the Geological Society of America. "But there are a lot of inaccuracies in the statements we are seeing, and we have to temper that with real data" (Broad 2007). In an e-mail exchange about the critics, Gore said his work made "the most important and salient points" about climate change, if not "some nuances and distinctions" scientists might want. "The degree of scientific consensus on global warming has never been stronger," he said, adding, "I am trying to communicate the essence of it in the lay language that I understand" (Broad 2007).

"He's a very polarizing figure in the science community," said Roger A. Pielke Jr., an environmental scientist at the University of Colorado. "Very quickly, these discussions turn from the issue to the person, and become a referendum on Mr. Gore" (Broad 2007). Paul Reiter, a global warming contrarian who directs the insects and infectious diseases unit at the Pasteur Institute in Paris, faulted Gore for alleging that global warming spreads malaria. "For 12 years, my colleagues and I have protested against the unsubstantiated claims," Reiter wrote in the *International Herald-Tribune*. "We have done the studies and challenged the alarmists, but they continue to ignore the facts" (Broad 2007).

Gore does have support from scientists who believe his work is basically sound. During December 2006, Gore spoke in San Francisco at the American Geophysical Union's annual meeting to an audience of several thousand scientists. "He has credibility in this community," said Tim Killeen, the group's president and director of the National Center for Atmospheric Research, a top group studying climate change. "There's no question he's read a lot and is able to respond in a very effective way" (Broad 2007).

James E. Hansen, director of NASA's Goddard Institute for Space Studies and a frequent adviser to Gore, said, "Al does an exceptionally good job of seeing the forest for the trees," adding that Gore often did so "better than scientists" (Broad 2007). Hansen, however, noted some imperfections and technical flaws in Gore's work. For example, he faulted Gore's simplistic association of strengthening hurricanes with warming seawater (many other factors play a role), especially following devastating storms (including Katrina and Rita) in 2005. "We need to be more careful in describing the hurricane story than he is," Hansen said. "On the other hand," Hansen continued, "he has the bottom line right: most storms, at least those driven by the latent heat of vaporization, will tend to be stronger, or have the potential to be stronger, in a warmer climate" (Broad 2007).

Gore defended his work as fundamentally accurate. "Of course," he said, "there will always be questions around the edges of the science, and we have to rely upon the scientific community to continue to ask and to challenge and to answer those questions." He said, "not every single adviser" agreed with him on every point, "but we do agree on the fundamentals—that warming is real and caused by humans" (Broad 2007).

Michael Oppenheimer, a professor of geosciences and international affairs at Princeton University who advised Mr. Gore on his book and documentary, said that reasonable scientists disagreed on the role of rising temperatures as a cause of malaria (and other issues). In general, he said, Gore has distinguished himself with integrity. "On balance, he did quite well—a credible and entertaining job on a difficult subject," Oppenheimer said. "For that, he deserves a lot of credit. If you rake him over the coals, you're going to find people who disagree. But in terms of the big picture, he got it right" (Broad 2007).

Advocacy of Renewable Energy

Gore was not deterred by his critics' stridency. He has said that enough solar energy falls on Earth's surface every 40 minutes to meet 100 percent of the entire world's energy needs for a full year. "Tapping just a small portion of this solar energy could provide all of the electricity America uses" (Gore 2008). "Today I challenge our nation to commit to producing 100 percent of our electricity from renewable energy and truly clean carbon-free sources within 10 years," Gore said. "This goal is achievable, affordable, and transformative. It represents a challenge to all Americans—in every walk of life: to our political leaders, entrepreneurs, innovators, engineers, and to every citizen" (Gore 2008).

Gore developed his proposal in some detail:

To be sure, reaching the goal of 100 percent renewable and truly clean electricity within 10 years will require us to overcome many obstacles. At present, for example, we do not have a unified national grid that is sufficiently advanced to link the areas where the sun shines and the wind blows to the cities in the East and the West that need the electricity. Our national electric grid is critical infrastructure, as vital to the health and security of our economy as our highways and telecommunication networks. Today, our grids are antiquated, fragile, and vulnerable to cascading failure. Power outages and defects in the current grid system cost US businesses more than $120 dollars a year. It has to be upgraded anyway. (Gore 2008)

"The political system, like the environment, is nonlinear," Gore said. "As this pattern [of global warming] becomes ever more clear, there will be a rising public demand for action" (Revkin 2006, May 22). The film concludes with Gore stating that political will is the one element missing in the fight against global warming. "In America, political will is a renewable resource," he said. "This isn't a political film," he said of *An Inconvenient Truth*. "Global warming isn't a political issue. It's about the survival of the planet. Nobody is going to care who won or lost any election when the Earth is uninhabitable" (Svetsky 2006, 32).

Further Reading

Broad, William J. "From a Rapt Audience, a Call to Cool the Hype." *The New York Times*, March 13, 2007. http://www.nytimes.com/2007/03/13/science/13gore.html.

Denby, David. "Review: An Inconvenient Truth." *The New Yorker*, June 19, 2006, 23.

Dionne, E. J., Jr. "Gore's Energy Oomph." *Washington Post*, July 18, 2008. http://www.washingtonpost.com/wp-dyn/content/article/2008/07/17/AR2008071701840_pf.html

Gibbs, Walter, and Sarah Lyall. "Gore Shares Peace Prize for Climate Change Work." *The New York Times*, October 13, 2007, A1, A13.

Gore, Al. "A Generational Challenge to Repower America." July 17, 2008. http://www.realclearpolitics.com/articles/2008/07/al_gores_energy_speech.html.

Kolbert, Elizabeth. "Testing the Climate." *The New Yorker*, December 24 and 31, 2007, 43–44.

Revkin, Andrew C. "'An Inconvenient Truth': Al Gore's Fight against Global Warming." New York *Times*, May 22, 2006. http://www.nytimes.com/2006/05/22/movies/22gore.html.

Revkin, Andrew C. "Gore Calls for Immediate Freeze on Heat-Trapping Gas Emissions." *The New York Times*, September 19, 2006. http://www.nytimes.com/2006/09/19/washington/19gore.html.

Svetsky, Benjamin. "How Al Gore Tamed Hollywood." *Entertainment Weekly*, July 21, 2006, 26–32.

HANSEN, JAMES E. (MARCH 29, 1941–)

If the American president's cabinet were to include a secretary of climate security, James E. Hansen might just be the best nominee. Hansen, who was the longtime director of NASA's Goddard Institute for Space Sciences in New York City, is too much of a scientist—and not enough of a politician—to ever accept such a job, however. Regarding global warming, however, he may be our Paul Revere. Hansen has been an international leader in both the scientific and political worlds since global warming became an important public issue during the 1980s.

As author Mark Bowen wrote in *Thin Ice: Unlocking the Secrets of Climate in the World's Highest Mountains* (2006),

> As testament to [Hansen's] fine character, he never accused his colleagues of pettiness. In the 30 years he has been working on the greenhouse issue, he has been the object of endless scorn and personal attack, yet he has never responded in kind. He has always focused resolutely on the science and the facts and, moreover, gone out of his way to point out the shortcomings in his own arguments. (Bowen 2006, 159)

Hansen enjoys reading fiction with heroes who flout conviction and are persecuted for sticking to their principles. He also maintains an extensive e-mail list of people to whom he sends early drafts of his papers and requests critical review.

Early Life

The fifth of seven children (he has four older sisters), Hansen was born in a farmhouse and raised in Denison, Iowa, about 60 miles northeast of Omaha. His father, a tenant farmer, moved to Denison when Jim was four years old and took up work as a bartender; his mother worked as a waitress. James was known for his intelligence as a young man, even when he was not being overly studious. A lifelong New York Yankees fan, he played second base in the Babe Ruth league.

With a scholarship and money saved from his *Omaha World-Herald* paper route, Hansen attended the University of Iowa and graduated summa cum laude in 1963 after majoring in mathematics and physics. He later earned a master's degree in astronomy.

The University of Iowa was an exciting place to study astronomy; the department had its own satellite, and its chairman was James Van Allen, the man who discovered the Earth-girdling radiation belts named after him. Hansen decided to specialize in the atmosphere of Venus at a time when scientists were discovering that the planet's super-hothouse atmosphere (with temperatures above 850°F) was 95 percent carbon dioxide. He earned a doctorate in 1967 with a dissertation on Venus. "I was so shy and unconfident that when I had an opportunity to take a course under Professor Van Allen, I avoided it because I didn't want him to realize how ignorant I was," Hansen told an audience at his alma mater in 2004 (Johansen 2006, 27).

Career as a Scientist

His doctorate completed, Hansen went to work at NASA's Goddard Institute for Space Studies (GISS), which is affiliated with Columbia University in New York City. On leave for 1969–70 at the Leiden Observatory in the Netherlands, Hansen met his wife, Anniek. During 1976, Hansen worked as principal investigator on the Pioneer Venus orbiter when a Harvard postdoctoral researcher asked him to help calculate the greenhouse effect of human-generated emissions in Earth's atmosphere. Soon Hansen was captivated by potential global warming on Earth.

As he worked at GISS, the institution became more involved in Earth studies, a trend that accelerated after his appointment as director in 1981. Also in 1981, he and several colleagues were the first to use the term "global warming" in a scientific context. Following an important article in *Science* (Hansen et al. 1981, 957–966), Reagan administration functionaries withdrew GISS funding from the Department of Energy.

As has often been the case, Hansen elicited conflict among scientists as well as among politicians as described by Justin Gillis in *The New York Times:*

> The divide is characteristic of the strong reactions that Dr. Hansen has elicited playing dual roles in the debate over climate change and how to combat it. As the head of the Goddard Institute for Space Studies . . . , he is one of NASA's principal climate scientists and the primary custodian of its records of the Earth's temperature. Yet he has also become an activist who marches in protests to demand new government policies on energy and climate. The latter role—he has been arrested four times at demonstrations, always while on leave from his government job—has made him a hero to the political left, and particularly to college students involved in climate activism. But it has discomfited some of his fellow researchers, who fear that his political activities may be sowing unnecessary doubts about his scientific findings and climate science in general. (Hansen 2012)

As he grew older, Hansen became more adamant in his opposition to coal mining and power production as the lynchpin in a damaging degree of human-provoked warming. He came to advocate civil disobedience and was himself arrested several times. Hansen was among 31 protesters arrested for obstructing officers and impeding traffic during a protest mountaintop mining in June 2009 as they tried to enter grounds of Massey Energy, the largest company conducting mountaintop mining in West Virginia. But several hundred miners and relatives and other mining advocates blocked the entrance. He also was arrested several times at the White House for civil disobedience against the Keystone XL pipeline. Hansen senses development of a mass political movement led by young people, to which he gives full support. "At my age," he said, "I am not worried about having an arrest record" (Gillis 2013).

Hansen's retirement from NASA in April 2013 at age 72, after 46 years at GISS summoned a spate of media speculations about his activism and his legacy. "Few other figures in modern science have straddled—and for that matter blurred—the boundaries between science, policy, and advocacy quite like the homespun but outspoken climatologist," wrote Eli Kintisch in *Science* (2013, 540).

Hansen's retirement from NASA did not mean that he was quitting climate science and activism. However, it enabled him to take part in lawsuits against the government, something that had been difficult when he was a NASA employee. From his office in a converted barn at his home in rural Pennsylvania, he has "done the most important science on the most important question that there ever was," said climate activist and author Bill McKibben (Gillis 2013). "Jim has a real track record of being right before you can actually prove he's right with statistics," said Raymond T. Pierrehumbert, a University of Chicago planetary scientist.

Further Reading

Bowen, Mark. *Thin Ice: Unlocking the Secrets of Climate in the World's Highest Mountains.* New York: Henry Holt, 2006.

Gillis, Justin. "Study Finds More of Earth Is Hotter and Says Global Warming Is at Work." *The New York Times*, August 6, 2012. http://www.nytimes.com/2012/08/07/science/earth /extreme-heat-is-covering-more-of-the-earth-a-study-says.html.

Gillis, Justin. "Climate Maverick to Retire from NASA." *The New York Times*, April 1, 2013. http://www.nytimes.com/2013/04/02/science/james-e-hansen-retiring-from-nasa-to -fight-global-warming.html.

Hansen, James E. "Climate Change Is Here—and Worse Than We Thought." *Washington Post*, August 3, 2012. http://www.washingtonpost.com/opinions/climate-change-is-here—and -worse-than-we-thought/2012/08/03/6ae604c2-dd90-11e1-8e43-4a3c4375504a_print .html.

Hansen, Janes E., et al. "Climate Impact of Increasing Atmospheric Carbon Dioxide," *Science* 213 (1981), 957–966.

Johansen, Bruce E. "The Paul Revere of Global Warming." *The Progressive*, August 2006, 26–28.

Johansen, Bruce E. "Hansen Battles Climatic Self-Deception World-Wide." *The Nebraska Report*, October 2008, 7.

Kintisch, Eli. "Hansen's Retirement from NASA Spurs Look at His Legacy." *Science* 340 (May 3, 2013): 540–541.

Revkin, Andrew C. "NASA's Goals Delete Mention of Home Planet." *The New York Times*, July 22, 2006, A1, A10.

INHOFE, JAMES M. (1934–)

James Mountain Inhofe, senior Oklahoma senator in the U.S. Senate and chair of the Senate Committee on Environment and Public Works (2003–2007 and again beginning in 2015) may be the world's most prominent denier of climate change, as well as the science supporting it.

Early Life and Career

Inhofe was born November 17, 1934, in Des Moines, Iowa, the son of Perry Dyson Inhofe and Blanche (née Mountain). The family moved to Tulsa, Oklahoma, when James was a child. He attended Tulsa Central High School, graduating in 1953. He served in the U.S. Army between 1956 and 1958 and obtained a BA degree in economics from the University of Tulsa in 1973 at age 39. He had a checkered career as a businessman, including a leading role in a small firm, the Quaker Life Insurance Company, that later went into receivership and was liquidated in 1986.

Inhofe served in both the Oklahoma House of Representatives (1967–1969) and Senate (1969–1977). In 1974, he ran for governor of Oklahoma but lost decisively

(64 percent to 36 percent) to David Boren. He ran for the U.S. House first in 1976 and lost to James R. Jones, 54 percent to 46 percent. In 1978, he was elected mayor of Tulsa and served three two-year terms. Inhofe had purchased an abandoned building during the late 1970s and wanted to move a fire escape to an outside wall. A city official delayed his project. "I told him that I was going to run for mayor and fire him," Inhofe wrote in his book. "So I ran for mayor and I fired him" (Terris 2015). As mayor, Inhofe supported a tax increase, as well as solar and wind power. He also appointed Tulsa's first black city commissioner.

Ties to Fossil Fuels

As time went on, however, Inhofe shed his liberal inclinations as he accepted a growing amount of money from oil and gas interests. In 2008, a typical year, Inhofe received $446,900 in donations from the oil and gas industry in a state where a quarter of the jobs are tied to fossil fuel production. He also holds a lifetime service award from the Oklahoma Independent Petroleum Association. The oil and gas industry has remained Inhofe's top campaign funding source most of his political life. "Anytime someone asks me how much money I get from the oil industry, I always tell them the same thing," he said, smiling. "Not enough" (Terris 2015).

Between 1986 and 1994, Inhofe served four terms as representative from Oklahoma's first congressional district. Inhofe then assumed office in the U.S. Senate in 1994, winning the seat of Boren, who had resigned in the middle of his term to become president of the University of Oklahoma. Inhofe was reelected to a fourth full term in 2014 with 68 percent of the vote, the largest margin of his Senate career.

Inhofe married Kay Kirkpatrick, who was born to a staunchly Democratic family, and the couple had four children (and 15 grandchildren by 2014). By 2015, they had been married 55 years. Inhofe commutes home to Oklahoma by air most weekends when the Senate is in session. As a pilot, Inhofe had logged more than 11,000 flight hours as of 2014. He is also the only member of the House or Senate to fly an airplane around the world; he even re-created the original globe-girdling trip of famous aviator Wiley Post.

Inhofe has become steadily more right-wing as he has gotten older. In 2015, his official Senate biography pointed with pride to his ranking by the right-wing magazine *Human Events* as one of the Senate's "Top 10 Most Outstanding Conservative Senators." The publication ranked Inhofe number one, saying he is an "unabashed conservative" and that "he's unafraid to speak his mind" ("About Senator Inhofe" no date). As of 2013, *National Journal* ranked Inhofe among the five most conservative members of the Senate.

Blasting "Warmists" as Nazis

On environmental issues, Senator Inhofe gives no quarter. He is fond of comparing the environmentalist movement to the Nazis and the Environmental Protection Agency to the Gestapo. Despite the absolute nature of his opinions, Inhofe is known for his affable and sometimes self-deprecating good humor with political figures of diverse backgrounds. He is a good friend of Bernie Sanders, Vermont's socialist senator and former presidential candidate, as well as liberal Senator Barbara Boxer of California, who retired from Congress when her term ended in 2017.

On the stump, Inhofe is a tireless opponent of "big government," but at home he points with pride to his role in ending Oklahoma's role as a "donor state" (e.g., one that pays more in taxes than it gets in federal appropriations). Oklahoma "now receives more money than it sends to Washington in federal highway funding." He also points to his role as "a strong supporter of Oklahoma's defense industry and military community" with as much federal money as possible ("The Arena Profile" no date). His opposition to wasteful spending in the other 49 states has earned Senator Inhofe the National Taxpayers Association's "Friends of the Taxpayer" award and the "Hero of the Taxpayer Award" from Americans for Tax Reform. He also enjoys an A+ rating from the National Rifle Association.

Confronted with the fact that nearly all climate scientists support the case that humanity is having an impact on global warming, Inhofe calls on his conception of deity and his professed distrust of government-sponsored research, and says with ritualistic repetition that that his opponents' case is "predicated on fear rather than science." Inhofe authored *The Greatest Hoax: How the Global Warming Conspiracy Threatens Our Future* (2012). To Inhofe, climate change science invokes a conspiracy to overregulate private business.

Reviews of Inhofe's book have been starkly split. By January 2015, 53 of 151 reviewers on Amazon.com gave Inhofe's book five out of five stars. Eighty, however, gave him one star, the lowest rating. One reviewer, who signed as "Clear to Take Off," said the book was

> utterly despicable [and] ignorant. The senator should be ashamed, held on charges for crimes against humanity and crimes against the planet. Just ask any Katrina, Sandy victim, Texas or California farmer or family suffering from 80 percent exceptional drought. . . . The "book" is a sewer of foul lies, falsehoods, misrepresentations, and foul propaganda from beginning to end. . . . Where are the publications in *Nature, Science, Proceedings of the National Academy of Sciences* [?]. . . . This book is a hoax.

On Amazon.com on December 24, 2014, however, "Linda" wrote, "James Inhofe is truly an American Hero! He laid out precisely what he did to defeat the climate change politicians. Also taught me how government works or not. Makes me wonder why the left keeps pushing the climate change agenda. . . ."

"He likes to say that we may just be in a cycle of relative warming, and that as recently as the 1970s we were being warned of a looming ice age," Ben Terris wrote in the *Washington Post* (2015). "Plus, as he writes in his book, 'God is still up there, and He promised to maintain the seasons and that cold and heat would never cease as long as the earth remains.'" To Inhofe, God created resources for humans to use, and the senator invokes the Bible's book of Genesis, which says that Adam and Eve were called forth from the Garden of Eden to multiply and subdue the Earth. "The arrogance of people to think that we, human beings, would be able to change what

He is doing in the climate is to me outrageous," he said in a 2012 radio interview (Terris 2015).

Further Reading

"About Senator Inhofe." U.S. Senate Official Biography. No date. http://www.inhofe.senate .gov/biography. Accessed January 10, 2015.

"The Arena Profile: Senator James Inhofe." Politico.com. No date. Accessed January 11, 2015. http://www.politico.com/arena/bio/sen_james_inhofe.html.

"Clear to Take Off." Review of Inhofe, *The Greatest Hoax*, Amazon.com. September 4, 2014. http://www.amazon.com/The-Greatest-Hoax-Conspiracy-Threatens/dp/1936488493.

Inhofe, James M. *The Greatest Hoax: How the Global Warming Conspiracy Threatens Our Future.* New York: WND Books, 2012.

"Linda." Review of Inhofe, *The Greatest Hoax*, Amazon.com. December 24, 2014. http://www .amazon.com/The-Greatest-Hoax-Conspiracy-Threatens/product-reviews/1936488493 /ref=cm_cr_dp_synop?ie=UTF8&showViewpoints=0&sortBy=bySubmissionDateDesce nding#R3PEM77NQ445GW (no longer available).

Terris, Ben. "Jim Inhofe Is a Small-Plane-Flying, Global-Warming-Denying Senator. And Now He's Got a Gavel." *Washington Post*, January 9, 2015. http://www.washingtonpost .com/lifestyle/style/the-senates-top-climate-change-denier-is-flying-high-with-a -committee-chairmanship/2015/01/07/bf625fe0-9365-11e4-ba53-a477d66580ed _story.html.

KEELING, CHARLES D. (APRIL 20, 1928–JUNE 20, 2005)

Until the late 1950s, scientists had no reliable records of carbon dioxide and other greenhouse gas levels in the atmosphere, nor did they even have an instrument that could accurately measure the amount of CO_2 in the air. To measure such concentrations in the atmosphere, someone first had to construct a machine to obtain readings in parts per million. Charles D. Keeling worked on his "manometer" for a year using an old blueprint at the California Institute of Technology in Pasadena. His first readings on the roof of a Caltech laboratory showed an atmospheric concentration of 310 parts per million. Furthermore, his calculations indicated that the level was continuing to rise. It reached 385 ppm by 2008 and 400 ppm by 2015. The historic graph of Earth's atmospheric carbon dioxide level came to be called the "Keeling curve."

Keeling took his manometer on family vacations to record concentrations in several different places. What he found contradicted scientific assumptions of the time, which held that carbon dioxide levels would vary widely, depending on local sources of the gas. Instead, he found that carbon dioxide readings were very similar in different places. "I decided all the data in the literature was wrong," Keeling later told William K. Stevens, a science reporter for *The New York Times* (Stevens, 1999, 140). After Keeling and his associates recorded data for more than a year, he discovered that the readings on his manometer were rising steadily year by year.

Keeling also was discovering that atmospheric carbon dioxide levels vary annually, with lower readings in spring and summer (when plants are transpiring oxygen on the large land masses of the Northern Hemisphere), and higher levels during fall and winter when many plants in the same regions are dormant and more carbon is being released through vegetative decay. This annual cycle can vary by as much as 3 percent in the Northern Hemisphere and 1 percent in the Southern Hemisphere, where the dominance of much larger oceans mutes seasonal cycles and restricts seasonal variability.

Until Keeling showed that CO_2 levels were rising, most scientists who studied carbon dioxide levels in the atmosphere had assumed that the oceans absorbed all of the extra carbon injected into the air by human activities. The readings of Keeling and his associates showed that human activity was steadily raising the proportion of carbon dioxide in the atmosphere more quickly than the oceans or other "sinks" of the gas could absorb it.

Further Reading

Keeling, Charles D., and Timothy P. Whorf. "The 1,800-year Oceanic Tidal Cycle: A Possible Cause of Rapid Climate Change." *Proceedings of the National Academy of Sciences of the United States of America* 97(8) (April 11, 2000): 3814–3819.

Stevens, William K. *The Change in the Weather: People, Weather, and the Science of Climate.* New York: Delacorte Press, 1999.

LOMBORG, BJORN (JANUARY 6, 1965–)

Bjorn Lomborg, a Danish economist who wrote *The Skeptical Environmentalist* (2001) and *Cool It: The Skeptical Environmentalist's Guide to Global Warming* (2007), became an unusual figure in the climate change debate who marketed himself as *both* a climate contrarian *and* an environmentalist. Following the publication of *The Skeptical Environmentalist, TIME* magazine put Lomborg on its list of 100 most influential people in 2004. In 2015, he was director of the Copenhagen (Denmark) Consensus Center. Lomborg also has been named one of *Foreign Policy* magazine's top 100 global thinkers.

Lomborg also has written in the *Wall Street Journal, The New York Times,* the *Washington Post, USA Today,* and *The Guardian* (United Kingdom), among other outlets.

Life and Career

Lomborg is gay and a vegetarian. He is active in both as political causes. "Being a public gay is to my view a civic responsibility. It's important to show that the width of the gay world cannot be described by a tired stereotype, but goes from leather gays on parade-wagons to suit-and-tie yuppies on the direction floor, as well as everything in between" ("A OBLS Personer" 2007).

Born January 6, 1965, in Frederiksberg, Denmark, near Copenhagen, Lomborg earned a PhD in political science at the University of Copenhagen in 1994. He was an assistant and associate professor in the Department of Political Science at the University of Aarhus between 1994 and 2005. During the 1990s, he was active in Greenpeace but became critical of its views. His new analysis emphasized the cost and complexity of environmentalists' proposals to reduce atmospheric greenhouse gases. He suggested adaptation instead. His work has been severely criticized in *Nature* (Pimm and Harvey 2001, 149) and *Scientific American* (Schneider et al. 2002).

On Warming, Water, and Melting Glaciers

Lomborg dismisses fears that glaciers are melting too quickly, arguing that we should expect them to dissolve when Earth exited the Little Ice Age in the middle of the 19th century. He asserts that Mount Kilimanjaro's shrinking snows are a result of a drying climate, not warming temperatures (this assertion has been widely advertised by many contrarians), and that some of the glaciers in the Hindu Kush have been growing slowly, not shrinking. Lomborg ignores the *reason* why some higher-elevation glaciers are growing; as temperatures rise, heavier, moisture-laden snow falls higher on mountainsides. This is also a symptom of warming, the same reason why some of the heaviest lake-effect snows downwind of the Great Lakes have occurred during the last decade.

Likewise, on the subject of water, Lomborg writes, "Global warming will mean more precipitation and more water availability for more people" (Lomborg 2007, 111). As is so often the case, Lomborg is superficially correct because warmer air on average holds more moisture. The phrase "for more people" (and for what purpose) is problematic. Lomborg admits to some problems with access to water in a warmer world but stops there. Outside Lomborg's world, it is widely believed that additional water often will fall in useless (or devastating) deluges such as the 37-inch flood that inundated Mumbai (Bombay) on one day and night during the 2005 Indian monsoon. Lomborg also ignores the fact that intensifying drought aggravated by heat can occur even in a world in which average precipitation rises.

Debates Regarding Lomborg's Work

According to a British author who took a close look at a sample of Lomborg's sources, *Cool It!* flunks some of the most rudimentary tests of scholarly verification. The following critique of *Cool It!* was prepared by Chris Goodall, author of *How to Live a Low-Carbon Life: The Individual's Guide to Stopping Climate Change*, winner of the 2007 Clarion prize for nonfiction.

"Bjorn Lomborg has many strengths, but accuracy is not one of them," Goodall said (no date). "I looked at 27 footnoted sources. In 10 cases, he has taken a sentence from the source and copied it into his text without using quotes. He has strengthened the statement on three of these. In two instances, he has actually copied a sentence from a source that is not the one specified in the footnote" (Goodall no date). Goodall analyzed only footnotes 328 to 358 in the English edition of *Cool It!*, which is a tiny fraction of the total, and concluded that in this section of Lomborg's book, more than two-thirds of Lomborg's attributions contained at least one factual error or plagiarized another person's work.

Lomborg advises global warming advocates to temper their assertions, but he is no slouch with the rhetorical hot brand. The bombastic quotient of his own work is hardly temperate. The many people who do not agree with Lomborg are said by him to be "jumping on the bandwagon of catastrophe," "sexing up the ramifications of global warming," and "exploiting fear of disaster," among other things (Lomborg 2007, 131). Lomborg also has an annoying habit of dropping names off of incendiary quotes that beg for precise citation. At one point, for example, he quotes "one climate scientist" who has wondered whether "we have created a monster" (Lomborg 2007, 129). A trip to the page references produces a reference to a newspaper story in the *Houston Chronicle*, not to the quoted but unnamed scientist.

Sometimes the weather wars get downright visceral. On page 143 of *Cool It!*, Lomborg writes that British author and climate activist Mark Lynas threw a pie in his face to promote a book while Lomborg was appearing at an Oxford bookshop in Lynas's home town. He used Lynas as an example of prowarming rhetorical extremism, quoting him as comparing climate contrarians to deniers of the Jewish Holocaust who will be held to account someday in Nuremburg-style trials.

Lynas said of this:

Publicity stunt it may have been, but it wasn't in order to publicise my book (which only came out four years later), it was to challenge his. In any case, I was younger then, and a little more hotheaded. It probably backfired against me because the episode then became part of Lomborg's narrative that "environmentalists don't listen, they just attack," which I should have foreseen. Still, a lot of people heard about it, and I guess it demonstrated the level of opposition Lomborg's book sparked—I was even congratulated by a government official . . . in Tuvalu, a couple of years later! Also, Bjorn Lomborg and I have actually done a face-to-face debate subsequent to that episode in the Science Museum's Dana Center in London. Neither of us mentioned the pie incident then, so I'm surprised he does now. (Lynas 2007)

Lomborg wrote in a 2013 op-ed for *The New York Times*, by which time he was directing the Copenhagen Consensus Center, described in his piece as "a nonprofit group focused on cost-effective solutions to global problems." In this piece, Lomborg was bearing the banner of the world's poor, the 1.2 billion people who have little or no electricity and cook with dung and wood. "The Poor Need Cheap Fossil Fuels," he wrote, at least until more ecologically compatible alternatives exist.

Further Reading

Goodall, Chris. *How to Live a Low-Carbon Life: The Individual's Guide to Stopping Climate Change*. London: Earthscan Publications, 2007.

Goodall, Chris. Analysis of 27 footnoted sources in Bjorn Lomborg's *Cool It!* No date. http://www.carboncommentary.com/2007/10/15/29 (no longer available). Accessed October 19, 2007.

Lomborg, Bjorn. *The Skeptical Environmentalist*. Cambridge, UK: Cambridge University Press, 2001.

Lomborg, Bjorn. *Cool It!: The Skeptical Environmentalist's Guide to Global Warming*. New York: Knopf (Random House), 2007.

Lomborg, Bjorn. "The Poor Need Cheap Fossil Fuels." *The New York Times*, December 3, 2013. http://www.nytimes.com/2013/12/04/opinion/the-poor-need-cheap-fossil-fuels.html.

Lynas, Mark. Personal communication, October 19, 2007.
"A OBLS Personer: Bjorn Lomborg." Danmarks Radio. June 12, 2007. http://www.dr.dk/obls
   /personer-lomborg.html (no longer available).
Pimm, Stuart, and Jeff Harvey. "No Need to Worry about the Future." *Nature* 414 (November
   8, 2001): 149.
Schneider, Stephen, et al." Misleading Math about the Earth." *Scientific American*, January
   2002. http://www.scientificamerican.com/article/misleading-math-about-the/.

McKIBBEN, BILL (1960–)

Author, activist, and worldwide agitator for a sustainable Earth, Bill McKibben has been the driving force behind the organization of the global climate change movement that came to be called 350.org. The name was acquired in 2008 after he asked James E. Hansen, then director of NASA's Goddard Institute for Space Studies, what the maximum sustainable level of atmospheric carbon dioxide would be. His answer: 350 parts per million. The fight against climate change is "as morally compulsory as the battles for civil rights or against totalitarianism," McKibben has written (Rustin 2010).

"If humanity wishes to preserve a planet similar to that on which civilization developed and to which life on Earth is adapted, paleoclimate evidence and ongoing climate change suggest that CO_2 will need to be reduced from its current 385 ppm [400 in 2015] to at most 350 ppm, but likely less than that," Hansen had written (Hansen et al. 2008). The name was acquired shortly before that level passed 400 ppm, lending urgency to the campaign. His first book on the subject, *The End of Nature* (1989), had been published in in 24 languages as of 2015.

McKibben has been Schumann Distinguished Scholar in Environmental Studies at Middlebury College in Vermont, as well as a fellow of the American Academy of Arts and Sciences. He won the Gandhi Prize in 2013, holds the Thomas Merton Prize, and has been granted honorary degrees from 18 colleges and universities (as of 2015). The journal *Foreign Policy* included McKibben on a list of the world's 100 most important thinkers. McKibben also won a Guggenheim Fellowship in 1983.

As an activist, McKibben has become well acquainted with the Washington, D.C., city jail, having been arrested several times protesting the Keystone XL pipeline (with James Hansen and many others). His 350.org specializes in coordinated demonstrations across borders on common themes. In 2009, for example, 5,200 demonstrations were held in 181 countries (and communicated via the Internet). A year later, a global work party was held (on October 10, 2010—10/10/10) with more than 7,000 people in 188 countries again linked by Internet. Participants in the rallies posted 30,000 photographs. In 2011 and 2012, 350.org helped coordinate demonstrations in Washington, D.C., against the Keystone XL pipeline that filled the local jail with hundreds of people, including McKibben and Hansen.

McKibben calls global warming the

first civilization-scale challenge that humans have yet faced. Newly emerging science shows that we have underestimated the scale and urgency of the crisis. Everything frozen on Earth is melting fast, for instance, threatening to produce an inhospitable planet in the decades ahead and an unbearable one in the lifetime of those being born. Political rhetoric needs to reflect the stark fact that this is an emergency. (McKibben 2006).

"Writers," he says, "almost by definition are people who aren't so good at communicating in other ways. What I would like to do is spend my time in this small study, writing." Instead, he spends most of his life travelling (Rustin 2010). McKibben's role has made him a world traveler whose carbon footprint is "crazily high," even though most of 350.org's work is done by local people with computer and phone links.

Early Life and Career

William Ernest McKibben was born December 8, 1960, in Palo Alto, California. He attended Harvard University and married writer Sue Halpern. His daughter, Sophie McKibben, was born in 1993. His family lives in Vermont, making a home in the mountains above Lake Champlain.

McKibben was raised in Lexington, Massachusetts, near Boston, through high school. His father was a business editor at the *Boston Globe* and a Vietnam war veteran who had protested the war and been arrested. "The Vietnam Veterans Against the War wanted to camp out on Lexington Green. The police told them they couldn't and so hundreds of townspeople joined them and were arrested, and my father was among them," McKibben recalled later (Rustin 2010). McKibben's father came from the Pacific Northwest, to which the family returned to vacation each year for "treasured times" in a cabin lacking electricity on the slopes of Mount Rainier (Rustin 2010).

The younger McKibben wrote for Lexington's newspaper and took part in competitive debates at a statewide level. He edited the *Harvard Crimson*, the university's student newspaper, after he began attendance there as a student in 1978. He decided to concentrate on environmental issues while attending Harvard, from which he graduated in 1982. After that, McKibben wrote for *The New Yorker* during much of the 1990s, as he studied the philosophy of nonviolent protest. Writing a story on the lives of homeless people, McKibben decided to spend time living on the streets, where he met Susan Halpern, author and homeless advocate. The two later married.

After McKibben resigned from *The New Yorker*, he moved to the Adirondack range in upstate New York to freelance, becoming more intensely involved in the politics and science of climate change. In the meantime, he developed a broad list of freelance contacts, including *The New York Times*, *The Atlantic*, *Mother Jones*, *Harper's*, *Orion*, the *Washington Post*, *Rolling Stone*, *Granta*, and many others. His first book, *The End of Nature* (1989) was serialized in *The New Yorker*. In addition to addressing climate change, McKibben has written a wide variety of books and articles on many subjects ranging from frugal living to training for athletic competition and endurance events as well as solo hiking. He also edited an anthology of environmental and nature writing, *American Earth*, in the United States since the early 19th century.

Life as an Activist

Traveling the world, has McKibben stressed the central role of coal in raising carbon dioxide emissions. In Omaha, he protested with Nebraskans for Peace at the headquarters of United Pacific, one of the world's largest coal-hauling railroads. McKibben opened the protest by stressing that "coal is the dirtiest of fossil fuels, emitting more carbon dioxide than natural gas or oil" as a small number of Tea Party demonstrators waved signs in favor of coal power (Ruff 2010). McKibben and environmentalist Wendell Berry called for massive protest at the Capitol Power Plant in Washington, D.C., in March 2009.

McKibben is an expert at putting issues in human terms. For example: motor vehicles in the United States account for 40 percent of the country's oil consumption, and 10 percent of the world's (Kolbert 2007, 88, 90). According to McKibben, "The average American car, driven the average American distance—ten thousand miles—in an average . . . year releases its own weight in carbon into the atmosphere" (McKibben 1989, 6). He also aims to personalize the situation in other countries, uniting a worldwide constituency.

Bill McKibben also skilled at weaving patterns from worldwide climatic news. He paired tornadoes in Joplin, Missouri, and Tuscaloosa, Alabama, in 2011, for example, with a record wave of wildfires across Oklahoma and Texas that brought drought, deluge, and destruction. Many evenings the lead story on the evening television news is the weather. "It's far smarter to repeat to yourself the comforting mantra that no single weather event can ever be directly tied to climate change," McKibben wrote in the *Washington Post*:

> There have been tornadoes before, and floods—that's the important thing. Just be careful to make sure you don't let yourself wonder why all these record-breaking events are happening in such proximity—that is, why there have been unprecedented mega-floods in Australia, New Zealand and Pakistan in the past year. Why it's just now that the Arctic has melted for the first time in thousands of years. No, better to focus on the immediate casualties, watch the videotape from the store cameras as the shelves are blown over. Look at the news anchorman standing in his waders in the rising river as the water approaches his chest. (McKibben 2011)

German philosopher Friedrich Nietzsche once said, "The Earth is a beautiful place, but it has a pox called man" (Browne 1998). McKibben's thesis in *The End of Nature* (1989) is similar: "We have built a greenhouse, a human creation, where once bloomed a sweet and wild garden" (McKibben 1989, 91). "If industrial civilization is ending nature, it is not utter silliness to talk about ending—or, at least, transforming—industrial civilization," he wrote (1989, 186). The "inertia of affluence, the push of poverty, [and] the soaring population" make McKibben pessimistic about humankind's chances of averting a general ruination of Earth under greenhouse conditions (McKibben 1989, 204).

Further Reading

Browne, Malcolm W. "Will Humans Overwhelm the Earth? The Debate Goes On." *New York Times*, December 8, 1998, n.p.

Hansen, James E., et al. "Target Atmospheric CO_2: Where Should Humanity Aim?" *Open Atmospheric Science Journal*, 2 (2008): 217–231.

Kolbert, Elizabeth. "Running on Fumes." *The New Yorker*, November 5, 2007, 87–90.

McKibben, Bill. *The End of Nature*. New York: Random House, 1989.

McKibben, Bill. "Welcome to the Climate Crisis: How to Tell Whether a Candidate Is Serious about Combating Global Warming." *Washington Post*, May 27, 2006, A25. http://www.washingtonpost.com/wp-dyn/content/article/2006/05/26/AR2006052601549_pf.html.

McKibben, Bill. "A Link between Climate Change and Joplin Tornadoes? Never!" *Washington Post*, May 23, 2011. http://www.washingtonpost.com/opinions/a-link-between-climate-change-and-joplin-tornadoes-never/2011/05/23/AFrVC49G_print.html.

Ruff, Joe. "U[nion] P[acific] Target of Global Warming Protest." *Omaha World-Herald*, June 17, 2010. http://www.omaha.com/article/20100617/MONEY/100619732 (no longer available).

Rustin, Susanna. "A Life in Writing: Bill McKibben." *The Guardian* (U.K.), December 3, 2010. http://www.theguardian.com/books/2010/dec/06/bill-mckibben-interview.

REVELLE, ROGER (1909–1991)

As part of the International Geophysical Year in 1957, Roger Revelle and Hans Suess warned, "Human beings are now carrying out a large-scale geophysical experiment of a kind that could not have happened in the past nor be reproduced in the future. Within a few centuries we are returning to the atmosphere and oceans the concentrated organic carbon stored in sedimentary rocks over hundreds of millions of years" (Christianson 1999, 155–156).

Revelle would become known to the world some years later as mentor of a graduate student, Albert Gore, who was elected vice president of the United States in 1992 (a year after Revelle died). The same year, Gore published a book, *Earth in the Balance*, that argued for mitigation of the greenhouse effect. Gore went on to lose a close contest for the U.S. presidency in 2000. After his electoral loss, he became a worldwide publicist regarding global warming, winning a Nobel Prize for Peace in 2007 and an Academy Award for his documentary film *An Inconvenient Truth*.

"Grandfather" of the Greenhouse Effect

Revelle, who became known as the "grandfather" of the greenhouse effect, was born in Seattle on March 7, 1909, into a family with both Huguenot (on his father's side) and Irish ethnicity (on his mother's side). His parents—William Roger Revelle, an attorney and schoolteacher, and Ella Robena Dougan Revelle, also a schoolteacher—both had graduated from the University of Washington.

After he was admitted to Pomona College at age 16, Revelle entertained thoughts of a career in journalism. However, under the influence of charismatic professor Alfred Woodford, Revelle became interested in geology and, after receiving his BA in 1929, spent a year of additional study with Woodford. Revelle entered graduate studies at the University of California–Berkeley in 1930 under the tutelage of geologist George Louderback, who stimulated his interest in marine sedimentation. In 1936 Revelle, completed his doctorate at Berkeley.

Called to active duty in the U.S. Navy six months before the attack on Pearl Harbor, Revelle was assigned to oceanographic research applied to wartime needs as an officer in the U.S. Navy. Near the end of World War II, he was assigned to Joint Task Force One, the military command supervising the first postwar atomic test on Bikini Atoll in Operation Crossroads. Revelle organized the Crossroads scientific program, which included a study of the diffusion of radionuclides in the atoll, as well as radioactivity's impact on marine life. What he learned through these studies and others made Revelle a lifelong opponent of nuclear weapons. In 1951, Revelle became director of the Scripps Institute of Oceanography in La Jolla, California.

Career

In 1964, Revelle accepted an appointment as Richard Saltonstall Professor of Population Policy at Harvard University, where he served as Director of the Center for Population Studies until 1974. He held that endowed chair at Harvard until 1978. Revelle also was sympathetic to the book *Silent Spring* by Rachel Carson, published in 1962, which argued that ecosystems were being disrupted by the use of halogenated hydrocarbon pesticides such as DDT.

Revelle played a key role in the 1970 creation of the Scientific Committee on Problems of the Environment of the International Council of Scientific Unions (ICSU). He also suggested the objective for the ICSU's International Geosphere-Biosphere Program in 1986: "To describe and understand the interaction of the great global physical, chemical, and biological systems regulating planet Earth's favorable environment for life, and the influence of human activity on that environment" (Malone et al. no date).

Revelle served as president of the American Association for the Advancement of Science during 1974. In 1976, he returned to the University of California–San Diego to become a professor of science and public policy. He received the National Medal of Science in 1991 "for his pioneering work in the areas of carbon dioxide and climate modification, oceanographic exploration presaging plate tectonics, and the biological effects of radiation in the marine environment, and studies for population growth and global food supplies." To a reporter asking why he received the medal, Revelle said, "I got it for being the grandfather of the greenhouse effect" (Malone et al. no date).

Further Reading

Christianson, Gale E. *Greenhouse: The 200-Year Story of Global Warming*. New York: Walker & Co., 1999.

Malone, Thomas F., Edward D. Goldberg, and Walter H. Munk. "Roger Randall Dougan Revelle, 1909–1991." Reading Room. No date. http://www.nap.edu/readingroom/books /biomems/rrevelle.html.

Revelle, R., and H. E. Suess. "Carbon Dioxide Exchange between Atmosphere and Ocean and the Question of an Increase of Atmospheric CO_2 during the Past Decades." *Tellus* 9 (1957): 18–27.

WATT-CLOUTIER, SHEILA (1953–)

As international chair of the Inuit Circumpolar Conference (ICC) from 2002 to 2006, Sheila Watt-Cloutier brought Inuit struggles against persistent organic pollutants and global warming to the world. As president of the ICC in Canada and vice president of the ICC before 2002, she was a key figure in negotiating the Stockholm Convention (2001) that banned synthetic pollutants such as polychlorinated biphenyls (PCBs) and dioxins from lower latitudes because of the threat they pose to the lives of all living things in the Arctic, including her indigenous people. Forms of dioxin and PCBs have become so prevalent in the Arctic that they have been biomagnified up the food chain to the point where Inuit mothers cannot breast-feed their babies without endangering their health.

Formed in 1977, the Inuit Circumpolar Conference represents the common interests of roughly 155,000 Inuit in northern Canada, Greenland, Alaska, and Russia. As ICC head, Watt-Cloutier ranged the world on jets trying to protect her people, many of whom are only one generation off the land, from the effluvia of the industrialized world. Her goal, as she told a world conference on climate change in Milan, Italy, in December 2003, is to "bring a human face to these proceedings." She travels the world injecting "life," as she phrases it, into international negotiations with the delicacy of a diplomat and the verve of a social activist (Watt-Cloutier 2003).

Personal Life

Watt-Cloutier was born in Kuujjuaq, northern Quebec, December 2, 1953, to Daisy Watt (1921–2002), who was a well-regarded Native elder, as well as a musician, interpreter, and healer. Charlie Watt, a brother of Sheila's, served in the Canadian federal Senate during the 1980s. After being raised traditionally and often traveling by dog sled, Watt-Cloutier was taken from her home to attend boarding schools from age 10 in Churchill, Manitoba, and Nova Scotia. She later attended Montreal's McGill University during the middle 1970s, specializing in education, human development, and counseling. She also displayed her mother's gift for interpretation in Inuktitut at Ungava Hospital, the beginnings of a 15-year career in Inuit health care.

In 1992, Watt-Cloutier critiqued northern Quebec's educational system in a report for the Nunavik Educational Task Force titled *Siatunirmut: The Pathway to Wisdom*. She also worked in land claims negotiations that led to establishment of Nunavut as a semisovereign Inuit Canadian province. Her new career as an Inuit political leader and environmental advocate began during 1995 after she was elected to chair Canada's division of the ICC, to which she was reelected in 1998. She also contributed to academic and ecological studies on the Arctic.

Watt-Cloutier's own life illustrates how the Inuit struggle to maintain some semblance of tradition in a swiftly changing, melting, and now frequently polluted ice world. Watt-Cloutier watches global warming change the climate from her home overlooking Frobisher Bay in Iqaluit on Baffin Island. The town's name means "fishing place," an indication of the importance of sustenance from the sea in the traditional Inuit diet. Nunavut means "our home" in the Inuktitut language. Nunavut, a territory four times the size of France, has a population of roughly 25,000, 85 percent of whom are Inuit.

Watt-Cloutier also has been active in promoting sustainable development, retention of traditional ecological knowledge, and other forms of education. In 2007, she was nominated for the Nobel Prize for Peace for her ability to extend what were largely defined as environmental issues to threats to Inuit culture, health, and traditional ways of life. She also played a leading role in suing the United States for negligence regarding global warming in international legal tribunals, including the Organization of American States.

Watt-Cloutier became well known around the world as an adroit diplomat and charismatic speaker. Sometimes, she brought diplomats to tears as she described the effects of persistent organic pollutants (POPs) and global warming on Inuit peoples. She also extended cooperation among Inuits around the Arctic from Canada to Greenland, Alaska, and Russia.

By 2007, Watt-Cloutier had two grown children and a grandson. Displaying her grandmother's musical talents. Watt-Cloutier's daughter is a traditional Inuit throat singer and dancer; her son is a pilot who became the youngest pilot hired by Inuit Air (Kafarowski 2007).

"We Feel Like an Endangered Species"

The toxicological and climatic bills for modern industry at lower latitudes are being left on the Inuits' table in Nunavut. Native people whose diets consist largely of sea animals (whales, polar bears, fish, and seals) have been consuming a concentrated toxic chemical cocktail. Abnormally high levels of dioxins and other industrial chemicals are being detected in Inuit mothers' breast milk.

"As we put our babies to our breasts we are feeding them a noxious, toxic cocktail," said Watt-Cloutier. "When women have to think twice about breast-feeding their babies, surely that must be a wake-up call to the world" (Johansen 2000, 27). More than mother's milk is at stake here. "Greenland has no trees, no grass, no fertile soil," wrote Marla Cone in *Silent Snow* (2005), "which means no cows, no pigs, no chickens, no grains, no vegetables, no fruit. In fact, there is little need for the word 'green' in Greenland. The ocean is its food basket" (Cone 2004).

A tourist with no interest in environmental toxicology might find the Inuits' Arctic homeland as pristine as ever during its long, snow-swept winters. Many Inuit still guide sleds onto the pack ice surrounding their Arctic island homelands to hunt polar bears and seals. Such a scene may seem unspoiled—until one realizes that the body fats in both polar bears and seals are laced with dioxins and PCBs. Geographically, the Arctic could not be in a worse position for toxic pollution because a ring of industry in Russia, Europe, and North America pours out pollutants that are carried northward on prevailing winds. Arctic ecosystems and indigenous communities are particularly at risk because POPs biomagnify—that is, they increase in potency by several orders of magnitude—with each step up the food chain.

As Watt-Cloutier noted,

Imagine for a moment the shock of the Inuit as they discovered that what has nourished them for generations, physically and spiritually, is now poisoning them. Some Inuit now question whether they should eat locally gathered food; others ask whether it is safe to breast-feed their infants. What sort of public outcry and government action would there be if the same levels of POPs found among Inuit were found in women in Toronto, Montreal, or Vancouver as a result of eating poultry or beef? (Watt-Cloutier and Fenge 2000)

Persistent organic pollutants have been linked to cancer, birth defects, and other neurological, reproductive, and immune system damage in people and animals. At high levels, these chemicals also damage the central nervous systems of human beings and other animals. Many of them also act as endocrine disrupters, causing deformities in sex organs as well as long-term dysfunction of reproductive systems. These chemicals also can interfere with the function of the brain and endocrine system by penetrating the placental barrier between mother and unborn child, scrambling the instructions of the naturally produced chemical messengers. The latter tell a fetus how to develop in the womb and postnatally through puberty. If such interference occurs, the immune, nervous, and reproductive systems may not develop according to the contents of the genes inherited by the embryo.

"At times," said Cloutier, "We feel like an endangered species. Our resilience and Inuit spirit and of course the wisdom of this great land that we work so hard to protect gives us back the energy to keep going" (Watt-Cloutier 2001).

The Inuit Confront Climate Change

In addition to some of the planet's highest levels of toxicity, the Inuit have been forced to confront some of Earth's most rapid rates of global warming. Climate change arrived swiftly and dramatically in the Arctic during the 1990s, including such surprises as warm summer days, winter freezing rain, thunderstorms, and invasions of Inuit villages by heretofore unknown insects and birds. Yellowjacket wasps have been sighted for the first time, for example, on northern Baffin Island.

The Inuit Circumpolar Conference has assembled a human-rights case against the United States (specifically the George W. Bush administration) because global warming is threatening the Inuit way of life. The ICC has invited the Washington-based Inter-American Commission on Human Rights (a body of the Organization of American States) to visit the Arctic to witness the human effects of rapid warming. Introducing the legal action, Watt-Cloutier said, "We want to show that we are not powerless victims. These are drastic times for our people and require drastic measures" (Brown 2003). The Inuit say that by repudiating the Kyoto Protocol and refusing to cut carbon dioxide emissions in the United States, which make up 25 percent of the world's total, the White House imperils their way of life (Brown 2003). A hearing was held on the ICC petition on March 1, 2007.

Addressing a U.S. Senate Commerce Committee hearing on global warming August 15, 2004, Watt-Cloutier described the disorienting effect of rapid change on the Inuit: "The Earth is literally melting. If we can reverse the emissions of greenhouse gases in time to save the Arctic, then we can spare untold suffering." She continued: "Protect the Arctic and you will save the planet. Use us as your early-warning system. Use the Inuit story as a vehicle to reconnect us all so that we can understand the people and the planet are one" (Pegg 2004). The Inuits' ancient connection to their hunting culture may disappear within her grandson's lifetime, Watt-Cloutier said. "My Arctic homeland is now the health barometer for the planet" (Pegg 2004).

How quickly is the Arctic climate changing? On February 27, 2006, Watt-Cloutier described a heretofore unheard of event in Iqaluit: rain, lightning, thunder, and mud in *February*. The temperature had risen to 6°C to 8°C, the average for June, winds had reached 55 miles an hour, and the town was paralyzed in a sea of dirty slush. "Much of the snow has melted on the back of my house and all the roads are already slushy and messy. All planes coming up from the south were cancelled because the runways were icy from the rain," she wrote. "Unfortunately the predictions of the Arctic Climate Impact Assessment are unfolding before my very eyes" (Watt-Cloutier 2006). Iqaluit hunters expressed concern for the caribou, which would go hungry once freezing temperatures returned and encased the lichen, their food source, in a crust of ice. Instead of waiting until spring, the usual hunting season, Inuit were taking caribou in early March, believing that they would be too skinny later to be useful as food.

"Snow Age" to "Space Age"

As Watt-Cloutier remarked on one occasion,

> We have gone from the "snow age" to the "space age" in one generation. I was born on the land in Nunavik, northern Quebec and we were still traveling by dog team when I was sent to school in southern Canada at the age of ten. Notwithstanding significant economic, and social changes Inuit remain connected with the land and in tune with its rhythms and cycles.
>
> Inuit live in two worlds—the traditional and the modern. To this day our hunting based culture remains viable and important. We depend on seals, whale, walrus and other marine mammals to sustain our nutritional, cultural and spiritual well being. (Watt-Cloutier 2004)

Further Reading

Brown, Paul. "Inuit Blame Bush for Impending Extinction." *The Guardian* (U.K.), December 13, 2003. http://www.smh.com.au/articles/2003/12/12/1071125653832.html.

Cone, Marla. "Dozens of Words for Snow, None for Pollution." *Mother Jones*, January–February 2004. http://www.hartford-hwp.com/archives/27b/059.html.

Cone, Marla. *Silent Snow: The Slow Poisoning of the Arctic.* New York: Grove Press, 2005.

Johansen, Bruce E. "Pristine No More: The Arctic, Where Mother's Milk Is Toxic." *The Progressive,* December 2000, 27–29.

Johansen, Bruce E. "Arctic Heat Wave." *The Progressive,* October 2001, 18–20.

Kafarowski, Joanna. "Cloutier, Shiela-Watt." *Encyclopedia of the Arctic.* London: Routledge, 2007. http://www.routledge-ny.com/ref/arctic/watt.html (no longer available). Accessed November 10, 2007.

Pegg, J. R. "The Earth Is Melting, Arctic Native Leader Warns." Environmental News Service, September 16, 2004. http://www.ens-newswire.com/ens/sep2004/2004-09-16-10.asp.

Watt-Cloutier, Sheila. "Honouring Our Past, Creating Our Future: Education in Northern and Remote Communities." In Lynne Davis, Louise Lahache, and Marlene Castellano, eds., *Aboriginal Education: Fulfilling the Promise.* Vancouver: University of British Columbia Press, 2000.

Watt-Cloutier, Sheila. Personal communication, March 28, 2001.

Watt-Cloutier, Sheila. "Speech Notes for Sheila Watt-Cloutier, Chair, Inuit Circumpolar Conference." Conference of Parties to the United Nations Framework Convention on Climate Change, Milan, Italy, December 10, 2003. http://www.inuitcircumpolar.com/index.php?ID=242&Lang=En (no longer available).

Watt-Cloutier, Sheila. "Canada and Inuit: Addressing Global Environmental Challenges, Remarks by Sheila Watt-Cloutier, Chair, Inuit Circumpolar Conference at the Inaugural Environmental Protection Service Inuit Speaker Series." Ottawa, Ontario, January 16, 2004. http://www.inuitcircumpolar.com/index.php?ID=250&Lang=En

Watt-Cloutier, Sheila. Personal communication, February 27, 2006.

Watt-Cloutier, Sheila and Terry Fenge. "Poisoned by Progress: Will Next Week's Negotiations in Bonn Succeed in Banning the Chemicals That Inuit Are Eating?" Press release, Inuit Circumpolar Conference, March 14, 2000. http://www.inuitcircumpolar.com/index.php?ID=250&Lang=En (no longer available).

TERRAIN

OVERVIEW

Changes in the hydrological cycle wrought by rising temperatures, described in Part I ("Climate and Weather") are having major effects on weather patterns over land masses worldwide. Although a warming atmosphere generally increases the amount of moisture available in the atmosphere, this is not always the case. Enduring droughts (also described in Part I) also have intensified. These have scorched an entire continent (Australia) and damaged areas that historically have been wet such as the Amazon River valley of South America. Always a desert continent, Australia has become even drier and hotter in recent years, and people there have come to realize that climate change is the culprit—but evidence had to convince them.

The Indian Ocean monsoon affects the daily lives of hundreds of millions of people on the Indian subcontinent. In the past, a switch in upper-level winds caused abnormally heavy rain or drought and played a role in floods, droughts, and famines that have killed hundreds of thousands of people. With half of India's 1.2 billion people working in agriculture, a drought during monsoon season can have a major effect on million of peoples' lives. The monsoon rains of the Indian subcontinent and eastern Africa also have become more erratic, playing a role in floods some years and droughts in others. Human practices, including deforestation, also play a role in spreading droughts that turn lands once subject to occasional drought into deserts.

In areas such as western North America, warming accelerates the breeding cycle of pine beetles, killing large swaths of forests. Pine-beetle infestations have been encouraged by several factors: a warming trend that allows the beetles to multiply more quickly and reach higher altitudes, a drought that deprives trees of the sap they would have used to keep the beetles under control, and years of fire suppression that increased the amount of elderly wood susceptible to attack. Across the American West, the area burned each year has increased significantly over the past several decades, a trend that scientists attribute both to warming and drying and to a century of wildfire suppression and other human activities.

All of these patterns are likely to intensify as temperatures continue to rise in coming years. As with many processes related to climate change, various feedbacks come into play that reinforce each other. A rain forest that had been a major "sink" (absorber) of carbon dioxide (CO_2) can become a net source of the gas once it has been subject to drought for an extended period. Areas subject to human development, including deforestation (as in parts of the Amazon valley) compound this problem. Trees are roughly 50 percent carbon. Alive, they remove carbon dioxide from the atmosphere and replace it with oxygen. Dying and dead trees (or those being burned as fuel) become sources of and not absorbers of carbon dioxide and other greenhouse gases.

By 2007, approximately one-fifth of the human-caused greenhouse gases released into the atmosphere came from deforestation as carbon stored in trees entered the air. Indonesia has been clearing more forests than any other country. In some areas, such as the province of Riau (on the island of Sumatra), more than half the forests have been felled in a decade, many for palm oil plantations.

The region of Mato Grosso ("Great Forest" in Portuguese) in the Amazon, formerly rain forest, is becoming a dryland savanna. The destruction of the forest reinforces itself because rain clouds form more easily above moist forests. Deforestation also degrades soil quality. Because most of the Amazon's nutrients come from decaying vegetation, removing the forests removes the nutrients. The deforestation of the Matos Grasso is being aggravated by logging; some 17 percent of this region's forest already has been cleared

Africa has been plagued with expanding deserts, as the Sahara pushes the populations of Morocco, Tunisia, and Algeria northward toward the Mediterranean. At the southern edge of the Sahara, in countries from Senegal and Mauritania in the west to Sudan, Ethiopia, and Somalia in the east, the demands of growing human populations and livestock are converting land into desert. Iran also is battling its own desert. Mohammad Jarian, who directs Iran's antidesertification organization, reported in 2002 that sandstorms had buried 124 villages in the southeastern province of Sistan and Baluchistan. Here, as elsewhere, people are losing their livelihoods on lands that have become too hot, too dry—and, by turns, sometimes too wet—in a world of diverging extremes.

AMAZON, THE
Heat and Drought: Observations

Urbanization and industrialization continues in the Amazon valley, large parts of which are no longer rain forest. It has become Brazil's wild northwest, the fastest growing region in terms of human presence in South America, the "world's last great settlement frontier" in the words of Brian J. Godfrey, a geography professor at Vassar College and coauthor of *Rainforest Cities* (Romero 2012). All of this has implications for the world's carbon cycle as both climate change and human agency convert a longtime carbon dioxide sink into a net source of greenhouse gases. Deforestation in the Amazon valley already is one of Earth's largest sources of greenhouse gas emissions.

Even with the rate of deforestation in the Amazon declining as of late 2015, Survival International reported late that year, "Wildfires are raging through the Brazilian Amazon, destroying vast areas of forest on the eastern fringes of the 'earth's lungs.' The fires are reportedly being started by illegal loggers, in retaliation for tribal peoples' [the Awá tribe's] efforts to defend their territories and keep the invaders out. They threaten one of the few remaining areas of pre-Amazon forest in Brazil, the last environment of its type in the world" ("COP21" 2015). More than half of the Awá's land have been destroyed by fire.

The Parauapebas area, for example, "has changed in a generation from an obscure frontier settlement with gold miners and gunfights to a sprawling urban area with an air-conditioned shopping mall, gated communities and a dealership selling Chevy pickup trucks" (Romero 2012).

> Indeed, the streets of Parauapebas pulse with vitality. People shout to be heard along Rua 24 de Março, a traffic-clogged thoroughfare reverberating with the buzzing of motorcycle taxis, Pentecostal preachers bellowing warnings of sin and car stereos blaring electromelody, the thumping electronic music style popular in this part of the Amazon. (Romero 2012)

Manaus's population has grown to more than 1.7 million. By 2015, the population of the area as a whole had surpassed 25 million. Of 19 Brazilian cities that have doubled in population in a decade, 10 are in the Amazon. "More population leads to more deforestation," said Philip M. Fearnside, a researcher at the National Institute for Amazon Research (Romero 2012).

Simon Romero of *The New York Times* described the explosion of human settlement:

> Sinop, a city of 111,000 people in Mato Grosso State, grew about 50 percent in the past decade as soybean farmers expanded operations there. Fiscal incentives for manufacturing promote growth in Manaus and satellite towns like Manacapuru and Rio Preto da Eva. Logging still provides the lifeblood for growing towns along BR-163, an important Amazon highway now being paved. (Romero 2012)

The construction of hydroelectric dams is also drawing workers to the Amazon valley, as are open-pit iron ore mines near Pará, Parauapebas, and other locations that are trying to meet heavy demand in several other nations, most notably China. According to one observer,

> Already the Amazon desert is emerging in Rondonia. As far as the eye can see, there is only red, cracked ground. The red dust fills the air and sky itself turns red. Ten years ago this area was jungle, then it was slashed and burned to form farmland. But the jungle soil is notoriously poor, so it was used for pasture. Once overgrazed it became desert. Now even the landless poor who created it have moved on. They have moved west, to begin the process deeper in the land that used to be the Amazon rain forest. (Brandenburg and Paxton 1999, 225)

The Mato Grosso region, formerly a rain forest, is becoming a dryland savanna, as the destruction of forest reinforces the development of desert because rain clouds form more easily above moist forests. Deforestation also degrades soil quality. Because most of the Amazon's nutrients come from decaying vegetation, removing the forest removes the nutrients, according to Yadvinder Malhi of the University of Oxford. The deforestation of the Matos Grasso is being aggravated by logging, and already some 17 percent of this region's forest has been cleared (Brahic 2009).

According to Oliver L. Phillips and colleagues, a drying Amazon forest may "accelerate climate change through carbon losses and changed surface energy balances" (2009, 1344). The scientists used records from several long-term monitoring plots across Amazonia to evaluate response of the forest to the intense 2005 drought, which they considered to be a possible analog of future events. The forest in these areas lost biomass, "reversing a large long-term carbon sink, with the greatest impacts observed where the dry season was unusually intense. . . . Amazon forests therefore appear vulnerable to increasing moisture stress, with the potential for large carbon losses to exert feedback on climate change" (Phillips et al. 2009, 1344).

In the Amazon basin, the government has recognized lush forests as assets because they are sources of carbon sequestration as well as harvestable products. Members of the police force known as *Forca Verde* ("Green Police") now pursue men who burn logs into charcoal in ragged kilns, a "gritty, hellish business," as well as deforestation traffickers who sell high-end hardwoods for $1,000 or more per tree (Schapiro 2014, 77, 81). As Jennifer Balch wrote in *Nature* (2014, 41), "Aircraft have captured the 'breath' of the Amazon forest—carbon emissions over the Amazon basin. The findings raise concerns about the effects of future drought and call for a reassessment of how fire is used in the region."

The Kamayurá Lose Their Fish

Members of the Kamayurá tribe, who live in the Amazon's Xingu National Park, say that hotter, drier weather has imperiled the fish that are at the basis of their diet and played a role in deforestation and wildfires that have threatened their homeland for the first time in memory. "Us old monkeys can take the hunger, but the little ones suffer—they're always asking for fish," said Kotok, the tribe's chief (Rosenthal 2009). Fishermen haul in nearly empty nets while people swim in water that was once rife with piranhas. Some Kamayurá children have been substituting ants for fish that they eat on flatbread made from cassava flour.

The Kamayurás' homeland in the formerly isolated National Park has become a forested island in a sea of farms and ranches, while deforestation decreases the amount of rainfall, making rainy seasons less dependable. Elisabeth Rosenthal reported in *The New York Times*:

> Last year, for the first time, the beach on the lake that abuts the village was not covered by water in the rainy season, rendering useless the tribe's method of catching turtles by putting food in holes that would fill up, luring the animals. The tribe's agriculture has suffered, too. For centuries, the Kamayurá planted their summer crops when a certain star appeared on the horizon.

"When it appeared, everyone celebrated because it was the sign to start plant-ing cassava since the rain and wind would come," Chief Kotok recalled. But starting seven or eight seasons ago, the star's appearance was no longer fol-lowed by rain, an ominous divergence, forcing the tribe to adjust its sched-ule. It has been an ever-shifting game of trial and error since. Last year, families had to plant their cassava four times—it died in September, October and November because there was not enough moisture in the ground. It was not until December that the planting took. The corn also failed, said Mapulu, the chief's sister. "It sprouted and withered away," she said. (Rosenthal 2009)

Once too wet to burn, the forest has become dangerously flammable. During 2007, fire burned several thousand acres in the park. "The whole Xingu was burning—it stung our lungs and our eyes," Chief Kotok said. "We had nowhere to escape. We suffered along with the animals" (Rosenthal 2009).

Severe Drought Plagues Amazon Valley

During 2005, a severe drought spread through the Amazon valley at the same time that new evidence was being assembled, which indicated that damage from log-ging had been more than had been reported. "We think this [additional logging] adds 25 percent more carbon dioxide to the atmosphere" from the Amazon than previously estimated, said Michael Keller, an ecologist with the U.S. Department of Agriculture's Forest Service and coauthor of an Amazon logging inventory published in *Science* (Naik and Samor 2005; Asner et al. 2005, 480–481).

This study differed from others that measured only the clear-cutting of large for-est areas. The study by Asner and colleagues included these measures of deforesta-tion and added trees cut selectively, while much of the surrounding forest was left standing in five Brazilian states (Mato Grosso, Para, Rondonia, Roraima, and Acre) that account for more than 90 percent of deforestation in the Brazilian Amazon (Asner et al. 2005, 480). In addition, the Amazon valley's worst drought in some 40 years was causing several tributaries to evaporate, probably contributing even more carbon dioxide via wildfires.

A report by the United Nations Intergovernmental Panel on Climate Change issued in April 2007 projected that drying would continue in the Amazon valley. "By mid-century, increases in temperature and associated decreases in soil water are projected to lead to gradual replacement of tropical forest by savanna in east-ern Amazonia." The report anticipated that "crop productivity is projected to decrease for even small local temperature increases" in tropical areas, "which would increase risk of hunger" (Rohter 2007).

By 2008, the drought in some areas was the worst since record keeping began a century previously. This drought may be only an early indication of a new weather regime in the Amazon valley, which holds nearly a quarter of the world's freshwa-ter (Giles 2006, 726). The Amazon valley could be caught in a double vise as the world warms, as rising Atlantic Ocean temperatures combine with El Niño events to provoke more frequent droughts. El Niño events tend to reverse the air circulation over the Amazon from east–west to west–east, setting up drying downslope winds.

Amazon: Scientific Context

The Amazon is highly sensitive to drought, according to a 30-year study published March 6, 2009, in *Science*. The study provides the first scientific evidence that a drying environment causes increasing carbon loss as it kills trees in tropical forests ("Amazon Carbon" 2009). "For years the Amazon forest has been helping to slow down climate change. But relying on this subsidy from nature is extremely dangerous," said Professor Oliver Phillips, from the University of Leeds and the lead author of the research. "If the earth's carbon sinks slow or go into reverse, as our results show is possible, carbon dioxide levels will rise even faster" ("Amazon Carbon" 2009).

"The study, a global collaboration between more than 40 institutions, was based on the unusual 2005 drought in the Amazon. This gave scientists a glimpse into the region's future climate, in which a warming tropical North Atlantic may cause hotter and more intense dry seasons" ("Amazon Carbon" 2009). The 2005 drought sharply reversed decades of carbon absorption, in which Amazonia helped slow climate change. In an average year today, the Amazon forest absorbs nearly 2 billion tons of carbon dioxide. The drought caused a loss of more than 3 billion tons. The total impact of the drought—5 billion extra tons of carbon dioxide in the atmosphere—exceeds the annual emissions of Europe and Japan combined ("Amazon Carbon" 2009).

A lush tropical rain forest augments precipitation, so deforestation can enhance drought. As D. V. Spracklen and colleagues wrote in *Nature* (2012, 282):

> Vegetation affects precipitation patterns by mediating moisture, energy and trace-gas fluxes between the surface and atmosphere. When forests are replaced by pasture or crops, evapotranspiration of moisture from soil and vegetation is often diminished, leading to reduced atmospheric humidity and potentially suppressing precipitation. Climate models predict that large-scale tropical deforestation causes reduced regional precipitation. . . . We find that for more than 60 percent of the tropical land surface (latitudes 30° south to 30° north), air that has passed over extensive vegetation in the preceding few days produces at least twice as much rain as air that has passed over little vegetation. . . . We combine these empirical relationships with current trends of Amazonian deforestation to estimate reductions of 12 and 21 percent in wet-season and dry-season precipitation respectively across the Amazon basin by 2050, due to less-efficient moisture recycling.

What is more, drought in the Amazon and other rain forests combines with rapid climate change, habitat loss, and attacks by invasive species to reduce biodiversity in what used to be the richest habitats on land anywhere on Earth. Thiago Rangel wrote in *Science*: "These human-induced processes may have boosted the background rate of species extinction by 100 to 1,000 times. However, species do not go extinct immediately when their habitat shrinks, climate changes beyond their tolerance limit, or an invasive species spreads. It may take several generations after an initial impact before the last individual of a species is gone" (Rangel 2012, 162).

Capacity as Carbon Sink Declines

The Amazon's capacity to serve as a carbon sink has declined. As R. J. W. Brienen and colleagues wrote in *Nature* (2015, 344), "Rates of net increase in above-ground biomass declined by one-third during the past decade compared to the 1990s." This is important because until recently forests, especially those in the Amazon valley, have been soaking up much of the carbon dioxide produced by human activities. Signs are that this important carbon sink may have reached its maximum capacity (Hedin 2015, 295). "While this analysis confirms that Amazon forests have acted as a long-term net biomass sink," Brienen et al. (2015) wrote, "We find a long-term decreasing trend of carbon accumulation. This is a consequence of growth rate increases leveling off recently, while biomass mortality persistently increased throughout, leading to a shortening of carbon residence times. Potential drivers for the mortality increase include greater climate variability, and feedbacks of faster growth on mortality, resulting in shortened tree longevity" (2015, 344).

The study reported that the Amazon valley's uptake of carbon dioxide peaked at roughly 2 billion tons a year during the 1990s and has fallen by about half in two decades. Initially, trees responded to rising carbon dioxide levels with accelerating growth. However, with time, "the growth stimulus feeds through the system, causing trees to live faster and so die younger," said Oliver L. Phillips, a tropical ecologist at the University of Leeds, Great Britain, a leader of the research team (Gillis 2015).

Two "100-Year Droughts" in Five Years

In 2010, the Amazon valley experienced its second "100-year drought" in five years. "This is what's quite alarming—that we've seen these two very unusual events," said Simon Lewis, a University of Leeds (Great Britain) forest ecologist. "And those two unusual events are consistent with those predictions that suggest that the Amazon may be severely impacted over the next few decades by these droughts" (Joyce 2011).

Writing in *Science*, Lewis and colleagues said the droughts have been aggravated by the northward movement of warm water in the Atlantic Ocean. That shift carries moisture north and away from the Amazon.

As a moisture-starved Amazon basin's capacity to absorb carbon dioxide declines, burning trees also inject carbon dioxide into the air. If the forest gets dry enough, air can get into the vessels that carry water through a tree—kind of like air bubbles in a fuel line—and a tree dies. If enough trees die, that too could affect the atmosphere. "As these dead trees rot and release their carbon in their trunks and roots into the atmosphere as carbon dioxide, then we see it probably turning into a source of carbon emissions," Lewis said (Joyce 2011). Drought also increases the probability of fires, adding more CO_2. Areas already experiencing intense deforestation (usually the ones with the most human development) face the greatest risk; areas in the northwest region of the valley, which have the least human development, seem least likely to experience debilitating drought.

The potential for drought seems to operate on several levels. Some are local, such as the inhibition of transpiration. Others, including changes in the El Niño and La Niña cycle, occur outside the valley but influence weather and climate there (Malhi et al. 2008, 169–172). Drought also tends to feed on itself: a series of dry seasons desiccates trees, which recycle less rainfall and intensify the cycle. Human activities (clearing land for farming and pasture, especially using fire) often makes things even worse. Deforestation leaves behind combustible material that makes future fires more intense and quicker to spread. "A tipping point may be reached," according to one study, "[w]hen grasses can establish in the forest understory, providing a renewable source of fuel for repeated burns" (Malhi et al. 2008, 171).

Human activity is pushing the Amazon valley toward drought in many ways, of which fossil fuel effluvia is only one. Rising sea surface temperatures in the Atlantic Ocean change circulation patterns in the atmosphere in ways that can induce drier weather in parts of the valley because changes in climate patterns from more frequent El Niño episodes (which also may be provoked by rising temperatures) do the same. Land-use changes associated with the spread of logging, ranching, and farming also cause parts of the shrinking rain forest to dry out as world demand surges for tropical timber, soybeans, and free-range beef, among other products. Roads expand to link production centers in the Amazon valley to ports on South America's Pacific coast, allowing exports to the rapidly developing economies of China, India, and other parts of Asia. Demand for "green" ethanol meanwhile causes more Amazon forest to be cleared for sugarcane fields. In an area where 25 percent to 59 percent of rainfall is "recycled" by the forest, once roughly 30 percent to 40 percent of the rain forest goes, the area as a whole could pass a threshold, a "tipping point," into a drier climate on a long-range basis (Malhi et al. 2008, 169).

Mônica Carneiro Alves Senna and colleagues at the Federal University of Viçosa, Brazil, have used computer models to simulate how the Amazon could recover from various degrees of deforestation. Their models ranged from obliteration of the entire forest to 20-percent removal, which is close to present-day conditions (Brahic 2009; Senna, Costa, and Pires 2009).

"Cascading Positive Feedbacks"

"The Amazon is a kind of canary-in-a-coal-mine situation," said Daniel C. Nepstad, a senior scientist at the Woods Hole Research Center in Massachusetts and the Amazon Institute of Ecological Research in Belem, Brazil (Rohter 2005). "A warmer Atlantic not only helps give more energy to hurricanes, it also aids in evaporating air," said Luiz Gylvan Meira, a climate specialist at the Institute for Advanced Studies at the University of São Paulo. "But when that air rises over the oceans in one region, it eventually has to come down somewhere else, thousands of miles away. In this case, it came down in the western Amazon, blocking the formation of clouds that would bring rain to the headwaters of the rivers that feed the Amazon" (Rohter 2005).

Per Cox of the Center for Ecology and Hydrology in Winfrith, United Kingdom, anticipates that increased drought frequency could devastate about 65 percent of the Amazon's forest cover by the end of the 21st century (Cox et al. 2004, 137). Nepstad has been simulating drought in the Amazon valley since 1998. In a paper he was preparing for *Ecology* during 2006, Nepstad estimated that that reduced wood formation and tree mass killed largely by drought in the Amazon valley added half a billion tons of carbon dioxide to the atmosphere during the 2005 drought— yet another example of cascading positive feedbacks that intensify carbon-related warming (Giles 2006, 727). For comparative purposes, note that the Kyoto Protocol's goal as negotiated in 1997 was to remove twice that amount, 1 billion tons, from human-generated emissions.

Nepstad's fear is that a drier Amazon valley "will be invaded by highly flammable grasses," adding still more carbon dioxide to the atmosphere (Giles 2006, 727). In the state of Acre in western Brazil, parched trees turned to tinder during the 2005 drought, and the number of forest fires recorded during 2005 tripled to nearly 1,500 by September compared with a year earlier. The resulting smoke, which may itself have intensified the drought by impeding the formation of storm clouds, was so thick on some days that residents took to wearing masks outdoors (Rohter 2005). "Because droughts remain registered in the soil for up to four years, the situation is still very critical and precarious, and will remain so," Nepstad said. Where there are "forests already teetering on the edge," he added, the prospect of "massive tree mortality and greater susceptibility to fire" must be considered (Rohter 2005).

The Amazon Valley: No Longer the "Lungs of the World"

Given deforestation and other land-use changes in the Amazon valley, the area no longer functions generally as "the lungs of the world." The area acts as a carbon sink (absorber) only when temperature and precipitation patterns are healthy for the forest. In years of unusual drought (such as during the El Niño episodes of the middle and late 1990s and that of 2015–2016), unusually hot, dry weather turns the Amazon into a net producer of carbon dioxide. Usually, the Amazon absorbs about 700 million tons of carbon dioxide each year, but in the El Niño year of 1995 the Amazon forest *added* 200 million tons of carbon dioxide to the atmosphere. According to A. Lindroth, A. Grelle, and A. S. Moren, writing in *Nature*, a decrease in soil moisture can turn a forest that consumes carbon dioxide into one that adds the gas to the atmosphere. The authors assert that this variation occurs in temperate-zone forests as well as those of the tropics (Lindroth, Grelle, and Moren 1998).

In Indonesia and Brazil, new landowners have been migrating into the tropical forests year by year, altering the landscape (and Earth's carbon balance) by felling trees and burning what they have cut. The rain forest canopy, which had heretofore produced a surplus of oxygen, is being replaced by cattle, human beings, and motor vehicles. A survey conducted by the United Nations in 1982 estimated that tropical forests were being felled at a rate of 70,000 square miles (180,000 square kilometers) per year. The report estimated that 300 million subsistence farmers around the world, most of them in the tropics, were turning an area of forest the size of Denmark into a net producer of carbon every three months (Bates and Project Plenty 1990, 92). In 1987, 25,000 square miles of Brazilian rain forest were felled; a year later, the area that was deforested doubled. In the 1980s, lowland tropical forests in Indonesia, the Philippines, Malaysia, and parts of West Africa shrank quickly as well.

According to the Brazilian government's own figures, the amount of Amazonian forest lost to fire doubled between 1997 and 1998. The Woods Hole Research Center in Massachusetts said that 7,800 square miles of Amazonian rain forest caught fire in 1997. Approximately 80 percent of the fires were set on purpose, according to the United Nations; 42 million acres of rain forest (an area the size of Florida) were being lost per year by the 1990s.

In recent years, Brazil has been reducing forest loss but in an uneven fashion.

Between 1995 and 2005, forest loss in the Brazilian Amazon averaged 19,500 square kilometers per year—roughly the area of Israel. By 2013, that rate had been cut by 70 percent, even as beef and soya production continued to grow. A combination of measures was applied: corporate commitments coupled with strong laws, satellite surveillance and robust enforcement, restrictions on access to credit for farms and ranches in counties with high deforestation, the creation of protected areas and indigenous reserves, and improvements in land tenure and governance. Brazil's federal government worked closely with the beef and soya industries, NGOs [nongovernmental organizations] and international partners. In 2008, for example, Norway committed US$1 billion to Brazil because it wanted to demonstrate practical new ways to protect forests globally. Even so, Brazil's progress is fragile—deforestation in the Amazon has increased over the past 18 months [2014–2015]. (Victor and Leape 2015, 440)

By the 1990s, according to R. A. Houghton et al., "This large area of tropical forest [was] nearly balanced with respect to carbon" (Houghton et al. 2000, 301).

The combined effects of deforestation, abandonment, logging, and fire may thus yield sources of carbon. . . . These fluxes are similar in magnitude (but opposite in sign) to the sink calculated recently for natural ecosystems in the region. Taken together, the sources (from land-use change and fire) and the sinks (in natural forests) suggest that the net flux of carbon between Brazilian Amazonia and the atmosphere may be nearly zero, on average. (Houghton et al. 2000, 303–304)

Threats to the last remnants of Amazonian forests sometimes provoke action. According to Stephen Schwartzman, "China, not a world leader in green consciousness, last year [1998] banned all logging in its few remaining natural forests after disastrous flooding wreaked havoc along heavily populated rivers. In so doing, China hoped to save remnants of forest cover on the upper headwaters. But so much forest is already gone that it may not make much difference" (Schwartzman 1999, 61).

Deforestation has taken the trees from the world's tropical forests to some notable venues. Vintage ships of the British Royal Navy, for example, were refurbished during the 1990s with large amounts of mahogany harvested illegally from an indigenous reserve in the Amazon River basin. The harvest raised a ruckus among English environmentalists, including Friends of the Earth, as the defense ministry's mahogany purchase was questioned on national television. More than 7,000 cubic feet of mahogany was cut and moved by 10 to 12 trucks a day from the Kayapo indigenous reserve in Para state to the town of Redencao, home of the Juary logging company, harvester of the lumber. Each log was harvested illegally under Brazilian law. Despite the illegality of the harvest (aided by slack enforcement) the mahogany was purchased for the British ministry of defense by an English supplier, Parker Kislingbury.

Humanity Shapes the Rainforest

A rain forest generates its own humidity and shapes its own weather—up to a point. Roughly half of a pristine rain forest's precipitation is generated by the forest itself. Change the climate, and the forest's character changes. As humanity has invaded the Amazon River valley and spread urban areas and cleared forests, the land has generally dried, which aggravates drought. These changes are compounded in the case of the Amazon River valley by broader global trends such as more frequent El Niño conditions that change prevailing winds from east to west (which generates moisture off the Atlantic Ocean) to west to east (which wrings out Pacific Ocean moisture over the Andes mountains). The net result is a much drier Amazon that produces much less oxygen and more carbon dioxide.

Fewer plants produce less moisture and less rain. With less ground cover, more water runs off barren soil, eroding topsoil and leaving less nourishment for trees. As a result, large areas of the Amazon River valley that were once heavily forested have become savanna, a mixture of malnourished trees, shrubs, and grasslands. Human-wrought changes thus reinforce each other. As a tropical forest declines, its weather and climate generally become drier and hotter and more likely to suffer fire that savages what vegetation remains.

Deforestation debilitates the forest because once an area is denuded, local soil conditions (and sometimes weather) change in such a way as to make the regrowth of trees more difficult. In many areas where deforestation has been extensive, previously forested areas have been replaced by sparse grasses, stunted shrubs, and bare eroded earth. An area that has long been regarded as "the lungs of the world" is gradually becoming (like the human beings who are felling trees and building cities), a greenhouse gas producer. Fire (both natural and human-set) now adds considerably to the atmosphere's overload in the Amazon as well as in other tropical forests in Southeast Asia, equatorial Africa, Mexico, and Indonesia.

In February 1999, massive floods inundated São Paulo, the largest city in South America, and deforestation was a major cause. In 2014 and 2015, the same area experienced record drought, with stringent restrictions on water use. Fifteen years later, São Paulo experienced a historic drought. Also in 1999, deforestation helped provoke flooding in Caracas, Venezuela, which is surrounded by denuded mountains. The increasing intensity of deluges provoked by a warming climate also plays a major role in deadly flooding. The same combination of circumstances played out in the Peruvian area of the Amazon River valley in 2017. Deforestation also aggravated death and destruction in Central America by land-falling hurricanes.

Fires strip the forest cover and thus make future fires more likely in the Amazon valley and in many other areas where once-lush tropical forests are in peril. In some of them (Indonesia is a prime example) peat smolders underground below the forest canopy, pouring carbon dioxide into the atmosphere. As fire destroys the forest, wildlife is killed or flees its dying habitat. Humanity's appetite for timber and farmland destroys the forest, producing air pollution (in the form of smoke) and a great deal of excess carbon dioxide that may remain in the atmosphere for centuries, warming the atmosphere.

The very vocabulary now being used to describe the Amazon indicates just how endangered the rainforest has become. The Amazon River valley is now being described as "world's last great settlement frontier." Ten of 19 Brazilian cities that doubled in population by 2016 are in the Amazon valley, "providing economic opportunities to Brazilians struggling to emerge from poverty" (Romero, 2012). The area is compared to the U.S. West in the heyday of Manifest Destiny. One Amazon city, Manaus, is Brazil's fastest growing urban area.

As Simon Romero reported in Planetizen, "The torrid expansion of rain forest cities is visible in places like Parauapebas, which has changed in a generation from an obscure frontier settlement with gold miners and gunfights to a sprawling urban area with an air-conditioned shopping mall, gated communities, and a dealership selling Chevy pickup trucks" (Romero, 2012). The main source of employment in Parauapebas is an open-pit iron ore mine that provides thousands of jobs.

The Hadley Centre for Climate Prediction and Research, part of the United Kingdom's meteorological office (Met Office), has set out to survey the world and forecast how much of its quickly declining tropical forests will switch from carbon "sinks" to carbon sources during the 21st century. Its research is sobering, indicating just how influential human activity has become in pushing the world's atmosphere toward an unsustainable degree of warming. Humanity's role is not just a personal matter of consuming fossil fuels and producing a personal budget of greenhouse gases. Every time a tree falls in the Amazon valley (or anywhere else), the world's atmosphere changes. Usually, this means production of more carbon dioxide. As human populations and affluence rise, the world's forests and their ability to consume carbon dioxide decline.

Deforestation sometimes places additional debits on the world account of greenhouse gases. When tropical forests are cut and slashed, for example, the resulting piles of dead wood attract termites, whose populations explode as they feast on a food supply that has suddenly increased. Termites are extremely efficient at digesting plant mass and turning it into methane, a greenhouse gas that remains in the atmosphere only two weeks compared with carbon dioxide's residency of a century or more.

Hadley Centre projections, now almost two decades old (released in 1998), anticipated that large parts of the Amazon valley could become desiccated by 2050, an event that is forecast to force humanity and every other living thing to endure an unstoppable greenhouse effect. Land temperatures will rise by an average 10°F over the next century. In addition to intensifying aridity in parts of tropical South America, large parts of tropical Africa also are expected to become desiccated by 2050, according to the Hadley Centre researchers. These climate changes will force roughly 30 million people to face intense hunger, the report asserts. In the two decades since the Hadley Centre made these projections, humanity has followed the prediction. The population of Manaus, Brazil's Amazon metropolis, for example, has doubled in two decades to some 2.5 million.

Fire in the Amazon Valley

"The scale of the problem is mind-boggling," said Daniel C. Nepstad of the Woods Hole Research Center, who has been surveying deforestation in the Amazon valley. Nepstad estimated that 400,000 square kilometers of rain forest, an area 20 times the size of Massachusetts, became vulnerable to fire during 1998, compared to an average of 15,000 square kilometers a year that had been cleared and burned during the previous few years. "Once burned," wrote Nepstad, "Amazonian forests become more susceptible to future burning. Wildlife is killed or dispersed (except for some species that thrive on disturbance). Timber, forest medicines and forest fruits are destroyed, and a large part of the carbon contained in the trees is gradually released into the atmosphere" (Nepstad 1998).

A major problem, according to Nepstad, "is that the rain forest is available in abundance and is therefore very cheap. When forest is cheap and labor and capital are scarce, it is the forest itself that becomes the fertilizer, the pesticide, the herbicide, the plow" (Nepstad 1998). Nepstad suggested a "new model of rural development . . . in Amazonia, which restricts access to most (about 70 percent) of the region's forests, and which creates a situation of land scarcity, higher land prices, and higher investments in agricultural production systems" (Nepstad 1998). Nepstad believes that people will value the forest only when it becomes an economic good in human financial terms. People will quit burning the forests when they can earn more money by sustaining it. Approximately half of the area burned each year in Amazonia is accidental, according to Nepstad, who recommends prohibition of burning whenever fires could easily spread out of control.

Destruction of the Amazon rain forest accelerated toward the end of the 20th century. According to studies by Nepstad, the forest was being destroyed two to three times more quickly than previous estimates. Nepstad conducted his research with scientists at the Institute of Environmental Research in Belem. They measured forest losses at 1,104 sample points from a light plane and on the ground. They also interviewed more than 1,500 mill operators and landholders in the Amazon region. Nepstad and his associates assert that satellite imagery, on which most previous estimates have been based, "often fails to distinguish between pristine forest and burned or cut forest land newly covered with fast-growing brush" ("Amazon Destruction" 1999).

Research published during October 1998 by the Hadley Centre for Climate Prediction and Research, which is part of the United Kingdom's meteorological office (Met Office), anticipates that large areas of "closed" tropical forests will die as the world warms during the next century. Instead of soaking up pollution, these forests will dump more carbon dioxide into the atmosphere than all of the world's power stations and cars have produced during the past 30 years. "Closed" tropical forests, according to the Hadley Centre report, contain trees covering a large proportion of the ground where grass does not form a continuous litter on the forest floor. Forests of this type declined in Africa from 289,700 thousand hectares in 1980 to 241,500 in 1990. In Latin America, tropical forests declined from 825,000 to 753,000 hectares during the same period; in Asia, tropical forest coverage fell from 334,500 hectares in 1980 to 287,500 in 1990 (National Academy of Sciences 1991, 9).

Deforestation sometimes places additional debits on the world account of greenhouse gases. When tropical forests are cut and slashed, for example, the resulting piles of dead wood attract termites, whose populations explode as they feast on a food supply that has suddenly increased. Termites are extremely efficient at digesting carbon and turning it into methane, a greenhouse gas (McKibben 1989, 17).

Amazon Valley Desiccated by 2050?

The Hadley Centre released a report in 1998 that predicted that large parts of the Amazon valley could become desiccated by 2050, an event that the center said may threaten the world with an unstoppable greenhouse effect. The Hadley report asserts that land temperatures will rise by an average 10°F over the next century. In addition to intensifying aridity in parts of tropical South America, large parts of tropical Africa also are expected to become desiccated by 2050, according to the Hadley Centre researchers. These climate changes will force roughly 30 million people to face intense hunger, the report asserts.

By 2050 or so, according to one study, "a decrease in annual rainfall of up to 500 millimeters in some key areas, combined with temperature rises of up to [7°C], will begin to kill tropical forests, creating deserts" (Nuttall 1998). At this point, the study says, these forests will become carbon sources rather than absorbers (sinks) as their remains burn or rot. "Up to 2050, the land surface takes up carbon through an increased growing season, which is being measured at the moment. But when we move to 2050, the tropical dieback releases so much carbon to the atmosphere it causes the concentrations to increase. This may enhance the buildup of carbon dioxide in the atmosphere" (Nuttall 1998).

Further Reading

"Amazon Carbon Sink Threatened by Drought." NASA Earth Observatory. March 5, 2009. http://earthobservatory.nasa.gov/Newsroom/view.php?id=37493&src=eoa (no longer available).

"Amazon Destruction More Rapid Than Expected." April 10, 1999. http://www.gsreport.com /articles/art000099.html (no longer available).

Asner, Gregory P., et al. "Selective Logging in the Brazilian Amazon." *Science* 310 (October 21, 2005): 480–481.

Balch, Jennifer. "Atmospheric Science: Drought and Fire Change Sink to Source." *Nature* 506 (February 6, 2014): 41–42.

Bates, Albert K., and Project Plenty. *Climate in Crisis: The Greenhouse Effect and What We Can Do.* Summertown, TN: The Book Publishing Co., 1990.

Brahic, Catherine. "Parts of Amazon Close to Tipping Point." *New Scientist.* March 5, 2009. http://www.newscientist.com/article/dn16708-parts-of-amazon-close-to-tipping-point .html.

Brandenburg, John E., and Monica Rix Paxson. *Dead Mars, Dying Earth.* Freedom, CA: The Crossing Press, 1999.

Brienen, R. J. W., et al. "Long-Term Decline of the Amazon Carbon Sink." *Nature* 519 (March 19, 2015): 344–348.

"COP 21: Amazon Fire Destroying Rare Forest Home of Uncontacted Tribe." Survival International press release by e-mail, December 1, 2015, by e-mail. http://www .survivalinternational.org/news/11033.

Cox, P. M., et al. "Amazonian Forest Dieback under Climate-Carbon Cycle Projections for the 21st Century." *Theoretical and Applied Climatology* 78 (2004): 137–156.

Giles, Jim. "The Outlook for Amazonia Is Dry." *Nature* 442 (August 17, 2006): 726–727.

Gillis, Justin. "A Climate Change Safety Net Frays." *The New York Times*, March 24, 2015, D3.

Hedin, Lars O. "Biogeochemistry: Signs of Saturation in the Tropical Carbon Sink." *Nature* 519 (March 19, 2015): 295–296.

Houghton, R. A., et al. "Annual Fluxes of Carbon from Deforestation and Regrowth in the Brazilian Amazon." *Nature* 403 (January 20, 2000): 301–304.

Joyce, Christopher. "'Alarming' Amazon Droughts May Have Global Fallout." National Public Radio, February 7, 2011. http://www.npr.org/2011/02/07/133462608/alarming -amazon-droughts-may-have-global-fallout?ft=1&f=1007.

Lindroth, A., A. Grelle, and A. S. Moren. "Long-Term Measurements of Boreal Forest Carbon Balance Reveal Large Temperature Sensitivity." *Global Change Biology* 4 (April 1998): 443–450.

Malhi, Yadvinder, et al. "Climate Change, Deforestation, and the Fate of the Amazon." *Science* 319 (January 11, 2008): 169–172.

McKibben, Bill. *The End of Nature.* New York: Random House, 1989.

Naik, Gautam, and Geraldo Samor. "Drought Spotlights Extent of Damage in Amazon Basin." *Wall Street Journal*, October 21, 2005, A12.

National Academy of Sciences. *Policy Implications of Greenhouse Warming.* Washington, DC: National Academy Press, 1991.

Nepstad, Daniel C. "Report from the Amazon: May, 1998." Fairmouth, MA: Woods Hole Research Center. http://terra.whrc.org/science/tropfor/fire/report2.htm (no longer available).

Nuttall, Nick. "Global Warming 'Will Turn Rainforests into Deserts.'" *The Times* (London). November 3, 1998. http://bonanza.lter.uaf.edu/~davev/nrm304/glbxnews.htm (no longer available).

Phillips, Oliver L., et al. "Drought Sensitivity of the Amazon Rainforest." *Science* 323 (March 6, 2009): 1344–1347.

Rangel, Thiago F. "Amazonian Extinction Debts." *Science* 337 (July 13, 2012): 162–163.

Rohter, Larry. "A Record Amazon Drought, and Fear of Wider Ills." *The New York Times*, December 11, 2005. http://www.nytimes.com/2005/12/11/international/americas/11ama zon.html.

Rohter, Larry. "Brazil, Alarmed, Reconsiders Policy on Climate Change." *The New York Times*, July 31, 2007. http://www.nytimes.com/2007/07/31/world/americas/31amazon.html.

Romero, Simon. "Swallowing Rain Forest, Cities Surge in Amazon." *The New York Times*, November 24, 2012. http://www.nytimes.com/2012/11/25/world/americas/swallowing -rain-forest-brazilian-cities-surge-in-amazon.html.

Rosenthal, Elisabeth. "An Amazon Culture Withers as Food Dries Up." *The New York Times*, July 25, 2009. http://www.nytimes.com/2009/07/25/science/earth/25tribe.html.

Schapiro, Mark. *Carbon Shock: A Tale of Risk and Calculus on the Front Lines of the Disrupted Global Economy; How Carbon Is Changing the Cost of Everything.* White River Junction, VT: Chelsea Green Publishing, 2014.

Schwartzman, Stephen. "Reigniting the Rainforest: Fires, Development and Deforestation." *Native Americas* 16(3–4) (Fall–Winter 1999): 60–63.

Senna, Mônica Carneiro Alves, Marcos Heil Costa, and Gabrielle Ferreira Pires. "Vegetation-Atmosphere-Soil Nutrient Feedbacks in the Amazon for Different Deforestation Scenarios." *Journal of Geophysical Research*, February 18, 2009. doi: 10.1029/2008JD010401.

Spracklen D. V., S. R. Arnold, and C. M. Taylor. "Observations of Increased Tropical Rainfall Preceded by Air Passage over Forests." *Nature* 489 (September 13, 2012): 282–285.

Victor, David G., and James P. Leape. "Global Climate Agreement: After the Talks." *Nature* 527 (November, 26, 2015): 439–441.

See also: Australia, Heat and Drought in; Deforestation; Drought, United States; Drought, Worldwide; Drought and Deluge, Scientific Analysis of

AUSTRALIA, HEAT AND DROUGHT IN

Always a desert continent (except in its tropical north), Australia has become even drier and hotter in recent years. Some Australians have long suspected that global warming is the culprit, but mounting evidence has convinced most of them. Reports of heat, drought, and devastating wildfires often dominate both national and international news. In one example of many wildfires during February 2015, two large conflagrations, both ignited by lightning, threatened towns in southwestern Australia—one east of Northcliffe and another more than 60 miles to the northeast near Boddington ("Bushfires Menace" 2015).

Summer during the intense El Niño year of 2015 arrived in Australia with record late spring heat (104°F in Sydney). Fire ravaged 580 square miles of wheat fields and killed farmers at work in the south of Western Australia, near Esperance, a town of 14,000 some 450 miles southeast of Perth. Furious flames were driven by a scorching 50 mile-per-hour wind at temperatures more than 100°F. Farmers were beginning their wheat harvest as exceedingly dry fields, record heat, and strong winds fed the flames.

October 2015—the calendrical equivalent of April north of the equator—was Australia's hottest month on record. "It is going to be a horror summer," said Trevor Tasker, a firefighter and regional emergency services inspector from Western

Australia. "I've never seen conditions like this. "By the time we knew that fire was alight, it was unstoppable," said Tasker, a firefighter and regional emergency services inspector. "There was nothing anyone could do but get out of the way and let it unleash its fury." One farmer was singed inside his car as he raced to warn others of the wall of flames headed their way (Innis 2015). The townships of Salmon Gums, Grass Patch, and Norseman, all near Perth on Australia's southwest coast, recorded temperatures around 107°F on Tuesday, November 17, according to Neil Bennett, a weather forecaster based in Perth at the Bureau of Meteorology. "This is the earliest we have seen the temperatures this high," Bennett said. "But we have had a drying trend going back 30 to 40 years" (Innis 2015).

Southwestern Australia saw intensifying heat and drought into 2016. On January 6, a lightning-ignited fire in Lane Poole Reserve enveloped Yarloop, a town about 70 miles south of Perth, destroying 128 homes and 41 other structures, including bridges and community buildings, according to the Department of Fire and Emergency Services. Two Yarloop residents were killed and four firefighters were injured. Within five days, the fire had burned 276 square miles ("Bushfire Devastates" 2016).

Heat made outdoor sporting events impossible near Melbourne during mid-January 2014, when a hot wind from the interior brought temperatures of 111°F to the Australian Open tennis matches, nearly shutting them down. Press reports described tennis great Maria Sharapova as being happiest off the sizzling court, "underneath the shade of an umbrella, an ice vest draped over the back of her neck" (Bishop 2014).

The year 2013 started and ended with record heat and was Australia's hottest on record to that time. In addition, one-tenth of Australia's reporting stations experienced record heat between January 1 and 4, 2014. The highest air temperature during that period was in Moomba, Queensland, at 49.3°C (120.7°F) on January 2 ("Heat Wave" 2014).

As reporter Matt Siegel of *The New York Times* wrote from Sydney (Siegel 2013, January 9), "Four months of record-breaking temperatures stretching back to September of 2012 have combined over the past week with widespread drought conditions and strong winds to create what the government had labeled 'catastrophic' fire conditions along the heavily populated eastern and southeastern coasts of the country, where much of the population is centered." Meteorologists in Australia added two new colors to temperature charts and added a range to 54°C (129°F) to accommodate highs not seen before. "If you look at yesterday [January 8, 2013], at Australia as a whole, it was the hottest day in our records going back to 1911," said David Jones, manager of climate monitoring prediction at the Bureau of Meteorology. "From this national perspective, one might say this is the largest heat event in the country's recorded history" (Siegel 2013, March 4).

"Drought-Proof" Reservoirs Running Low

Australia has long experience with drought. During the 1950s and 1960s, it built reservoirs that were supposed to protect against recurring droughts. These gave the country the highest storage capacity per capita in the world. Combined with hundreds of miles of irrigation conduits, the reservoirs were said at the time to make Australia "drought-proof." However, by mid-2007, Melbourne's water storage was 28 percent of capacity, Sydney's 37 percent, and Perth's 15 percent. In May, rains described as a once-in-30-year phenomenon hit the Hunter valley north of Sydney but did little to help. The land was so dry that the brief torrential rains were quickly absorbed (Nowak 2007, 10).

Six years of drought aggravated by warming temperatures from 2002 to 2008 reduced Australia's rice crop by 98 percent and played a major role in rapidly rising worldwide prices, including a doubling an three months during the first half of 2008. The price rise contributed to food riots as far away as Haiti, and rice shortages "spurred panicked hoarding in Hong Kong and the Philippines, and set off violent protests in countries including Cameroon, Egypt, Ethiopia, Haiti, Indonesia, Italy, Ivory Coast, Mauritania, the Philippines, Thailand, Uzbekistan and Yemen" (Bradsher 2008). Police suppressed violent demonstrations in Dakar, Senegal, on March 30.

During the summers of 2002 and 2003, wildfires pushed by raging hot, dry winds from Australia's interior seared parts of Canberra, charring hundreds of homes, killing four people, and forcing thousands to flee the area. "I have seen a lot of bush fire scenes in Australia . . . but this is by far the worst," Australian Prime Minister John Howard said ("Australia Assesses" 2003). Flames spread through undergrowth and exploded as they hit oil-rich eucalyptus trees. The 2002–2003 drought knocked 1 percent off Australia's gross domestic product and cost $6.8 billion in exports. It reduced the size of Australia's cattle herd by 5 percent and its sheep flock by 10 percent (Macken 2004). Recovery was impeded during 2004 by continuing drought.

Members of the Australian Greens urged the country's federal government to expand an inquiry into the devastating bushfires to consider global warming. Bob Brown, an Australian senator and member of the Greens, said that new research indicates record daytime temperatures and unprecedented rates of water evaporation made the current drought the worst on record. The extreme dryness of vegetation arising from global warming was what made this fire season so devastating, he said ("Greens Want" 2003).

Even as drought aggravated wildfires in Australia, some parts of the country were dealing with deluges. In Sydney, Tomihiro Taniguchi, vice chairman of the Intergovernmental Panel on Climate Change (IPCC), said that recent Australian flooding and similar weather conditions around the world were further evidence that the impact of global warming was beginning to be felt. Although he admitted it was difficult to directly link deluges in New South Wales to global warming, he added, "We can say with high confidence that the likelihood of flooding will increase in the future" (Maynard 2001). The higher temperatures create greater seawater evaporation, which in turn produces more precipitation in coastal areas.

"Climate change is potentially the biggest risk to Australian agriculture," said Ben Fargher, chief executive of the National Farmers' Federation in Australia (Bradsher 2008). Many Australian farmers have switched from rice to wine grapes, which require much less water and produce a pretax profit of about $2,000 an acre compared to rice at about $240 an acre. In the meantime, strains of rice are being developed that bloom during the earlier and cooler hours of the day. Rice that blooms late in the day is vulnerable to warming.

With irrigation, cultivation of cotton prospered in Australia's Outback. By 2007, however, with Australia suffering record heat and its worst drought on record, production of cotton, with its thirst for water, had fallen sharply. In the *Wall Street Journal*, Patrick Barta wrote that in the town of Wee Waa (Australia's self-described "cotton capital," population 2,000, and some 250 miles northwest of Sydney), "The Cotton Fields Motel that once was busy with seasonal workers now struggles to fill rooms. Elsewhere in the flat basin [of Australia's largest river system, the Murray–Darling], kangaroos hop along dry levees, and the end of giant water-transport pipes poke out over empty reservoirs" (Barta 2007, A1).

Australia's cotton production fell by two-thirds between 2001 and 2007 (Barta 2007, A12). Some fields were down to 2 percent of usual, and reservoirs as large as 240,000 acres had completely dried up. In 2008, the drought in the Murray–Darling river basin continued, following a record dry June, even after parts of Australia received occasional torrential rains during a strong La Niña that briefly countered some effects of the drought.

The Heat Wave of 2009

The scorching summer of 2015 had Australians recalling another El Niño summer: searing drought and heat waves during the summer of 2008–2009. According to an Associated Press account published in *The New York Times*, "Towering flames razed entire towns in southeastern Australia and burned fleeing residents in their cars as the death toll rose [to more than 180], making it the country's deadliest fire disaster" ("Death Toll" 2009). Several thousand homes were destroyed in early February as temperatures rose above 100°F for several days in Melbourne and Sydney; temperatures as high as 117°F were reported in the scorched interior of Victoria state, the highest temperatures on record.

Thousands of volunteer firefighters and Australian army troops battled the wind-whipped flames. Entire small villages (including Marysville and several hamlets in the Kinglake district 50 mikes north of Melbourne) were wiped out by racing flames. In Kinglake, according to the same Associated Press report:

> Just five houses out of about 40 remained standing. . . . Street after street was lined by smoldering wrecks of homes; roofs collapsed inward, iron roof sheets twisted from the heat . . . burned-out hulks of cars dotted roads. . . . [T]he landscape was blackened as far as the eye could see. Entire forests were reduced to leafless, charred trunks, farmland to ashes. ("Death Toll" 2009)

During the fires, on the morning of January 29, 2009, in the northern suburbs of Adelaide, around 3 a.m., strong northwesterly winds mixed hot air aloft to the surface. At RAAF Edinburgh, a regional airport, the temperature rose to 41.7°C (106°F) at 3:04 a.m., an event without recorded precedent in southern Australia.

On February 7, 2009, the temperature in Melbourne hit 115.5°F (and as high as 120°F in nearby areas) with relative humidity as low as 6 percent. The next day, more than 600 fires blew up on stiff northwesterly winds, the worst outbreak in the continent's recorded history. Some "fire fronts" reached 300 feet in height, sweeping through and searing small towns, killing 173 people. The fires moved so quickly that they caught some people from behind in their cars as they tried to flee (Kenneally 2009, 46–53). Also during mid-November 2009, Australia suffered a spring heat wave that set record highs characteristic of midsummer and pushed fire danger to "catastrophic" in some regions of South Australia and New South Wales (NSW). Adelaide had its highest springtime temperatures on record (35°C, 95°F) for eight days in a row, and temperatures reached 109°F in Sydney on one day.

By 2009, the Murray–Darling basin (in the continent's southeastern quadrant) was becoming extremely dry. Hundreds of thousands of river red gum trees, the world's largest such forest, died suddenly. Some former wetlands, now encrusted with cracking dried earth where swamps once stood, have been reacting with air to form sulfuric acid because they no longer have periodic flushing. These are signs of an epic drought beyond the cycles that usually affect the area (Draper 2009, 56). On September 23, 2009, Australia's worst dust storm in 70 years produced an orange dawn in Sydney. Dust clouds from Australia's interior, which was suffering its worst drought on record, spread over much of eastern Australia. Visibility was low enough to close Sydney's airport.

Floods Replace Drought for a Year

With the advent of a strong La Niña cycle in the Pacific Ocean during 2010, Australia's wheat belt's three-year drought broke dramatically. Record-setting precipitation drenched sections of Victoria and New South Wales in August, followed by the wettest September on record in Australia as a whole. By the end of December 2010, 200,000 people had been stranded by floods in parts of northeastern Australia's province of Queensland, an area the size of France and Germany combined. In Rockhampton near the coast, the Fitzroy River reached a level not seen since 1918 (Goodman 2010). Rain was falling at four times its average rate, producing the worst floods in nearly a century. In mid-January 2011, floodwaters were surging toward Brisbane, Australia's third-largest city with 2 million people. The floodwaters killed 10 people, including five children in Toowoomba and other sections of the Lockyer valley. On January 10, buildings west of Brisbane were ripped from the ground and many cars were swept away. On February 2, 2011, flooding was topped off by Cyclone Yasi, the strongest tropical storm to hit Australia in a century, a category 5 with wind gusts up to 200 miles per hour.

After the floods submerged large parts of Brisbane, the rebuilding effort was compared to what would follow a war. More than 30,000 homes and businesses were inundated in Australia's northeast at the peak of the flood as the area remained under water for several days while the Brisbane and Bremer Rivers crested. "We now face a reconstruction task of postwar proportions," state Premier Anna Bligh said, choking back tears (Belford and Foley 2010). The floods also nearly paralyzed Queensland's suppliers of coking coal, an ingredient used in steel production.

Political Context of Australian Heat and Drought

Australia was the only major industrial country other than the United States that long refused to ratify the Kyoto Protocol. Under Prime Minister John Howard, Australia's conservative government was slow to recognize the economic perils of global warming. Largely because 85 percent of the country's electricity was generated from coal and because it hosts a large number of energy-intensive industries such as aluminum smelting and steel manufacturing, Australia generated more greenhouse gases per capita in 2004 than any other country, 26.3 tons, except for Luxembourg at 28.1 tons (the comparable U.S. figure was 24.1 tons) (Wiseman 2007). The Howard government's inability to realize the role of climate change in the devastation of Australia's economy probably played a major role in the country's 2007 elections, when the government was replaced by the Labor Party and Kevin Rudd, who signed the Kyoto Protocol as his first official act after he was sworn into office December 3, 2007. Howard had been quite direct with his disregard of global warming. During September 2006, former U.S. presidential candidate Al Gore had come to the country, and Howard had refused to meet with him.

By the time Howard faced voters, many Australian urban areas were facing severe water shortages. Inflows behind Sydney's dams from 1991 to 2006 were 71 percent less than their averages from 1948 to1990 (Pincock 2007, 336). By 2006, polls in Australia indicated that a large proportion of people there ranked a dysfunctional climate as the number-one danger in their lives, ahead of such right-wing mainstays as international terrorism. Howard had been reelected three times (1998, 2001, and 2004) as the drought intensified, but by 2007 his string had run out despite his pledge in June 2007 to implement a cap-and-trade emissions-reduction scheme.

Record Heat and Walls of Flame: Atmospheric Dynamics

While northern subtropical Australia, which usually has high rainfall, has been getting wetter, the more arid south and west have been drier for a combination of reasons: more frequent El Niños, which bring low rainfall and usually peak in the summer; higher temperatures, which increase evaporation of what little rain does fall; an active Indian Ocean dipole, which reduces crucial spring rains in the agricultural areas of southeastern Australia; and changes in the southern annular mode, which has been preventing rain-bearing storms from reaching southern parts of the continent in winter (Nowak 2007, 10). These last effects may be part of the reorganization of atmospheric circulation—the Hadley cell—that some scientists attribute to global warming and may be intensifying drought in the western United States, southwestern China, and parts of Saharan Africa, notably Darfur.

Desalination Plants Built

Perth's first desalination plant was completed in 2006; a wind farm was constructed to provide the 24 megawatts required to operate the plant. Perth soon was drawing 17 percent of its water from the new facility. Sydney and Melbourne added desalination plants soon thereafter. Some industries, such as BHP Billiton's copper and uranium mines in the south, also may build their own desalination facilities. The plants require a great deal of energy to operate and produce a salty mush that will render whatever land or water is used for its disposal useless for any other purpose. Brisbane's government is considering recycling sewage after its main water supply runs dry (Nowak 2007, 11).

Desalination plants are being built not only in Australia but also in other affluent countries that can afford the up-front costs. After plant construction, water costs are approximately $3.50 per 1,000 gallons. Such plants are common in the Middle East, and Southern California has some smaller ones. The desalination plant in Perth is powered by wind from 48 turbines in the Emu Downs Wind Farm, some 100 miles to the north. These can produce as much as 80 megawatts of electricity a day, three times the plant's requirements. Similar plants are proposed for Sydney and the town of Tugun in Queensland (Mydans 2007).

Vulnerable Species

Australia's Queensland state may lose half of its wet tropical highland rain forest, including many of its rarest animals because of global warming. The state's emblem, the koala, is at risk because of rising carbon dioxide levels, which could strip the gum leaf (the koala's principal food) of its nutritional value. The same goes for many other vulnerable species. A report released February 4, 2002, by Climate Action Network Australia, "represents one of the most comprehensive pictures yet of the local ecological effects of global warming" (Ryan 2002). More than half of Australia's eucalyptus species are unlikely to survive an average temperature rise of 3°C, according to the report. In addition, higher carbon dioxide levels are expected to reduce carbohydrates and nitrogen in leaves, undermining food supplies for animals such as koalas (Ryan 2002).

Ian Hume, professor emeritus of biology at Sydney University, said that rising levels of carbon dioxide increase levels of "anti-nutrients" in the leaves, making the koalas' sole source of food toxic to them. Hume expects such problems to interfere with koala reproduction and reduce their populations substantially within 50 years. As it is, eucalyptus leaves contain little energy, and koalas conserve energy by sleeping as much as 20 hours a day. Populations already have been reduced in some parts of Australia by habitat loss, including the spread of farms and suburbs ("Scientist" 2008).

The study reviewed possible damage to Queensland's tropical areas, asserting that "90 Australian animal species, including a third of those on the endangered list, also [are] likely to suffer in the hotter, more extreme climate forecast this century" (Ryan 2002). David Hilbert, principal research scientist at the Tropical Forest Research Centre (which is run by the Commonwealth Scientific and Industrial Research Organisation or CSIRO), said even a 1°C temperature rise would devastate half of the rain forests of north Queensland's wet tropics.

"CSIRO atmospheric scientists are now predicting 2 to 5 degree C of warming by the end of the century, so a 1 degree C change is liable to happen within 30 to 50 years," Hilbert said (Ryan 2002). The Murray–Darling basin, for example, faces a 12- to 35-percent reduction in average flow by 2050 as southern regions grow hotter and drier, adding to the pressures on its overextracted and salt-affected rivers (Ryan 2002). The Great Barrier Reef also runs a risk of heat-related coral bleaching, which affected 16 percent of the world's reefs in 1998 (Ryan 2002). In addition, Australia's alpine ecosystems are expected to shrink to six high mountaintops with continued warming, with any remaining snowfields probably disappearing within 100 years, the report asserted.

In the meantime, ring-tailed possums were falling dead out of trees in Australia's far north Queensland because the climate had become too hot for them. Green ring-tailed possums cannot survive more than five hours at an air temperature above 30°C (O'Malley 2003, 14). The ring-tailed possums were not alone. The Cooperative Research Centre anticipates that half of all the unique mammals, reptiles, and birds in far north Queensland's rain forests could become extinct given a 3.5°C rise in temperature, which is at the midpoint of the IPCC's projected range for the end of the 21st century. Most of these animals are found only in the Wet Tropics World Heritage Area near Cairns at elevations above 600 meters (O'Malley 2003, 14). "It potentially takes just one day of six or seven hours of extreme heat . . . and you find dozens of dead possums on the ground," said Steve Williams, a tropical biologist at Australia's James Cook University (Braasch 2007, 83).

Drought to Become Chronic?

Drought may become chronic in Australia if modeling done by CSIRO proves accurate. According to Kevin Hennessey of the agency's Atmospheric Research Climate Impact Group, "There is a consistency between our modeling and the reality of Australia's weather. Our modeling suggests Australia will become warmer and drier in the future as a result of global warming. By 2030 most of Australia will be between 0.5 and 2 degrees warmer, and potentially 10 percent drier" (Macken 2004).

Anecdotal evidence of warmth and drought in Australian is supported by scientific study. Neville Nicholls of Australia's Bureau of Meteorology Research Centre (Melbourne) wrote in *Climatic Change*:

> Rainfall over nearly all of Australia during the cooler half of the year (May–October) was well below average in 2002. Mean maximum temperatures were very high during this period, as was evaporation. This would suggest that drought conditions (precipitation minus evaporation) were worse than in previous recent periods with similarly low rainfall (1982, 1994). Mean minimum temperatures were also much higher during the 2002 drought than in the 1982 and 1994 droughts. The relatively warm temperatures in 2002 were partly the result of a continued warming evident in Australia since the middle of the twentieth century. The possibility that the enhanced greenhouse effect is increasing the severity of Australian droughts, by raising temperatures and hence increasing evaporation, even if rainfall does not decrease, needs to be considered. (Nicholls 2004, 323)

Australia's decade-long drought, provoked by climate change and natural variability, also has been aggravated by deforestation, according to Clive McAlpine of the

University of Queensland in Brisbane. He and his colleagues used a climate model to simulate Australian climatic conditions from the 1950s to 2003 and compared them to ground conditions before European immigration began in 1788. The drought is most intense in the southeast regions of the continent, where less than 10 percent of original vegetative cover remains, which increases the number of days with temperatures above 35°C by 300 percent (six to 18 a year) and the number of dry days by a similar ratio, five to 15 ("Land Clearances" 2009).

By 2010, Australia's extended drought was being called "The Big Dry." The scientific journal *Nature* described work being done by Gavan McGrath and colleagues at the University of Western Australia:

> [They] analyzed satellite data from across the continent and found evidence of decreased water storage, rainfall and plant growth throughout the country between 2002 and 2010. In the southeast, the drought correlated with an irregular Indian Ocean circulation, whereas in the northwest it was associated with a decreased frequency of tropical cyclones. The authors say that the northwest drought coincided with and probably exacerbated the one in the southeast. The findings suggest that distinct climatic factors such as decadal cyclone trends and changes in ocean circulation can combine to create a continental-scale drought. ("The Extended Reach" 2012)

Human-Induced Fires

Some brushfires in Australia start with lightning, but others have been obviously human-caused such as a fire that burned more than 47,000 hectares in New South Wales. The State Mine fire started on Marrangaroo Army Range in the Blue Mountains of New South Wales near the city of Lithgow on October 16, according to the Environment News Service ("Australian Army" 2013). A spokesman for the Army confirmed that the Australian Defence Force had been responsible for starting a fire at the Cultana training area south of Port Augusta on Saturday as "part of an exercise." Australia's largest environmental group said the fires are partly the result of the "new climate reality" and that Australians must start cutting carbon pollution now. The South Australian Country Fire Service's Paul Sinclair added, "Sydney is, quite literally, on fire right now. Homes and cars are being lost as bushfires rage, fueled by a record-setting heat wave which NSW Fire Commissioner Greg Mullins says he's never seen the likes of so early in the year" ("Australian Army" 2013).

"Coincidentally, last week also saw the release of the Intergovernmental Panel on Climate Change's report unequivocally telling us if we don't start acting to rein in pollution, we can expect more of it," said Sinclair ("Australian Army" 2013). "In Australia, the problem we face is that we have been polluting unheeded for decades, and now we are beginning to pay the price," Sinclair said. "That price is a risk to our health and our well-being—access to clean drinking water, access to affordable food. The price is more frequent, more devastating bushfires, droughts, and floods" ("Australian Army" 2013).

Records Falling with Regularity

In 2013, Australia's Climate Commission, a federal government agency, associated extreme weather in that country, especially heat waves and droughts (both of which instigated many wildfires), with global warming. The fires scorched many heavily populated areas of New South Wales and Queensland, killed six people, and caused $2.43 billion in damage. The heat waves were followed by flooding rains. Although until now Australian climate scientists had been reluctant to connect these events with a more general pattern of climate change because Australia is prone to a drought–deluge cycle, the Climate Commission's "The Angry Summer" report argues that the "frequency and ferocity of recent extreme weather events indicate an acceleration that is unlikely to abate unless serious steps are taken to prevent further changes to the planet's environment. . . . I think one of the best ways of thinking about it is imagining that the base line has shifted," commission leader Tim Flannery told the Australian Broadcasting Corporation (Siegel 2013, March 4).

Weather records have been falling with a regularity unknown before, the report said. "Included were milestones like the hottest summer on record, the hottest day for Australia as a whole, and the hottest seven consecutive days ever recorded. To put it into perspective, in the 102 years since Australia began gathering national records, there have been 21 days when the country averaged a high of more than 102 degrees Fahrenheit (39 Celsius), and eight of them were in 2013" (Siegel 2013, March 4). In other words, a preponderance of evidence has established a pattern. Report author Will Steffen said, "The findings were consistent with an overall global acceleration of weather factors like rising temperatures and heavier rains attributed by scientists to human-caused climate change" (Siegel 2013, March 4).

As thousands of sheep and cattle died in fires during the record heat, climate scientists expressed alarm. "Those of us who spend our days trawling—and contributing to—the scientific literature on climate change are becoming increasingly gloomy about the future of human civilization," said Elizabeth Hanna, a researcher at Canberra's Australian National University. She told *The Sydney Morning Herald*. "We are well past the time of niceties, of avoiding the dire nature of what is unfolding, and politely trying not to scare the public" (Siegel 2013, March 4).

Further Reading

"Australian Army Started Bushfire Blazes, Climate also Blamed." Environment News Service, October 23, 2013. http://ens-newswire.com/2013/10/23/australian-army-started-bushfire-blazes-climate-also-blamed/.

"Australia Assesses Fire Damage in Capital." *Omaha World-Herald*, January 20, 2003, A4.

Barta, Patrick. "Parched Outback: In Australia, a Drought Spurs a Radical Remedy." *Wall Street Journal*, July 11, 2007, A1, A12.

Belford, Aubrey, and Meraiah Foley. "Australian Floods Rage through Brisbane." *The New York Times*, January 13, 2011. http://www.nytimes.com/2011/01/14/world/asia/14australia.html.

Bishop, Greg. "In Wilting Heat, Sharapova Escapes 2nd Round." *The New York Times*, January 16, 2014. http://www.nytimes.com/2014/01/17/sports/tennis/heat-wilts-players-and-crowd-at-australian-open.html.

Braasch, Gary. *Earth under Fire: How Global Warming Is Changing the World*. University of California Press, 2007.

Bradsher, Keith. "A Drought in Australia, a Global Shortage of Rice." *The New York Times*, April 17, 2008. http://www.nytimes.com/2008/04/17/business/worldbusiness/17warm .html.

"Bushfire Devastates Australian Town." NASA Earth Observatory, January 12, 2016. http://earthobservatory.nasa.gov/IOTD/view.php?id=87302&src=eoa-iotd.

"Bushfires Menace Towns in Western Australia." *NASA Earth Observatory*, February 6, 2015. http://earthobservatory.nasa.gov/IOTD/view.php?id=85225&src=eoa-iotd.

"Death Toll in Australian Bushfires Climbs to 84." Associated Press in *The New York Times*, February 9, 2009. http://www.nytimes.com/2009/02/09/world/asia/09australia.html.

Draper, Robert. "Australia's Dry Ruin." *National Geographic*, April 2009, 34–59.

"The Extended Reach of Australian Drought." *Nature* 483 (March 1, 2012): 8.

Goodman, J. David. "Australia Floods Show No Signs of Retreating." *The New York Times*, December 31, 2010. http://www.nytimes.com/2011/01/01/world/asia/01australia.html.

"Greens Want Global Warming Examined in Bushfire Inquiry." Australian Associated Press, January 21, 2003 (LEXIS).

"Heat Wave Stifles Australia." January 12, 2014. http://earthobservatory.nasa.gov/IOTD/view .php?id=82790&src=eoa-iotd.

Innis, Michelle. "Record Heat Puts Australia at Risk of Intense Fire Season." *The New York Times*, November 20, 2015. http://www.nytimes.com/2015/11/21/world/australia/australia -fires-record-temperatures.html.

Kenneally, Christine. "The Inferno." *The New Yorker*, October 26, 2009, 46–53.

"Land Clearances Turned up the Heat on Australian Climate." *New Scientist*, May 16, 2009. http://www.newscientist.com/article/mg20227084.700-land-clearances-turned-up-the -heat-on-australian-climate.html?DCMP=OTC-rss&nsref=environment.

Macken, Julie. "The Double-Whammy Drought." *Australian Financial Review*, May 4, 2004, 61.

Maynard, Roger. "Climate Change Bringing More Floods to Australia." *The Straits Times* (Singapore), March 14, 2001, 17.

Mydans, Seth. "Reports from Four Fronts in the War on Warming." *The New York Times*, April 3, 2007. http://www.nytimes.com/2007/04/03/science/earth/03clim.html.

Nicholls, Neville. "The Changing Nature of Australian Droughts." *Climatic Change* 63 (2004): 323–336.

Nowak, Rachel. "The Continent That Ran Dry." *The New Scientist*, June 16, 2007, 8–11.

O'Malley, Brenden. "Global Warming Puts Rainforest at Risk." *Cairns Courier-Mail* (Australia), July 24, 2003, 14.

Pincock, Steve. "Showdown in a Sun-Burnt Country." *Nature* 450 (November 15, 2007): 336–338.

Ryan, Siobhain. "National Icons Feel the Heat." *Courier Mail* (Australia), February 4, 2002, 1.

"Scientist: Warming Threatens Koalas." Associated Press in *Omaha World-Herald*, May 8, 2008, 7A.

Siegel, Matt. "Record Heat Fuels Widespread Fires in Australia." *The New York Times*, January 9, 2013. http://www.nytimes.com/2013/01/10/world/asia/record-heat-fuels-widespread -fires-in-australia.html.

Siegel, Matt. "Report Blames Climate Change for Extremes in Australia." *The New York Times*, March 4, 2013. http://www.nytimes.com/2013/03/05/world/asia/australian-government -blames-climate-change-for-angry-summer.html.

Wiseman, Paul. "Australia Pushes New Climate Plan." *USA Today*, September 6, 2007, 10A.

See also: Amazon, The; Drought, United States; Drought, Worldwide; Drought and Deluge, Scientific Analysis of; Heat Waves, Evidence of; Heat Waves, Forecasts of

DEFORESTATION

Forests are an important carbon "sink," or absorber. Worldwide, they absorb more than 25 percent of the carbon dioxide that human beings put into the atmosphere, or roughly the amount emitted by all cars and trucks and equal approximately to the volume absorbed by the oceans, which are growing acidic enough to imperil life forms in the sea with calcium-based shells. Dead (or dying) or burning forests no longer absorb carbon dioxide; they produce it. Humankind has been reducing forest cover for centuries, usually to make way for agriculture and cities. In our time, deforestation has been most intense in Indonesia and the Amazon valley of Brazil, but it has extended to many other areas with expanding populations as well such as China, India, and parts of sub-Saharan Africa.

Trees are roughly 50 percent carbon. When alive, they remove carbon dioxide from the atmosphere and replace it with oxygen. Trees that are dying, dead, or are being burned as fuel become sources, not sinks of carbon dioxide and other greenhouse gases. The destruction of forests around the world is no trivial matter for the atmosphere's carbon budget. Although the atmosphere contains some 750 billion tons of carbon dioxide, forests contain about 2,000 billion tons. Roughly 500 billion tons is stored in trees and shrubs and 1,500 billion tons in peat bogs, soil, and forest litter (Jardine 1994).

By around 2013, Earth had less than 5 million square miles of old-growth forest remaining. An estimated deforestation rate of 62,000 square miles a year, the amount being logged in recent years, could reduce that figure to zero within two human lifetimes. More than 50 percent of the world's forests have been destroyed in the past 100 years. An area equivalent to 57 soccer fields falls each minute (Webb 1998).

Drought Accelerates Deforestation Worldwide

Drought has been scarring landscapes worldwide, as Justin Gillis wrote in *The New York Times*:

> From the mountainous Southwest deep into Texas, wildfires raced across parched landscapes this summer, burning millions more acres. In Colorado, at least 15 percent of that state's spectacular aspen forests have gone into decline because of a lack of water. The devastation extends worldwide. The great euphorbia trees of southern Africa are succumbing to heat and water stress. So are the Atlas cedars of northern Algeria. Fires fed by hot, dry weather are killing enormous stretches of Siberian forest. Eucalyptus trees are succumbing on a large scale to a heat blast in Australia, and the Amazon recently suffered two "once a century" droughts just five years apart, killing many large trees. (Gillis 2011)

Deforestation Aggravates Drought

Drought may be reinforcing itself in the Amazon valley. Forests, especially large and dense ones, produce moisture that rises convectively into the atmosphere and returns as rain. In the Amazon region, a fully grown tree releases about 1,000 liters of water vapor per day. Trees also help provoke rain by releasing terpenes and isoprene, both volatile gases that help cause formation of rain droplets in clouds (Robbins 2015).

Clearing land for grazing and agriculture tends to reduce its moisture-producing capability even as deforestation releases additional carbon dioxide into the atmosphere. Above and beyond local effects, the reinforcing nature of droughts may influence other episodes of drying many thousands of miles away (Medvigy et al. 2013). "There's lots of evidence that changing the water cycle in the Amazon would have global consequences," said David Schmiel of NASA's Jet Propulsion Laboratory and author of *Climate and Ecosystems*. "It's a fairly robust notion" (Robbins 2015).

In 2015, Brazil was gripped by intense drought, and São Paulo rationed water. The arrival of an intense El Niño aggravated this pattern in much of Brazil as prevailing winds reversed from their usual east–west pattern to west–east. The east–west pattern picks up moisture from the Atlantic Ocean and runs upslope against the Andes Mountains, producing rain and snow on their eastern slopes and a dry downslope effect westward. This flow of moisture typically carries more water (as vapor) than the Amazon River itself. With El Niño, a west–east wind provokes heavy rain and floods on the western flank of the Andes and drying downslope winds to the east. Forest destruction impedes this "sky river."

During 2014, more than 2,000 square miles of Amazon rain forest fell to make way for agriculture and ranching. By 2015, one-fifth of this rain forest had been converted to farms and pastures, and another 20 percent was severely compromised (Robbins 2015). Antonio Donato Nobre, a climatologist in Brazil's National Institute for Space Research, "warned that if just 40 percent of the Amazon region is deforested there could be an abrupt large-scale shift to grasslands, which could substantially alter global weather patterns 'and cause a breakdown of the current climate system.' If deforestation continues, he has said, São Paulo will most likely 'dry up'" (Robbins, 2015).

Further Reading

Gillis, Justin. "The Threats to a Crucial Canopy." *The New York Times*, October 1, 2011. http://www.nytimes.com/2011/10/01/science/earth/01forest.html.

Jardine, Kevin. "The Carbon Bomb: Climate Change and the Fate of the Northern Boreal Forests." Ontario, Canada: Greenpeace International, 1994. http://dieoff.org/page129.htm.

Medvigy, David, et al. "Simulated Changes in Northwest U.S. Climate in Response to Amazon Deforestation." *Journal of Climate*, 26 (2013): 9115–9136.

Robbins, Jim. "Deforestation and Drought." *The New York Times*, October 9, 2015. http://www.nytimes.com/2015/10/11/opinion/sunday/deforestation-and-drought .html.

"The amount of area burning now in Siberia is just startling—individual years with 30 million acres burned [an area the size of Pennsylvania]," said Thomas W. Swetnam, an expert on forest history at the University of Arizona. "The big fires that are occurring in the American Southwest are extraordinary in terms of their severity, on time scales of thousands of years. If we were to continue at this rate through the century, you're looking at the loss of at least half the forest landscape of the Southwest," said Swetnam (Gillis 2011).

By 2007, approximately one-fifth of the human-caused greenhouse gases being released into the atmosphere came from deforestation, much of it as carbon that had been stored in trees. Indonesia has been clearing more forests than any other country. In some areas, such as the province of Riau (on the island of Sumatra), more than half the forests have been felled in a decade, many for palm oil plantations, and often for ethanol, a supposedly Earth-friendly fuel that replaces petroleum-based products such as gasoline. The burning and drying of Riau's peatlands, also to make way for palm oil plantations, similarly releases 1.8 billion tons of greenhouse gases a year (Gelling 2007).

Stephen Schwartzman, a senior scientist with the international program of the Environmental Defense Fund, sketches the scope of tropical deforestation:

> An area of forest bigger than Belgium, Holland, and Austria put together, or about 40 percent of California, was cut down and burned every year between 1980 and 1995, some 62,000 square miles per year. NASA's Landsat satellite photographs show that more than 200,000 square miles, an area about the size of France, has been cleared and burned in Brazil alone. All of this has happened since the 1970s. (Schwartzman 1999, 60)

Deforestation Accelerates over Time

Deforestation increased rapidly during the 1980s. Three million acres of forest were being lost per year in Indonesia, providing 70 percent of the world's plywood and 40 percent of its tropical hardwoods. By the early 1980s, roughly 1.2 percent of Indonesia's remaining tropical forest coverage was being felled each year. In 1988, 48,000 square miles of Amazon rain forest were burned to clear land for farming. Nearly overnight, most of the burned land was turned from a net producer of oxygen to a source of carbon dioxide and methane (Gordon and Suzuki 1991, 118–119).

Portugal's Wildfires

Portugal is being scorched by wildfires nearly every summer, a problem that it shares with many other countries, from the United States' Western states and

Alaska, to Canada, Russia, Australia, and elsewhere. In Portugal, the prevalence of fires, and their damage, has helped spark a country-wide conversation and action plan for dealing with the changing climate that includes a drive to substitute wind and solar power for fossil fuels.

The expansion of wildfires may be the most pervasive evidence of a warming climate. For example, By 2014, scientists were reporting that "The effects of human-induced climate change are being felt in every corner of the United States, with water growing scarcer in dry regions, torrential rains increasing in wet regions, heat waves becoming more common and more severe, wildfires growing worse, and forests dying under assault from heat-loving insects" (Gillis, May 6, 2014). Half of the United States Forest Service's budget by 2015 was going into fighting fire, which caused some who work there to complain that they were running a fire department, not a forest service.

Fires in Portugal during the summer of 2017 were the worst on record. On June 17, 2017, lightning ignited a deadly wildfire that spread across the mountainous areas of Pedrógão Grande, a city in central Portugal northeast of Lisbon. The raging fires were clearly visible to NASA satellites in space for several nights.

"Fires across Portugal's forested landscape during the warm, dry summer months are not uncommon," reported NASA's Earth Observatory. "In 2016, hundreds of fires raged on the mainland and also on the Portuguese island of Madeira" (Wildfires Light, 2017). The fires of 2017 were notable for a high death toll of 62 people because drought and heat were exceptionally intense, causing fires to spread so quickly that half of the people who died were caught from behind by the inferno and roasted alive in their cars as they tried to escape. The worst previous fire season was in 1966, when 25 people died, all of them Portuguese who were fighting the fires near Sintra.

Patrícia De Melo Moreira, a photographer for Agence France-Presse, was on the scene with firefighters when several blazes enveloped people fleeing in their cars., The flames were so hot that the asphalt highway under them "was completely destroyed—melted," she said (Minder, 2017).

The fire brought "a dimension of human tragedy that we cannot remember," Prime Minister António Costa said during a visit to the scorched area. The flames spread along four fronts at the same time with "great violence," said Jorge Gomes, the secretary of state for internal administration (Minder, 2017). In addition to burning people in their cars, several houses were engulfed as about 1,600 firefighters fought the flames, assisted by airplanes and helicopters. Low humidity, high temperatures, and a parched landscape combined to make the fire unusually violent and unpredictable. "We know fire behavior has changed and continues to change, yet we continue to be surprised every time, when we shouldn't be," said Don Whittemore, a former assistant fire chief in Colorado who has studied wildfire behavior. "The notion that firefighters will be able to put out, suppress or make safe a wildfire is becoming less and less of a reliable notion" (Minder, 2017).

From 1993 to 2013, Portugal, a relatively small nation, experienced more forest fires than France, Spain, Italy or Greece, according to a 2016 report from the European Environment Agency. Portugal sometimes reaches temperatures of more than

100 °F during the summer as hot, dry adiabatic (downslope) winds from the inland Iberian Peninsula dry forests and make them prone to fire. These winds are similar to the Santa Anas that create hazardous fire conditions in Southern California, and the offshore winds that sometimes blow from inland Australia to the coastal urban areas of Sydney and Melbourne. In Portugal, dense woodland and hilly terrain covered with pine and eucalyptus trees which contain oil that burns easily. During recent years, the forests have transitioned to pine and eucalyptus (which use useful commercially) from oak, which burn less easily. Portugal also has an unusually large number of fires that start due to arson or accidents.

The Pedrogao Grande fire of 2017 was not sparked by arson or accident, however, but by a "dry thunderstorm" that contained lightning but no rain that reached the ground. Police were able to find the tree that was struck (in the town of Escalos Fundeiros), starting the fire. Dry thunderstorms, which also may produce strong wind that spread the flames, also play a role in igniting fires in many other parts of the world. Fires in Portugal increased not only because of hot, dry weather, but because recession had caused cuts in forestry management. The World Wildlife Fund (WWF) in Portugal said that the government should improve forestry management, which "is more effective and financially more efficient than the huge mechanisms used every year to combat forest fires" (Minder, 2017).

Colorado's Aspens Endangered

If levels of greenhouse gases continue to accelerate, scientists warn that Colorado's iconic aspen forests may fall prey to drought by 2050. "In the fall," wrote Justin Gillis in *The New York Times* (2015), "stands of trembling aspens are among the most breathtaking sights in the West, turning hillsides an iridescent golden hue." "I think of aspens as a good canary-in-the-coal-mine tree," said William R. L. Anderegg, of Princeton University, whose study was published in March 2015 in *Nature Geoscience*. "They're a wet-loving tree in a dry landscape. They may be showing us how these forests are going to change pretty massively as that landscape gets drier still" (Gillis 2015). The aspens already have been reduced by periodic droughts. The aspens probably will survive only if greenhouse gases are brought under control quickly.

"In a high-emissions scenario," according to Anderegg's report, "climate models project that drought stress will exceed the observed mortality threshold in the southwestern United States by the 2050s. Our approach provides a powerful and tractable way of incorporating tree mortality into vegetation models to resolve uncertainty over the fate of forest ecosystems in a changing climate" (Anderegg et al. 2015). The study concentrates on biological processes that kill trees in droughts. As soil moisture declines, air bubbles infiltrate the microscopic tubes that carry water through a tree. "These air bubbles block the pipes and interrupt water transport, giving the tree a kind of heart attack, basically," Anderegg said (Gillis 2015).

Anderegg and colleagues developed a computer model that accurately predicts aspen mortality 75 percent of the time by using precipitation and temperatures expected with differing future greenhouse gas levels. Higher levels portend rising temperatures, less soil moisture, and higher rates of aspen die-off. The model also can be applied to tree species other than aspens and geographic areas beyond Colorado. Large stands of aspens died in "hot droughts" shortly after the year 2000.

Craig D. Allen, a forest expert with the U.S. Geological Survey, said that applying this model to other species beyond Colorado may be complicated by the wide "variability between species." For example, Allen said, "aspens have relatively shallow roots, limiting their ability to tap deep water in a drought, whereas other trees could be more resilient" (Gillis 2015).

Further Reading

Anderegg, William, et al. "Tree Mortality Predicted from Drought-Induced Vascular Damage." *Nature Geoscience*, March 30, 2015. doi:10.1038/ngeo2400.

Gelling, Peter. "Forest Loss in Sumatra Becomes a Global Issue." *The New York Times*, December 6, 2007. http://www.nytimes.com/2007/12/06/world/asia/06indo.html.

Gillis, Justin. "Climate Change Threatens to Kill Off More Aspen Forests by 2050s, Scientists Say." *The New York Times*, March 31, 2015. http://www.nytimes.com/2015/03/31/science/earth/climate-change-threatens-to-kill-off-more-aspen-forests-by-2050s-scientists-say.html.

Schwartzman, Stephen. "Reigniting the Rainforest: Fires, Development and Deforestation." *Native Americas* 16:3/4 (Fall/Winter 1999): 60–63.

Warming and Tropical Mountain Forests

Tropical mountain forests depend on predictable, frequent, and prolonged immersion in moisture-bearing clouds. Clearing upwind lowland forests alters surface energy budgets in ways that influence dry-season cloud fields (Lawton et al. 2001, 584). Cloud forests form where mountains force trade winds to rise above the condensation level, the point of orographic cloud formation. "We all thought we were doing a great job of protecting mountain forests [in Costa Rica]," said Robert O. Lawton, a tropical forest ecologist at the University of Alabama in Huntsville, Alabama. "Now we're seeing that deforestation outside our mountain range, out of our control, can have a big impact" (Yoon 2001).

The clearing of forests, sometimes many miles from the mountains, alters this pattern, often raising the elevation at which clouds form. Lawton and colleagues used Landsat and geostationary operational satellite images to measure such changes in the Monteverde cloud forests of Costa Rica. "Our simulations suggest that conversion of forest to pasture has a significant impact on cloud formation" (Lawton et al. 2001, 586). Patterns found in Costa Rica resemble those in other tropical areas, including parts of the Amazon valley. "These results suggest that current trends in tropical land use will force cloud forests upward, and they will thus decrease in area and become increasingly fragmented—and in many low mountains may disappear altogether" (Lawton et al. 2001, 587).

Deforestation's effects probably extend farther than most observers have heretofore believed. "Mountain forests . . . may be affected by what's happening some distance away," said Lawton. Each year approximately 81,000 square miles of tropical forests are cleared, according to Gary S. Hartshorn, president of the Organization for Tropical Studies, a consortium of rain forest researchers at Duke University (Polakovic 2001). Thus, the weather in the lush cloud forests of Costa Rica is changing because of land-use changes many miles away, including deforestation. As trees on Costa Rica's coastal plains are removed and replaced by farms, roads, and settlements, less moisture evaporates from soil and plants, in turn reducing clouds around forested peaks 65 miles away.

At risk is the Monteverde Cloud Forest Reserve, an ecosystem atop a Central American mountain spine, "a realm of moss and mist, the woodland in the clouds— a type of rain forest—[that] is home to more than 800 species of orchids and birds, as well as jaguar, ocelot, and the resplendent quetzal, a plumed bird sacred to the Mayans." The same area also is a watershed that supplies farms, towns, and hydropower plants in the lowlands (Polakovic 2001).

Hurricane Mitch and Deforestation

Five years before flooding from Hurricane Mitch devastated Honduras, J. Almendares, a Honduran medical doctor, warned readers of the British medical journal *Lancet* that deforestation was making the country more vulnerable than ever to deadly flooding. Almendares presented evidence that "desiccation and soil erosion caused by cattle grazing and sugarcane and cotton cultivation have altered the regional hydrological cycle" (Almendares and Sierra 1993, 1401). These changes have led to fewer rainy days but more intense downpours.

Almendares presented temperature statistics from one deforested area of Honduras indicating that the average ambient air temperature had risen 7.5°C between 1972 and 1990 (Almendares and Sierra 1993, 1403). In neighboring Nicaragua, during the final months of the Sandinista revolution (1979), similar temperature rises were reported in Managua after dictator Anastazio Somoza ordered the wholesale destruction of trees in the city to deny Sandinistas places to hide during gun battles.

Central America had 200,000 square miles of forest in 1900. By the 1980s, only 36,000 square miles survived, and the rate of deforestation was increasing, especially in the Miskito region of Nicaragua and Honduras. A 1998 report by United Nations agencies and nongovernmental organizations documented a regional deforestation rate of 958,360 acres (1,500 square miles) a year (Weinburg 1999, 51).

Honduras lost a third of its forests between 1964 and 1990. The country's forests continued to be felled at an annual rate of 80,000 hectares a year during the 1990s, a rate which, if sustained, would strip a quarter of remaining forested land per decade. In the meantime, people who can no longer wrest a living from denuded (or corporate-controlled) land have been moving to Honduran cities, where malaria has become endemic.

During October 1998, Hurricane Mitch made landfall on Nicaragua's Miskito coast, with winds as strong as 178 miles an hour, dropping as much as three feet of rain. Crossing the mountains, Mitch turned northwest, moving slowly through El Salvador, Guatemala, and southern Mexico. By the time the storm reached the sea again, it had killed 10,000 people and left nearly 3 million others homeless. In Honduras, thousands of people who survived the storm lost their jobs in devastated banana plantations.

The devastation wrought by Mitch raised questions in Central America not only about potential increases in hurricane severity from global warming but also about changes in land use across the region that make many areas more prone to severe flooding during heavy rains. The same questions were raised in Caracas, Venezuela, after devastating floods there in late 1999.

Bill Weinburg described the conditions that have made hurricane flooding so devastating in Central America:

> Fire and flood fuel each other in a vicious cycle. Landless peasants colonize the agricultural frontier, or clear-forested slopes for their *milpas* (fields). The more forest is destroyed, the more the hydrologic cycle is disrupted; with no canopy for transpiration, local rainfall and cloud cover decline; aridity makes the surviving forest vulnerable to wildfires. Then, when the rains do come, sweeping in from the Caribbean on the trade winds, there are no roots to hold the soil and absorb the water. Millennia [of] accumulated wealth of organic matter is swept from the mountainsides in deluges of mud. Tlaloc, the revered Nahua rain god who the Maya called Chac Mool, brings destruction instead of abundance. (Weinburg 1999, 51–52)

In Nicaragua, 2,000 people died in the municipality of Posoltega as 10 communities were buried in mudslides when the Casita volcano crater collapsed after several days of torrential rains. Three-quarters of a million people were left homeless in the area. Posoltega lies in an area that has been almost completely deforested. Hurricane Mitch's trail of death and damage in eastern Nicaragua also was intensified by deforestation of the upper Rio Coco watershed. More than 20 inches of rain caused the river to rise more than 60 feet within a few days.

Preservation of forests became an environmental issue in eastern Nicaragua following Hurricane Mitch's devastation. Local native peoples and ecologists united the same year to evict Korean-owned timber giant Solcarsa, which had won a government contract in a large area of the Miskito rain forest. According to Weinburg, at least 370,500 acres were being deforested annually in Nicaragua at the time of Hurricane Mitch. Nicaragua has lost 60 percent of its forest cover within the last two generations.

Hurricane Mitch left Nicaragua with an enduring legacy, as described by Weinburg:

> [I]n September 1999, almost a year after the disaster . . . President Aleman declared a national emergency over a plague of rats in . . . Nicaragua. Their numbers exploded from an overabundance of dead meat—both human and

animal—following the hurricane. The rats overran fields and homes, decimating crops. Poison had to be distributed in mass quantities to beat the infestation. (Weinburg 1999, 54)

Former Honduran President Rafael Callejas attributed the extreme death and damage wrought by Hurricane Mitch to "mudslides that were the result of uncontrolled deforestation and therefore could have been prevented" (Weinburg 1999, 54). Ecological activism, however, can be as dangerous in Honduras as the winds, rains, and mudslides of a major hurricane. A few weeks before Mitch made landfall, Carlos Luna, a local opponent of logging in the central mountains near Tegucigalpa, the country's capital and largest city, was gunned down by unknown assassins.

Deforestation and the Carbon Cycle

Trees are roughly 50 percent carbon. Alive, they remove carbon dioxide from the atmosphere and replace it with oxygen. Dying and dead trees (or being burned as fuel) become sources of and not absorbers ("sinks") of carbon dioxide and other greenhouse gases. The destruction of forests around the world is no trivial matter for the atmosphere's carbon budget. Although the atmosphere contains about 750 billion tons of carbon dioxide, forests contain about 2,000 billion tons. Roughly 500 billion tons is stored in trees and shrubs, and 1,500 billion tons is tied up in peat bogs, soil, and forest litter (Jardine 1994).

By 2007, approximately one-fifth of the human-caused greenhouse gases being released into the atmosphere came from deforestation, during which time carbon stored in trees enters the air. Indonesia has been clearing more forests than any other country. In areas such as the province of Riau (on the island of Sumatra), more than half the forests have been felled in a decade, many for palm oil plantations, and often for ethanol, a supposedly Earth-friendly fuel that replaces petroleum-based

Might Deforestation Exert a Cooling Effect?

Might deforestation exert a cooling effect? Xuhui Lee and colleagues have made such a case as described in *Nature* (2011, 384):

> Deforestation in mid- to high latitudes is hypothesized to have the potential to cool the Earth's surface by altering biophysical processes. In climate models of continental-scale land clearing, the cooling is triggered by increases in surface albedo and is reinforced by a land albedo–sea ice feedback. This feedback is crucial in the model predictions; without it other biophysical processes may overwhelm the albedo effect to generate warming instead. Ongoing land-use activities, such as land management for climate mitigation, are occurring at local scales (hectares) presumably too small to generate the

feedback, and it is not known whether the intrinsic biophysical mechanism on its own can change the surface temperature in a consistent manner. Night-time temperature changes unrelated to changes in surface albedo are an important contributor to the overall cooling effect. The observed latitudinal dependence is consistent with theoretical expectation of changes in energy loss from convection and radiation across latitudes in both the daytime and nighttime phase of the diurnal cycle, the latter of which remains uncertain in climate models. (Lee et al. 2011, 384)

Further Reading

Lee, Xuhui, et al. "Observed Increase in Local Cooling Effect of Deforestation at Higher Latitudes." *Nature* 479 (November 17, 2011): 384–387.

products like gasoline. The burning and drying of Riau's peatlands, also to make way for palm oil plantations, similarly releases some 1.8 billion tons of greenhouse gases a year (Gelling 2007).

As Indonesia's forests were being felled, government policy encouraged a homesteading program on Sumatra and Borneo for farmers moving from overcrowded Java. Indonesia was using some of the same government incentives Brazil was using to "develop" the Amazon valley. During the 1980s, hundreds of thousands of Brazilians moved to the Amazon valley in a fashion resembling the homesteading of the North American west a century earlier. Some of this migration was financed by international development agencies such as the World Bank.

Deforestation also has become a major problem (and political issue) in Mexico as well as in Central America. During the spring of 1999, fires raged out of control throughout the mountains of southern Mexico, sweeping through parts of the Chiapas highlands, as well as the Sierra Tarahumara of Chihuahua. The fires cloaked Mexican skies in an acrid haze from its borders with Guatemala and Texas. Among the causes of these fires were private timber operations on *ejido* (native-held) lands that compound the deforestation caused by campesinos clearing lands for their *milpas*. In October 1997, deforestation took a deadly toll when Hurricane Paulina hit the Sierra Madre del Sur in the Mexican states of Guerrero and Oaxaca. Mudslides cascading down denuded mountainsides left scores dead and thousands homeless, especially in Zapotec country.

Perhaps two-thirds of tropical deforestation is caused by slash-and-burn agriculture by small farmers and ranchers who are threatened by the spread of commercial farming and ranching enterprises. Driven out of more populous areas, subsistence farmers are forced to destroy large amounts of tropical forest to create new farmland. According to a study by the *Global Futures Bulletin* ("Much Deforestation" 1999), the pattern is the same in much of Africa, Asia, and Latin America. Increasing human population and pressure on the land is the ultimate cause of the deforestation by slash-and-burn agriculture, according to the report.

Deforestation Shapes Climate Worldwide

"The effects of tropical deforestation on climate go well beyond carbon," said Professor Deborah Lawrence. "It causes warming locally, regionally, and globally, and it changes rainfall by altering the movement of heat and water" (McSweeney 2014). Lawrence coauthored a worldwide study of tropical rainforest deforestation (Lawrence and Vandecar 2014) that indicates that deforestation, which raises global carbon dioxide emissions by 11 percent, changes climate in ways that put agricultural productivity at risk. Lawrence and Vandecar's study examined a wide range of studies on deforestation in the Amazon basin, Central Africa, and Southeast Asia. Their chief finding is that large-scale deforestation affects climate change worldwide over and above its more obvious effects on any given locale. These models suggest, for example, that "deforestation in the Amazon . . . can reduce rainfall over the [American] Midwest and even in northeast China. Deforestation in central Africa can cause a drop in rainfall in Southern Europe, and loss of trees in Southeast Asian can bring wetter conditions in Southern Europe and the Arabian Peninsula" (McSweeney 2014).

Robert McSweeney, writing for the Web site Carbon Brief (2014), explained the causal links: "First you have to bear in mind that rainforests cool the air above them by turning water from the soil into moisture in the air. Chop the trees down, and you remove the cooling effect from this additional moisture. The effect is so pronounced, the study finds, that if all the trees in the tropics were cut down global temperature could increase by as much as 0.7 degrees." Lawrence compared the effects of large-scale deforestation to boiling water: "Imagine steam rising off a pot of boiling water, hitting the ceiling in your kitchen and flowing outward, along the ceiling, out the door to your hallway" (McSweeney 2014).

As Lawrence and Vandecar wrote in *Nature Climate Change* (2014), "Future agricultural productivity in the tropics is at risk from a deforestation-induced increase in mean temperature and the associated heat extremes and from a decline in mean rainfall or rainfall frequency. . . . Negative impacts on agriculture could extend well beyond the tropics."

Similar localized weather changes have been observed in deforested parts of the Amazon basin. Scientists say that cloud forests in Madagascar, the Andes, and New Guinea also are at risk. According to the Hartshorn-led study, "These results suggest that current trends in tropical land use will force cloud forests upward and they will thus decrease in area and become increasingly fragmented and in many low mountains may disappear altogether" (Polakovic 2001, A1). "It's incredibly ominous that over such a distance deforestation can alter clouds in mountains. This is a very serious concern," said Hartshorn. "This is confirmation of what we have predicted for a long time," said Stanford University ecologist Gretchen Daily. "The implications are very serious for the tropics and other parts of the world" (Polakovic 2001, A1).

Using data collected from satellites and computer models, scientists examined how forest clearing along the Caribbean coastline, where more than 80 percent of lowland forests have been cleared for farms and towns. Such clearing influences weather downwind in the Cordillera de Tilaran, a mountain range in northern

Costa Rica. Evaporation from lowland vegetation is a principal source of moisture for the 4,000- to 5,000-foot mountains during the dry season of January to mid-May.

As Gary Polakovic explained in the *Los Angeles Times*:

The researchers found that the moisture content of the clouds over the mountains has declined by about half since intensive land clearing began in the 1950s. Also, the cleared land is warmer, pushing the base of clouds nearly a quarter of a mile higher on some days, meaning they pass over the mountain range dropping little moisture. In contrast, clouds were more abundant over forested lowlands just across the border in Nicaragua, where forest still blankets much of the coastal plain. (Polakovic 2001, A1)

Tropical rain forests are typically the last areas colonized by people, a final refuge of biodiversity in places where lowlands have been cleared and developed. "Many cloud forest organisms have literally nowhere to go," said Dr. Nalini M. Nadkarni, an ecologist at Evergreen State College in Olympia, Washington. "They're stuck on an island of cloud forest. If you remove the cloud, it's curtains for them" (Yoon 2001, F5).

Many watersheds fed by wet highland forests also are threatened. "We always knew that if you have a town and a mountain behind it, you protect the mountain forests to protect the water," said James O. Juvik, a tropical ecologist at the University of Hawaii at Hilo. "Even if you leave the cloud forest intact like good conservationists, if you clear the lowland forests, you can diminish the cloud forest and affect your water flow" (Yoon 2001, F5).

In Indonesia and Brazil, new landowners have been migrating into the tropical forests year by year, altering the landscape (and Earth's carbon balance) by felling and burning trees. The rain forest canopy that had once produced a surplus of oxygen has been replaced by cattle, human beings, and motor vehicles. A survey conducted by the United Nations in 1982 estimated that tropical forests were being felled at a rate of 70,000 square miles (180,000 square kilometers) per year. The report estimated that 300 million subsistence farmers around the world, most of them in the tropics, were turning an area of forest the size of Denmark into a net producer of carbon every three months (Bates and Project Plenty 1990, 92). In 1987, 25,000 square miles of Brazilian rain forest were felled; a year later, the area that was deforested doubled. In the 1980s, lowland tropical forests in Indonesia, the Philippines, Malaysia, and parts of West Africa shrank quickly as well.

According to the United Nations, 42 million acres of rain forest (an area the size of Florida) were being lost to fire every year by the 1990s, with 80 percent of the fires set on purpose. According to the Brazilian government's own figures, the amount of Amazonian forest lost to fire doubled between 1997 and 1998. The Woods Hole Research Center in Massachusetts said that 7,800 square miles of Amazonian rain forest caught fire in 1997.

By the 1990s, according to R. A. Houghton et al., "This large area of tropical forest is nearly balanced with respect to carbon" (Houghton et al. 2000, 301).

The combined effects of deforestation, abandonment, logging, and fire may thus yield sources of carbon. . . . These fluxes are similar in magnitude (but opposite in sign) to the sink calculated recently for natural ecosystems in the region. Taken together, the sources (from land-use change and fire) and the sinks (in natural forests) suggest that the net flux of carbon between Brazilian Amazonia and the atmosphere may be nearly zero, on average. (Houghton et al. 2000, 303–304)

Deforestation and Flooding

Approximately half of the rain that falls on a rain forest is produced by its own humidity. Deforestation over large areas influences regional climate because there are fewer plants to produce the moisture that returns to Earth as rain. More water runs off, carrying more topsoil, leaving less forest cover. Deforestation thus breaks a forest's hydrology cycle. As a tropical forest declines, regional weather becomes hotter, drier, and more prone to fire within surviving stands of trees. Increasing wildfires (such as those during the late 1990s in Mexico, Brazil, and Indonesia) add even more carbon to the atmosphere. By the late 1990s, the burning of tropical forests was contributing about 20 percent of the human-induced carbon dioxide buildup in the atmosphere. The burning of the Amazon rain forest alone contributes around a quarter of the worldwide total. In February 1999, deforestation was implicated as a major culprit when dozens of people died and hundreds of millions of dollars in property was destroyed in massive floods that shut down the industrial capital of South America, São Paulo. Flooding that year was also aggravated by deforestation and provoked unprecedented destruction in Caracas, Venezuela. Deforestation also intensified the death and destruction wrought in Central America by Hurricane Mitch.

According to Stephen Schwartzman,

The Woods Hole Research Center has found that for every acre cleared and burned in the Amazon, at least another acre burns in ground fires under the forest canopy or is degraded by selective logging (not picked up by the satellites). The frequency and extent of these ground fires skyrocket in El Niño events, which can then cause drought in some tropical forests. . . . [S]uch fires are likely to increase in frequency and intensity with global warming. (Schwartzman 1999, 63)

Deforestation is especially dangerous because once an area is denuded, local soil conditions (and sometimes weather) change in a way that makes tree regrowth more difficult. In many areas where deforestation has been extensive, previously forested areas have been replaced by sparse grasses, stunted shrubs, and bare eroded earth.

Further Reading

Almendares, J., and M. Sierra. "Critical Conditions: A Profile of Honduras." *Lancet* 342 (December 4, 1993): 1400–1403.

Bates, Albert K., and Project Plenty. *Climate in Crisis: The Greenhouse Effect and What We Can Do.* Summertown, TN: The Book Publishing Co., 1990.

Gelling, Peter. "Forest Loss in Sumatra Becomes a Global Issue." *The New York Times*, December 6, 2007. http://www.nytimes.com/2007/12/06/world/asia/06indo.html.

Gillis, Justin. "The Threats to a Crucial Canopy." *The New York Times*, October 1, 2011. http://www.nytimes.com/2011/10/01/science/earth/01forest.html.

Gillis, Justin. "Climate Change Study Finds U.S. Is Already Widely Affected." *New York Times,* May 6, 2014. http://www.nytimes.com/2014/05/07/science/earth/climate-change-report.html.

Gordon, Anita, and David Suzuki. *It's a Matter of Survival.* Cambridge, MA: Harvard University Press, 1991.

Houghton, R. A., et al. "Annual Fluxes of Carbon from Deforestation and Regrowth in the Brazilian Amazon." *Nature* 403 (January 20, 2000): 301–304.

Jardine, Kevin. "The Carbon Bomb: Climate Change and the Fate of the Northern Boreal Forests." Ontario, Canada: Greenpeace International, 1994. http://dieoff.org/page129.htm.

Lawrence, Deborah, and K. Vandecar. (2014) "Effects of Tropical Deforestation on Climate and Agriculture." *Nature Climate Change* (December 2014). http://www.nature.com/nclimate/journal/v5/n1/full/nclimate2430.html.

Lawton, R. O., et al. "Climatic Impact of Tropical Lowland Deforestation on Nearby Montane Cloud Forests." *Science* 294 (October 19, 2001): 584–587.

McSweeney, Robert. "Deforestation in the Tropics Affects Climate Around the World, Study Finds." *Carbon Brief* (December 18, 2014). https://www.carbonbrief.org/deforestation-in-the-tropics-affects-climate-around-the-world-study-finds.

Minder, Raphael. "Portugal Fires Kill More Than 60, Including Drivers Trapped in Cars." *New York Times*, June 18, 2017. https://www.nytimes.com/2017/06/18/world/europe/portugal-pedrogao-grande-forest-fires.html.

"Much Deforestation Driven by Population, Poverty." *Global Futures Bulletin* 84 (May 15, 1999). http://www.gsreport.com/articles/art000149.html.

Polakovic, Gary. "Deforestation Far Away Hurts Rain Forests, Study Says; Downing Trees on Costa Rica's Coastal Plains Inhibits Cloud Formation in Distant Peaks." *Los Angeles Times*, October 19, 2001, A1.

Romero, Simon. "The Alarming Urbanization of the Amazon." *Planetizen* (November 27, 2012). https://www.planetizen.com/node/59462.

Schwartzman, Stephen. "Reigniting the Rainforest: Fires, Development and Deforestation." *Native Americas* 16(3–4) (Fall–Winter 1999): 60–63.

Webb, Jason. "World Forests Said Vulnerable to Global Warming." Reuters. November 4, 1998. http://bonanza.lter.uaf.edu/~davev/nrm304/glbxnews.htm (no longer available).

Weinburg, Bill. "Hurricane Mitch, Indigenous Peoples, and Mesoamerica's Climate Disaster." *Native Americas* 16(3–4) (Fall–Winter 1999): 50–59. http://nativeamericas.aip.cornell.edu/fall99/fall99weinberg.html.

"Wildfires Light Up Portugal." NASA Earth Observatory, June 20, 2017. https://earthobservatory.nasa.gov/IOTD/view.php?id=90427&src=eoa-iotd.

Yoon, Carol Kaesuk. "Something Missing in Fragile Cloud Forest: The Clouds." *The New York Times*, November 20, 2001, F5.

See also: Australia, Heat and Drought in; Desertification; Drought, United States; Drought, Worldwide; Drought and Deluge, Scientific Analysis of

DESERTIFICATION

Climate change is playing a role in turning large amounts of fertile land to desert—enough within the next generation to create an "environmental crisis of global proportions"—as well as large-scale migrations, especially in parts of Africa and Central Asia, according to the United Nations. "The costs of desertification are large," said Zafar Adeel of the United Nations University in Bonn, Germany (Rosenthal 2007). "Already at the moment there are tens of millions of people on the move," Adeel said. "There's internal displacement. There's international migration. There are a number of causes. But by and large, in sub-Saharan Africa and Central Asia this movement is triggered by degradation of land," said Adeel (Rosenthal 2007). Overuse of limited water resources has been made worse by drought driven by climate change. In addition, populations are rising and many rivers are being diverted for irrigation and short-term gain.

"Today, those migrants who are escaping dry lands are mostly moving around far from the developed world," said Janos Bogardi of the U.N. University, a technical adviser on the report. "Those who end up on boats to Europe are the tip of an iceberg" (Rosenthal 2007). "The numbers we now find alarming may explode in an uncontrollable way," Bogardi said, eight years before Europe was inundated by a flood of refugees fleeing both war and drought in Syria, Iraq, Afghanistan, and parts of Africa. "Because if you look at land use now and dry land, there is the potential that we are nearing a tipping point" (Rosenthal 2007).

Approximately one-third of the land on Earth is now desert—almost 20 million square miles—and the percentage increases every year. The Intergovernmental Panel on Climate Change (IPCC) projects that rainfall could decrease by 15 percent to 20 percent in the Middle East and 25 percent in North Africa as weather becomes more extreme with global warming (Bilger 2011, 112).

Although warmer air holds more moisture, not everyone will see more precipitation in a globally warmed world. Many deserts already are expanding in a worldwide pattern influenced by atmospheric circulation patterns that meteorologists call *Hadley cells.*

Most deserts range between 20° and 40° north and south latitude. Although precipitation patterns are influenced by other factors (such as ready access or lack thereof to ocean-borne moisture), rainfall is strongly influenced by Hadley cells, which determine whether air generally rises or falls at certain latitudes. Rising air portends instability, low pressure, and storminess; descending air generally provokes high pressure and clear skies. In a warmer world, Hadley cells expand, which causes deserts to expand, a process that is already evident from news reports around the world (see Atmospheric Circulation).

Drought: The Most Pressing Problem

Drought is the most pressing problem caused by climate change but receives too little attention, according to Joseph Romm, writing in *Nature.* Rom makes the case that several climatic factors could produce an "American nightmare" of desertification (or, as he calls it, "dustbowlification") by 2050, of which the 2011 summer heat wave and drought in Texas was a foretaste. Drawing from a paper by Jonathan

Overpeck, Romm asserts, "We are at the dawn of the super interglacial drought" (Romm 2011, 450). As far back as 1990, scientists at the Goddard Institute for Space Studies projected that severe droughts that had roughly averaged once ever 20 years "could become an every-other-year-phenomenon by mid-century" (Romm 2011, 450).

In the United States, precipitation patterns are shifting to expand the dry subtropics to roughly a line from Kansas to California. Earlier snowmelt provides less water in the mountains to counter summer drought, which becomes more intense as dry ground allows the sun's energy to raise temperatures. What rain falls comes in deluges, and more runoff than drought relief is the result. These conditions look more likely in many midcontinental landmasses, not just the United States—southern Europe, southeast Asia, Brazil, and large parts of Australia and Africa. In the past six years, the Amazon has experienced two "100 year droughts" (Romm 2011, 451), and these have turned the area from a carbon dioxide sink to a source—as much as human emissions from the Untied States.

Romm asserts that these trends have ominous implications for food production. Already, sea level rise and saltwater intrusion are reducing output in river deltas such as the Nile and Ganges. At the same time, ocean acidification, warming, and overfishing are depleting food from the sea.

Deserts Spread in North China

Researchers based at Peking University and the Chinese Academy of Sciences disclosed in 2015 that lakes on the Mongolian plateau northwest of Beijing have been shrinking rapidly. Using several decades of satellite images, the researchers said that by 2015 lake surface area had decreased 30 percent in a quarter century because of warming temperatures and declining precipitation as well from increased mining and agricultural activities. Several lakes have completely dried up. Results were published in *Proceedings of the National Academy of Sciences* (Tao et al. 2015, 2281–2286).

Xinkai Lake, near the Russian border in northern Inner Mongolia, was sizable in 2001 but dry by 2006. Coal mining has been a major drain of water; the number of mines increased from 156 in 2000 to 865 in 2010. By 2014, Inner Mongolia was China's second-largest coal-producing area, as well as a major supplier of rare-earth minerals. Farmers also have been irrigating to an extent that reduces water levels. According to NASA's Earth Observatory ("Shrinking Lakes" 2015), quoting Tao et al., "The amount of irrigated cropland in Inner Mongolia has increased from 6,600 square kilometers in the late 1970s to 30,003 square kilometers in 2010."

Spreading Deserts in Africa's Sahel

Africa has been plagued with expanding deserts as the Sahara pushes the populations of Morocco, Tunisia, and Algeria northward toward the Mediterranean. At the southern edge of the Sahara—in countries from Senegal and Mauritania in the west to Sudan, Ethiopia, and Somalia in the east—demands of growing human populations and livestock numbers are converting land into desert. Iran also is battling desertification. Mohammad Jarian, who directs Iran's antidesertification organization, reported in 2002 that sandstorms had buried 124 villages in the southeastern province of Sistan and Baluchistan (Brown 2006).

At the end of the first 20 years of the 21st century, some 60 million people are expected to have abandoned the Sahelian region of northern Africa that borders the fringe of the Sahara Desert. According to former U.N. Secretary-General Kofi Annan in 2002, desertification is one of the world's most significant problems. In northeast Asia, for example, "dust and sandstorms have buried human settlements and forced schools and airports to shut down . . . while in the Americas, dry spells and sandstorms have alarmed farmers and raised the specter of another Dust Bowl, reminiscent of the 1930s." In southern Europe, "lands once green and rich in vegetation are barren and brown," Annan said ("Global Climate Shift" 2002).

Australian government researcher Leon Rotstayn has compiled evidence that air pollution probably has contributed to catastrophic drought in the Sahel. Sulfate aerosols, tiny atmospheric particles, have contributed to a global climate shift, he said. "The Sahelian drought may be due to a combination of natural variability and atmospheric aerosol," said Rotstayn. "Cleaner air in the future will mean greater rainfall in this region," he continued ("Global Climate Shift" 2002).

"Global climate change is not solely being caused by rising levels of greenhouse gases. Atmospheric pollution is also having an effect," said Rotstayn, who is affiliated with the Commonwealth Scientific and Industrial Research Organisation (CSIRO), the Australian government's climate-change research agency. Using global climate simulations, Rotstayn found that sulfate aerosols, which are concentrated mainly in the northern hemisphere, make cloud droplets smaller. This makes clouds brighter and longer-lasting so they reflect more sunlight into space, cooling Earth's surface below ("Global Climate Shift" 2002). As a result, the tropical rain belt, which migrates northward and southward with the seasonal movement of the sun, is weakened in the northern hemisphere and does not move as far north. This change has had a major impact on the Sahel, which has experienced devastating drought since the 1960s. Rainfall was 20 percent to 49 percent lower than in the first half of the 20th century, causing widespread famine and death ("Global Climate Shift" 2002).

Further Reading

Bilger, Burkhard. "The Great Oasis." *The New Yorker*, December 10 and 26, 2011, 110–121.

Brown, Lester R. "The Earth Is Shrinking." Environment News Service, November 20, 2006. http://www.ens-newswire.com.

"Global Climate Shift Feeds Spreading Deserts." Environment News Service, June 17, 2002. http://ens-news.com/ens/jun2002/2002-06-17-03.asp.

Romm, Joseph. "Desertification: The Next Dust Bowl." *Nature* 478 (October 27, 2011): 450–451.

Rosenthal, Elisabeth. "Likely Spread of Deserts to Fertile Land Requires Quick Response, U.N. Report Says." *The New York Times,* June 28, 2007. http://www.nytimes.com/2007/06/28/world/28deserts.html.

"Shrinking Lakes on the Mongolian Plateau." NASA Earth Observatory. April 8, 2015. http://earthobservatory.nasa.gov/IOTD/view.php?id=85665&src=eoa-iotd.

Tao, Shengli, et al. "Rapid Loss of Lakes on the Mongolian Plateau." *Proceedings of the National Academy of Sciences* 112(7) (2015): 2281–2286.

See also: Australia, Heat and Drought in; Drought, United States; Drought, Worldwide; Drought and Deluge, Scientific Analysis of

LAND USE

The growth of cities and industrial-scale agriculture may be responsible for some recent temperature increases across the United States, according to a study by scientists at the University of Maryland. Meteorologists Eugenia Kalnay and Ming Cai found evidence that a temperature increase of 0.50°C over 100 years may be attributed to changes in land use ("Half U.S." 2003).

Kalnay and Cai's research used records from 1,982 surface stations located below 500 meters in elevation in the 48 contiguous U.S. states during 50 years (1950–1999), as well as trends based on data from satellite and weather balloons. They reported the following:

> The most important anthropogenic influences on climate are the emission of greenhouse gases and changes in land use, such as urbanization and agriculture. But it has been difficult to separate these two influences because both tend to increase the daily mean surface temperature. . . . Our results suggest that half of the observed decrease in diurnal temperature range is due to urban and other land-use changes. Moreover, our estimate of 0.27 degrees C mean surface warming per century due to land-use changes is at least twice as high as previous estimates based on urbanization alone. (Kalnay and Cai 2003, 528–529)

Land-Use Changes: Rural and Urban

A comparison of urban and rural weather stations, without including agricultural effects, would underestimate the total impact of land-use changes, Kalnay and Cai asserted. The urban heat-island effect actually takes place mostly at night, the two scientists wrote, "when buildings and streets release the solar heating absorbed during the day" ("Half U.S." 2003). At the time of maximum temperature, they argue, the urban effect is one of slight cooling due to shading, aerosols, and thermal inertia differences between city and country that are not well understood currently ("Half U.S." 2003).

Urban sprawl, deforestation, and agricultural practices can alter temperatures and rainfall patterns in ways that sometimes augment the effects of increased greenhouse gases. "Our work suggests that the impacts of human-caused land-cover changes on climate are at least as important, and quite possibly more important than those of carbon dioxide," said Roger Pielke, an atmospheric scientist at Colorado State University. "Through land-cover changes over the last 300 years, we may have already altered the climate more than would occur . . . [from] the radiative effect of a doubling of carbon dioxide," added Pielke, who was lead author of a study published in the August 2002 issue of *Philosophical Transactions: Mathematical, Physical & Engineering Sciences* (Lazaroff 2002; Pielke 2002).

Pielke and his colleagues asserted that if carbon dioxide emissions continue to rise at recent rates, the level of carbon dioxide in the atmosphere will double from preindustrial levels by 2050. At the same time, land-surface uses will continue to change. According to a report by the Environment News Service (ENS), "Different land surfaces influence how the sun's energy is distributed back to the atmosphere. For example, if a rain forest is removed and replaced with crops, there is less transpiration, or evaporation of water from leaves. Less transpiration leads to warmer temperatures in that area" (Lazaroff 2002).

Land Use and Temporary Cooling

Land-use changes can produce temporary regional cooling as well as warming. If farmland is irrigated, for example, more water transpires and evaporates from moist soils, which cools and moistens the atmosphere, changing local precipitation and cloudiness. Forests may influence the climate in more complicated ways than previously thought, the authors contend. According to the ENS report, "In regions with heavy snowfall, reforestation or the growth of new forests would cause the land to reflect less sunlight, meaning that more heat would be absorbed. This could result in a net warming effect, even though the new trees would remove carbon dioxide from the atmosphere through photosynthesis during the growing season" (Lazaroff 2002). Reforestation also could increase transpiration in an area, putting more water vapor into the air. Water vapor in the troposphere, the lowest and densest part of Earth's atmosphere, is the biggest contributor to greenhouse gas warming, according to researchers.

Australian researchers also believe they have found strong evidence that clearing land may trigger major climatic changes. A Macquarie University team said that its findings indicate that the climate can respond suddenly and dramatically to centuries of environmental abuse. The researchers used one of Australia's most powerful supercomputers to model changing rainfall patterns after the mid-1970s in the southwestern corner of the state of Western Australia.

Some parts of the region have suffered declines of as much as 15 percent and 20 percent in winter rainfall, which threatens Perth's water supply. Research by Macquarie University atmospheric science researchers Andy Pitman, Neil Holbrook, and Gemma Narisma (working with Roger Pielke) suggests the clearing of land may be responsible for approximately half of the rain shortfall (Macey 2004). Pitman said the results of the modeling closely reflected the rainfall changes observed since the mid-1970s, suggesting that land clearing may exert five or 10 times the influence previously expected. "We didn't expect to find that," he said. "It scared the hell out of us" (Macey 2004).

Forests that have been cleared for farming once slowed moist winds blowing in from the Indian Ocean, Pitman said. "This slowing of the atmosphere causes turbulence, which in turn generates rainfall. Without the tree cover, the water in the atmosphere flows across the landscape and is deposited elsewhere. Our results suggest, and observations indicate, that it's falling farther inland, outside of the catchments that provide the Perth water supply. . . . It may be another argument why we shouldn't be logging old-growth forests" (Macey 2004). These findings, to be published in the *Journal of Geophysical Research*, also showed that climatic changes could appear decades or centuries after humans began interfering with the environment. "It may be that when the effects of deforestation suddenly exceed a threshold the climate is likely to respond in a dramatic way" (Macey 2004).

Further Reading

"Half U.S. Climate Warming Due to Land Use Changes." Environment News Service, May 28, 2003. http://ens-news.com/ens/may2003/2003-05-28-01.asp (no longer available).

Kalnay, Eugenia, and Ming Cai. "Impact of Urbanization and Land-Use Change on Climate." *Nature* 423 (May 29, 2003): 528–531.

Lazaroff, Cat. "Land Use Rivals Greenhouse Gases in Changing Climate." Environment News Service, October 2, 2002. http://ens-news.com/ens/oct2002/2002-10-02-06.asp (no longer available).

Macey, Richard. "Climate Change Link to Clearing." *Sydney Morning Herald*, June 29, 2004.

Pielke, Roger. "Land Use Changes and Climate Change." *Philosophical Transactions: Mathematical, Physical & Engineering Sciences* (Journal of the Royal Society of London), August 2002.

See also: Agriculture, Future of, Biodiversity; Drought, United States; Drought, Worldwide; Extinctions; Greenhouse Gas Emissions

PINE BEETLES

By 2015, several years of intense drought and rising temperatures had created perfect conditions for bark-beetle infestations across the U.S. West. These conditions had been ravaging forests there for more than a dozen years. This condition had been intensifying since the fall of 2002 when large swaths of evergreen forests in western Montana and the Idaho panhandle, as well as parts of California, Colorado, and Utah fell victim to unusually large infestations of these insects, including the Douglas fir bark beetle, spruce beetle, and mountain pine beetle. A warming climate distorts the natural cycle as heat and drought deliver a plague that devours forests from Alaska to New Mexico. The beetles have now moved as far east as New Jersey.

Before rising temperatures accelerated their breeding cycle during the 1990s, several hundred species of bark and pine beetles were part of the natural cycle in the United States and Canadian west. They helped break down dying trees, a few at a time. Some scientists argue that the beetles are a natural phenomenon and should not be interfered with. Diana L. Six, a professor of forest entomology and pathology at the University of Montana, is one of them. "Lodge pine [are] evolved to go with a stand-replacing event, such as fire or beetles, then regenerate really quickly," said Six, who has studied bark beetles for 16 years. "It's not the end of the forests, and they are not destroyed," said Six. She and some others assert that the beetles take out only older trees, causing the soil to replenish with nitrogen, with new growth replacing old. The insects recycle nutrients, said Ken Raffa, an entomologist at the University of Wisconsin "and they introduce microorganisms that further break down the wood" (Robbins 2009).

For the *Washington Post*, Doug Struck described how the beetles kill trees:

In an attack played out millions of times over, a female beetle no bigger than a rice grain finds an older lodgepole pine, its favored host, and drills inside the bark. There, it eats a channel straight up the tree, laying eggs as it goes. The tree fights back. It pumps sap toward the bug and the new larvae, enveloping them in a mass of the sticky substance. The tree then tries to eject its captives through a small, crusty chute in the bark.

Countering, the beetle sends out a pheromone call for reinforcements. More beetles arrive, mounting a mass attack. A fungus on the beetle, called the blue stain fungus, works into the living wood, strangling its water flow. The larvae begin eating at right angles to the original up-and-down channel, sometimes girdling the tree, crossing channels made by other beetles. The pine is doomed. As it slowly dies, the larvae remain protected over the winter. In spring, they burrow out of the bark and launch themselves into the wind to their next victims. (Struck 2006)

By late 2008, British Columbia, the westernmost Canadian province, had lost 33 million acres of high-altitude lodgepole pine forest. From the province's infested forests, unusually strong winds in 2006 pushed clouds of mountain pine beetles, a species of bark beetle, over the Continental Divide to northern Alberta, the neighboring province, where they had never before been observed. In years to come, the beetles could ride the winds eastward to the Great Lakes (Robbins 2008).

By the end of 2008, Montana, Wyoming, and Colorado each had lost more than 1 million acres of trees to the beetles, which were spreading in an exponential fashion, according to Clint Kyhl, director of a Forest Service incident management team in Laramie, Wyoming. Kyhl said that nearly all of Colorado's larger lodgepole pine trees—some 5 million acres—could die within five years "The Latin name [for the beetles] is *Dendroctunus*, which means tree killer," said Gregg DeNitto, a Forest Service entomologist in Missoula, Montana. "They are very effective" (Robbins 2008).

By 2008, so many trees were being killed by pine-bark beetles that the National Center for Atmospheric Research in Boulder, Colorado, was studying how the swaths of dead trees change the weather from southern Wyoming to northern New Mexico. Trees killed by the beetles may influence rainfall, temperatures, and smog. Preliminary computer modeling indicated that the alterations could amount to temporary temperature increases between 2°F and 4°F ("Pine Beetles" 2008).

By 2009, nearly every mature lodgepole pine in Colorado was dead across 5 million acres. Aspens, which grow with lodgepoles between 5,000 and 8,000 feet, also were dying, but no one knew why. All the dead trees were raising the risk of fires in the state. The area covered by dead aspens in Colorado grew from 30,000 acres in 2005 to 540,000 acres in 2008, 15 percent of the aspens in the state (Simon 2009).

In a letter to *The New York Times*, Bill McEwen wrote,

I reside in the semi-arid West, where scientists are just beginning to understand the enormous synergistic impact of global warming, atmospheric drying (drought), and the explosion in insect populations that is killing many of our forests. . . . On a recent vacation to the Northwest, I drive through Sun Valley, Idaho. Around Sun Valley and the nearby Salmon River Valley, entire mountainsides of forest are now being destroyed by out-of-control bark beetle infestations. (McEwen 2004)

Charles Petit described the spreading infestations also in *The New York Times*:

Like an army roaring out of the trenches, bark beetles overwhelm their next round of piney prey, emitting pheromones that draw more attackers to individual trees. Forest managers say defenses—quarantines, burning, and other methods—are ineffective over large areas The result, in British Columbia, is the largest forest insect blight ever seen in North America, and it seems nowhere near its peak. It covers a patch running about 400 miles north-south and 150 miles across. Officials expect 80 percent of British Columbia's mature lodgepoles to be dead by 2014. . . . Loggers are frantically cutting lifeless trees for lumber before they rot. Funguses carried by beetles stain the wood a blotchy blue. Lumber mills promote it as stylish "denim wood." In 2002, high winds carried the beetles through the Rockies and into the Alberta plain. They appear poised to sweep east to the Atlantic through Canada's jackpine boreal forest. (Petit 2007)

In Flagstaff, Arizona, near the world's largest contiguous ponderosa pine forest, Tom Whitham wondered how much more devastation the drought and beetles would cause and to what extent humans will contribute to it. "The thing that would make me really sad is if this was human caused," he said, glancing at the bare trees towering over his pickup truck. "If you lose a 200-year-old forest, you can't get it back" (Wagner 2004).

Deadly wildfires that burned at least 1,000 homes in Southern California during late October 2003 were aggravated not only by fierce Santa Ana winds and 100°F temperatures, drought, and exceptionally low humidity but also by the deaths of more than 1 million mature pine trees killed during the previous year by bark-beetle infestation. According to a Canadian governmental projection, pine-beetle infestation presents a "worst-case scenario" that could afflict 80 percent of British Columbia pine forests by 2020. "Mountain pine beetle outbreaks are stopped by severe winter weather or depletion of the host. The vast spatial extent of the outbreak implies that a weather-stopping event is unlikely," said the report (Baron 2004). By 2004, the beetle infestation already had spread through a broad swath of the B.C. interior, especially around the Prince George area, and extending south to the United States border.

By 2015, 44 million acres of pine forest (60 percent of mature pines), an area the size of Missouri, had been killed in British Columbia alone (Rosner 2015, 102–103). Three-quarters of mature whitebark pines in Yellowstone National Park had turned a signature ghostly white shade of death, provoking hunger for grizzly bears and some birds that eat their seeds. In a decade, foresters in Alberta had cut and burned a million infested pines in an attempt to slow the beetles' eastward migration (Rosner 2015, 101). According to one observer, "The vast tracts of Douglas fir that stood green and venerable for generations [east of Yellowstone National Park] are peppered and painted with swaths of rusty red and gray. For Douglas fir, those are the colors of death" (Stark 2002).

Some types of beetles that once propagated two generations a year have been reproducing three times. "This is all due to temperature," said Barbara Bentz, a research entomologist with the U.S. Forest Service who studies bark beetles. "Two or three degrees [of temperature] is enough to do it" (Wagner 2004). Outside Cody, Wyoming, an entire forest was killed by the drought and beetles. "It used to be a nice spruce forest," said Kurt Allen, a Forest Service entomologist. "It's gone now. You're not going to get those conditions back for 200 or 300 years. We're really not going to have what a lot of people would consider a forest" (Wagner 2004).

Warming not only increases many insects' reproductive energies but also compromises many trees' reproductive capacities. According to Kevin Jardine of Greenpeace International, rising temperatures can cause boreal (northern) trees' pollen and seed cones to develop too rapidly because of higher-than-usual spring temperatures, which leads to reproductive failure. Boreal seeds also germinate within a specific range of soil temperatures. For example, writes Jardine, black spruce seeds germinate between 15°C and 28°C. "If the soil temperature falls below 15 degrees," writes Jardine, "the processes that cause germination come to a halt. If soil temperatures rise above 28°C, bacteria and fungi can attack and consume seeds. The probability of germination also declines rapidly for higher temperatures" (Jardine 1994).

Warming, Insects, and Wildfires

Pine beetles are not the only insects favored by rising temperatures. Warming encourages leaf-eating as well as bark-consuming insects. John Couture and his colleagues at the University of Wisconsin–Madison found that in northern temperate forests the consumption of forest canopy by leaf-eating insects reaches levels that compromise their ability to act as carbon sinks ("Insects Feast" 2015, 133).

The infestations were being encouraged by several factors: a warming trend, which allows the beetles to multiply more quickly and reach higher altitudes; drought, which deprives trees of sap they would usually use to keep the beetles under control; and years of human fire suppression, which increased the amount of elderly wood susceptible to attack. Beetles had been attacking in "epic proportions, killing many stands of trees within a few weeks" (Stark 2002). By killing trees, bark beetles add more fodder for forest fires that become more likely and more widespread under increasingly warm, dry conditions fostered by rising temperatures. Without living tissue that holds water, dead trees burn much faster.

Absolute winter minimum temperatures in winter have risen between 6°F and 10°F in areas affected by pine beetles, according to Steve Running, an ecologist at the University of Montana. Coldest winter readings in the 1950s were 42°F to 47°F below zero. Since 2000, they have rarely dropped below −35°F. The lower the temperature, the lower the survival ratio of bark beetle larvae. As the growing season has become longer, said Running, precipitation on average has remained about the same, provoking more evaporation (and thus drought), which makes trees more vulnerable to beetle infestation (Robbins 2009).

Pine Beetles Reach the U.S. East Coast

Pine bark beetles are best known as a scourge in North America's western forests, but by 2013 they had reached New Jersey, via limited infestations in the U.S. South, as "Southern pine beetles." In the past, cold winters had limited their spread farther north when winters were colder. Winter nights at −8°F, which kill the beetles, have become rare in New Jersey and have allowed the insects to spread. "Scientists say it is a striking example of the way seemingly small climatic changes are disturbing the balance of nature," wrote Justin Gillis in *The New York Times* (2013). "They see these changes as a warning of the costly impact that is likely to come with continued high emissions of greenhouse gases." Because New Jersey's pinelands are flat, the devastation is visible only at a short distance from the ground, or from the air. "It's a tremendously serious issue, but it hasn't gotten anybody's attention," said State Senator Bob Smith, a Democrat from Piscataway and the chairman of the Environment and Energy Committee (Gillis 2013).

Having ravaged more than 30,000 acres in New Jersey between 2002 and 2015, pine beetles moved onto Long Island and were invading the remnants of pitch pine forests that once covered much of New England. As average winter temperatures rose, the beetles were reported in Connecticut and Massachusetts by 2015 and 2016, possibly headed northward into Canada. *The New York Times* reported, "Alarmed scientists first discovered the beetles last year [2015] along a front stretching more than 200 miles, from central Long Island to Cape Cod and Martha's Vineyard. . . . Many of the forests are already unhealthy, a result of overcrowding, making them especially susceptible to the pine beetle's attacks—boring through bark, laying eggs and spreading a crippling fungus" (Schlossberg 2016).

Matthew P. Ayres, a Dartmouth College biologist, said that if precautions were not taken, a widespread invasion could leave only a few pitch pines in the region. *The New York Times* quoted Ayres as saying that he could foresee the beetles possibly reaching the Great Lakes states and Canadian provinces, which are heavily forested with red and jack pines and could be vulnerable. From there, they could infiltrate much of Canada, spreading west to meet the mountain pine beetles that have been moving east into Alberta and Saskatchewan, leaving a ring of dead forests around the continent. "It's an example of something that's happening all over the world," Dr. Ayres said. "It's old pests in new places, and with that comes a whole new set of challenges." (Schlossberg 2016)

Pine Beetles in Canada

The mountain pine beetle has infested an area of British Columbia the size of Maine, devastating swaths of lodgepole pines and reshaping the future of the forest and the communities in it. The beetles are turning trees red and killing more of them than wildfires. "It's pretty gut-wrenching," said Allan Carroll, a research scientist at the Pacific Forestry Centre in Victoria, whose studies tracked a lockstep between warmer winters and the spread of the beetle. "People say climate change is something for our kids to worry about. No. It's now" (Struck 2006).

At the same time, cool temperatures along Alaska's south-central coast had kept the spruce-bark beetle under control. As temperatures warmed, however, the beetles have killed much of the tree cover across 3 million acres, one of the largest insect-caused forest devastations in North America's history. William K. Stevens of *The New York Times* described unprecedented destruction of boreal forests in Alaska by spruce bark beetles that have been whipped into a reproductive frenzy by a warming environment:

> Once these purple-gray stretches of tree skeletons were a green and vital part of the spruce-larch-aspen tapestry that makes up the taiga. . . . Today, in a stretch of 300 or 400 miles reaching westward from the Richardson Highway, north of Valdez, past Anchorage, and down through the Kenai Peninsula, armies of spruce bark beetles are destroying the spruce canopy or have already done so. Often the trees are red instead of gray—freshly killed but not yet desiccated. (Stevens 1999, 178)

Fires are being set in Alberta, along with traps, as thousands of trees are felled in an attempt to inhibit the beetles' spread across the Rocky Mountains from the West. "This is an all-out battle," said David Coutts, Alberta's minister of sustainable resource development. The Canadian Forest Service calls it the largest known insect infestation in North American history (Struck 2006). The United States is less vulnerable than Canada because its stands of lodgepole pine are not continuous and thus provide no easy path of infestation.

As the pine-beetle infestation spread in Western Canada, Werner A. Kurz, writing in *Nature,* found that mountain pine-beetle infestations have killed enough trees to convert large tracts of Canadian forests, especially in British Columbia, from carbon sinks to sources. By 2007, more than 32 million acres of forests were infested, and the affected area was spreading as warmer summers enable faster beetle reproduction and milder winters fail to kill them (Kurz et al. 2008, 987). "We are seeing this pine beetle do things that have never been recorded before," said Michael Pelchat, a forestry officer in Quesnel, as he followed moose tracks in the snow to examine a 100-year-old pine killed in one season by the beetle. "They are attacking younger trees, and attacking timber in altitudes they have never been before" (Struck 2006).

Until now, Canada's lodgepole pine had been protected from the beetles by early, cold winters. For two decades, however, winters have not been cold enough to kill the beetles. Scientists with the Canadian Forest Service say the average temperature of winters here has risen by more than 4°F in the last century. The same is now true in Alaska.

At British Columbia's Ministry of Forests and Range in Quesnel, forestry officer Pelchat saw the beetle expansion coming as "a silent forest fire" (Struck 2006).

Pelchat and his colleagues tried to stall the invasion, hoping for cold temperatures. They searched out and cut and burned beetle-ridden trees. They thinned forests. They set out traps. But the deep freeze never came. "We lost. They built up into an army and came across," Pelchat said. Surveys show the beetle has infested 21 million acres and killed 411 million cubic feet of trees—double the annual take by all the loggers in Canada. In seven years or sooner, the Forest Service predicts that kill will nearly triple and 80 percent of the pines in the central British Columbia forest will be dead (Struck 2006).

Spruce Bark Beetles in Alaska

On Alaska's Kenai peninsula, a forest nearly twice the size of Yellowstone National Park has been dying. "Century-old spruce trees stand silvered and cinnamon-colored as they bleed sap," from spruce bark-beetle infestations spurred by rising temperatures, wrote reporter Tim Egan of *The New York Times* (2002, June 16). Temperatures were in the middle 80s there by mid-May 2002 as park rangers worried that tinder-dry, warm conditions could provoke major wildfires. Over 15 years (1988–2003), 40 million spruce trees on the Kenai Peninsula died (Whitfield 2003, 338). The beetle infestations have reached Anchorage, where "[v]isitors flying into the city's airport cross islands covered with the bristling, white skeletons of dead trees that are easily visible through the plane windows" (Lynas 2004, 60).

Alaskan author Charles Wohlforth described a plague of bark beetles:

> On certain spring days in the mid-1990s, clouds of spruce bark beetles took flight among the big spruce trees around Kachemak Bay, 120 miles south of Anchorage. They could be seen from miles away, rolling down the Anchor River valley. People who witnessed the arrival sometimes felt like they were in a horror film, the air thick with beetles landing in their eyes and catching in their hair, and knew when it happened that their trees were destined to turn red and die. (Wohlforth 2004, 238)

The six-legged spruce beetle, which is only a quarter-inch long, takes to the air in the spring and looks for trees on which to feed. When beetles find a vulnerable group of trees, they send other beetles "a chemical message," said Ed Holsten, who studies insects for the Forest Service in Alaska. The insects then burrow under the bark, feeding on woody capillary tissue that the tree uses to transport nutrients. Healthy spruce trees produce chemicals (terpenes) that usually repel beetles. The chemicals cannot overwhelm a mass infestation of the type that has been taking place, however (Egan 2002, June 25). As a spruce dies, green needles turn red, and then silver or gray. According to Egan, "Ghostly stands of dead, silver-colored spruce—looking like black and white photographs of a forest—can be seen throughout south-central Alaska, particularly on the Kenai. Scientists estimate that 38 million spruce trees have died in Alaska in the current outbreak" (Egan 2002, June 25).

More than 4 million acres of white spruce trees on the Kenai peninsula were dead or dying by 2004 from beetle infestations. Beetles have been gnawing at spruce trees in Alaska for many thousands of years, but with rapid warming since the 1980s their populations have exploded (Egan 2002, June 25). A series of warmer-than-average years in Alaska allowed the spruce bark beetles to reproduce at twice their historic rate. "Hungry for the sweet lining beneath the bark," wrote Egan, "The beetles have swarmed over the stands of spruce, overwhelming the trees' normal defense mechanisms. . . . The dead spruce forest of Alaska may well be one of the world's most visible monuments to climate change." By 2002, nearly 95 percent of the spruce on the Kenai peninsula had been destroyed by the beetles, leaving a tinder-dry forest ripe for major wildfires that will further devastate the habitats of resident moose, bear, salmon, and other creatures (Egan 2002, June 25).

"The chief reason why the beetle outbreak has been the largest and the longest is that we have had an unprecedented run of warm summers," said Edward Berg, 62, a longtime student of the Kenai peninsula (Egan 2002, June 25). Berg has been piecing together the causes of the forest's demise. "His [Berg's] work is very convincing; I would even say unimpeachable," said Glenn Juday, a forest ecologist at the University of Alaska. "For the first time, I now think beetle infestation is related to climate change" (Egan 2002, June 25). Holsten said that he thinks climate change is only one reason for the beetle outbreak. The trees on the Kenai are old and ripe for beetle outbreaks, he said. If they had been logged or burned in fire, it might have kept the insects down, Holsten said (Egan 2002, June 25).

"It's very hard to live among the dead spruce; it's been a real kick in the teeth," said Berg. "We all love this beautiful forest" (Egan 2002, June 25).

Further Reading

Baron, Ethan. "Beetles Could Chew Up 80 Percent of B.C. Pine: Report: 'Worst-Case Scenario' by 2020 Blamed on Global Warming." *Ottawa Citizen*, September 12, 2004, A3.

Egan, Timothy Egan. "Now, in Alaska, Even the Permafrost Is Melting." *The New York Times*, June 16, 2002, A1.

Egan, Timothy. "On Hot Trail of Tiny Killer in Alaska." *The New York Times*, June 25, 2002, F1.

Gillis, Justin. "In New Jersey Pines, Trouble Arrives on Six Legs." *The New York Times*, December 1, 2013. http://www.nytimes.com/2013/12/02/science/earth/in-new-jersey-pines-trouble-arrives-on-six-legs.html.

"Insects Feast under High CO_2." *Nature* 519 (March 12, 2015): 133.

Jardine, Kevin. "The Carbon Bomb: Climate Change and the Fate of the Northern Boreal Forests." Ontario, Canada: Greenpeace International, 1994. http://dieoff.org/page129.htm.

Kurz, W. A., et al. "Pine Beetle and Forest Carbon Feedback to Climate Change." *Nature* 452 (April 24, 2008): 987–991.

Lynas, Mark. *High Tide: The Truth about Our Climate Crisis.* New York: Picador/St. Martins, 2004.

McEwen, Bill. "The West's Dying Forests" (Letter to the Editor). *The New York Times*, August 2, 2004, A16.

Petit, Charles. "In the Rockies, Pines Die and Bears Feel It." *The New York Times*, January 30, 2007. http://www.nytimes.com/2007/01/30/science/30bear.html.

"Pine Beetles Changing Rocky Mountain Air Quality, Weather," Environment News Service, October 1, 2008. http://www.ens-newswire.com/ens/oct2008/2008-10-01-091.asp (no longer available).

Robbins, Jim. "Bark Beetles Kill Millions of Acres of Trees in West." *The New York Times*, November 18, 2008. http://www.nytimes.com/2008/11/18/science/18trees.html.

Robbins, Jim. "Some See Beetle Attacks on Western Forests as a Natural Event." *The New York Times,* July 7, 2009, D3.

Rosner, Hillary. "The Bug' That's Eating the Woods." *National Geographic*, April 2015, 96–115.

Schlossberg, Tatiana. "Warmer Winter Brings Forest-Threatening Beetles North." *The New York Times*, March 18, 2016. http://www.nytimes.com/2016/03/22/science/southern-pine-beetles-new-england-forests.html.

Simon, Stephanie. "Aspen Trees Die across the West." *Wall Street Journal*, October 14, 2009, A3.

Stark, Mike. "Assault by Bark Beetles Transforming Forests; Vast Swaths of West Are Red, Gray, and Dying; Drought, Fire Suppression, and Global, Warming Are Blamed." *Billings Gazette* in *Los Angeles Times*, October 6, 2002, B1.

Stevens, William K. *The Change in the Weather: People, Weather, and the Science of Climate.* New York: Delacorte Press, 1999.

Struck, Doug. "'Rapid Warming' Spreads Havoc in Canada's Forests; Tiny Beetles Destroying Pines." Washington *Post*, March 1, 2006, A1. http://www.washingtonpost.com/wp-dyn/content/article/2006/02/28/AR2006022801772_pf.html.

Wagner, Angie. "Debate over Causes Aside, Warm Climate's Effects Striking in the West." Associated Press, April 27, 2004 (LEXIS).

Whitfield, John. "Alaska's Climate: Too Hot to Handle." *Nature* 425 (September 25, 2003): 338–339.

Wohlforth, Charles. *The Whale and the Supercomputer: On the Northern Front of Climate Change.* New York: Farrar, Strauss & Giroux, 2004.

See also: Biodiversity; Drought, United States; Drought, Worldwide; Drought and Deluge, Scientific Analysis of; Heat Waves, Evidence of; Heat Waves, Forecasts of; Wildfires, Worldwide

TAR SANDS

Exploitation of tar sands at the surface (some is also mined underground) requires a form of strip mining that scars Earth in ways that will not be quickly nor easily repaired. Opponents of their extraction assert that oil sands are a relatively new form of fossil fuel—the last thing Earth needs when carbon dioxide levels in the atmosphere have risen to more than 400 parts per million. This is more than 40 percent above peak preindustrial levels and damaging to climate, and it causes sea levels to rise and increases oceanic acidity. On a local level, many people worry about oil spills in fragile areas such as the Nebraska Sand Hills that could contaminate the Ogallala aquifer in an area where water is scarce. This aquifer supplies 78 percent of public water and 83 percent of irrigation water in Nebraska, almost a third of the irrigation water used in the United States.

The oil industry and supportive politicians argue that consumers will receive a secure source of vital energy from tar sands. Oil companies produce profits by removing carbon from the ground and selling it as products that ultimately inject carbon dioxide into the atmosphere as a by-product of both manufacturing and consumption. Tar sands advocates argue that the province of Alberta collects a carbon tax (it is very small) and that carbon emissions from oil sands mining have dropped about 25 percent since 1990 as refining become more efficient. Oil sands are still "dirtier" than conventional oil, however, because complex manufacturing is required to make products that are useful as fuel.

"If Canada proceeds [with oil sands development], and we do nothing, it will be game over for the climate," James Hansen, author of *Storms of My Grandchildren,* wrote in *The New York Times.*

> Canada's tar sands, deposits of sand saturated with bitumen, contain twice the amount of carbon dioxide emitted by global oil use in our entire history. If we were to fully exploit this new oil source, and continue to burn our conventional oil, gas and coal supplies, concentrations of carbon dioxide in the atmosphere eventually would reach levels higher than in the Pliocene era, more than 2.5 million years ago, when sea level was at least 50 feet higher than it is now. That level of heat-trapping gases would assure that the disintegration of the ice sheets would accelerate out of control. Sea levels would rise and destroy coastal cities. Global temperatures would become intolerable. Twenty to 50 percent of the planet's species would be driven to extinction. Civilization would be at risk. . . . If this sounds apocalyptic, it is.
>
> The concentration of carbon dioxide in the atmosphere has risen from 280 parts per million to slightly more than 400 ppm over the last 150 years, as of 2014. The tar sands contain enough carbon—240 gigatons—to add 120 ppm. Tar shale, similar to tar sands found mainly in the United States, contains at least an additional 300 gigatons of carbon. If we turn to these dirtiest of fuels, instead of finding ways to phase out our addiction to fossil fuels, there is no hope of keeping carbon concentrations below 500 ppm—a level that would, as earth's history shows, leave our children a climate system that is out of their control. (Hansen 2012)

According to NASA, the Alberta fields, which were first mined in 1967, are "the world's largest oil sands deposit, with a capacity to produce 174.5 billion barrels of oil—2.5 million barrels of oil per day for 186 years" ("Athabasca Oil Sands" 2011).

(The United States as a whole consumes 15 to 20 million barrels of oil per day.) Environmental activist Bill McKibben has called tar sands mining and the Keystone XL pipeline the "fuse to the biggest carbon bomb on the planet" (Tollefson 2013). "Saying that the tar sands are not necessarily worse than coal is like saying that drinking arsenic is not necessarily worse than drinking cyanide," said geophysicist Raymond Pierrehumbert of the University of Chicago. He says that fully developing the tar sands alone could—"even if we suddenly stopped burning coal"—warm the planet an additional 3.6°F by century's end, an amount climate scientists warn could be "catastrophic" (Koch 2014).

Tar Sands as Junk Energy

Tar sands are a mixture of clay and sand with bitumen, a thick, low-grade form of petroleum similar to asphalt. According to Thomas Homer-Dixon, who teaches global governance at the Balsillie School of International Affairs in Waterloo, Ontario, "Tar sands production is one of the world's most environmentally damaging activities. It wrecks vast areas of boreal forest through surface mining and subsurface production. It sucks up huge quantities of water from local rivers, turns it into toxic waste, and dumps the contaminated water into tailing ponds that now cover nearly 70 square miles" (Homer-Dixon 2013).

"Bitumen is junk energy," says Homer-Dixon. "A joule, or unit of energy, invested in extracting and processing bitumen returns only four to six joules in the form of crude oil. In contrast, conventional oil production in North America returns about 15 joules. Because almost all of the input energy in tar sands production comes from fossil fuels, the process generates significantly more carbon dioxide than conventional oil production" (Homer-Dixon 2013). According to NASA, oil sands refining produces the equivalent of 86 to 103 kilograms of carbon dioxide for every barrel of crude oil produced. By comparison, 27 to 58 kilograms of carbon dioxide are emitted in the conventional production of a barrel of crude oil.

By 2012, the mining and refining of tar sands in Canada consumed as much natural gas as that country uses for home heating (Kolbert 2007, 49). The gas is used to produce synthetic oil and by-products such as gasoline. Tar sands require about 15 percent to 40 percent more energy in its manufacturing compared to conventional crude oil; oil shales require about twice as much. Converting tar sands into oil costs about as much as oil at $30 a barrel, however. With oil pushing $100 a barrel in late 2014, tar sands were becoming highly profitable for many fossil fuel companies (Kolbert 2007, 49, 50).

Writing in *The New Yorker*, Ryan Lizza described the mining of tar sands:

Oil sand has the texture of soft asphalt; twenty per cent of it lies close to the surface, and the area is effectively strip-mined. The bitumen-rich sand is removed, mixed with water into a slurry, and spun in centrifuges until the oil is separated, leaving behind vast black tailings ponds that are hazardous to wildlife. The mining operations sprawl ruinously for miles. The remaining eighty per cent of the oil sands lie hundreds of feet down beneath a layer of hard rock. Steam is injected deep belowground until the oil naturally separates and is drawn out. The extra energy required to extract the sand makes it a more carbon-intensive fossil fuel—averaging seventeen percent more . . . than conventional oil. (2013, 42)

Carol Berry (2012) of the Indian Country Today Media Network described the environmental damage of tar sands mining:

If you can imagine the bleak landscape of the moon, you can envision the desolate, 54,000-square-mile tar sands of northern Alberta. . . . "It's literally a toxic wasteland—bare ground and black ponds and lakes—tailings ponds—with an awful smell," said Warner Nazile . . . [an] activist from British Columbia and member of the Wet'suwet'en First Nation. The mining is "despoiling an area roughly the size of England. . . . University of Alberta scientists "found indications that contamination from the tailings ponds was polluting a huge aquifer that ultimately flows into the Arctic Ocean," Nazile said. Two aboriginal communities downstream from the oil sands have experienced higher-than-average rates of cancer and other health problems, he added.

"For a vast stretch of western Canada's boreal forest, the fight over extracting bitumen has already been lost. The question is, how much more will we lose?" wrote Canadian journalist Andrew Nikiforuk (2014) who also is the author of *The Energy of Slaves: Oil and the New Servitude*. After 2000, intensive tar sand mining accelerated and almost 2 million acres of this forest have been cleared or degraded, according to Global Forest Watch. Canadian economist Jeff Rubin said, "When you're schlepping oil from sand, you're probably in the bottom of the ninth inning in the hydrocarbon economy" (Nikiforuk 2014).

After the forests have been removed, "the landscape is reduced to a treeless wasteland" (Nikiforuk 2014). The energy intensity of the bitumen harvest begins with drilling deep into frozen ground that is melted with water that has been heated to steam and pumped to the surface. Some of the bitumen that the Cree once heated to repair leaks in their canoes lies near the surface, from which it is removed by electric shovels the size of large buildings and then transported to mills that remove the sands in trucks carrying 400 tons at a time. Imagine their gas mileage. The toxic sludge that comes out of the mills is dumped into lakes.

Along the Athabasca River, more than a dozen of these enormous waste ponds hold back this industrial excrement. Pollutants in these lakes are leaking into groundwater and the Athabasca River. . . . Come the spring melt, these pollutants rush into the Athabasca River. A growing ring of mercury contamination surrounds the project. (Nikiforuk 2014)

Further Reading

"Athabasca Oil Sands." NASA Earth Observatory. November 30, 2011. http://earthobservatory
.nasa.gov/IOTD/view.php?id=76559&src=eoa-iotd.

Berry, Carol. "Alberta Oil Sands Up Close: Gunshot Sounds, Dead Birds, a Moonscape."
Indian Country Today Media Network, February 2, 2012. http://indiancountrytoday-
medianetwork.com/article/alberta-oil-sands-up-close%3A-gunshot-sounds%2C-dead
-birds%2C-a-moonscape-95444 (no longer available).

Hansen, James E. "Game Over for the Climate." *The New York Times,* May 9, 2012. http://
www.nytimes.com/2012/05/10/opinion/game-over-for-the-climate.html.

Homer-Dixon, Thomas. "The Tar Sands Disaster." *The New York Times*, March 31, 2013.
http://www.nytimes.com/2013/04/01/opinion/the-tar-sands-disaster.html.

Koch, Wendy. "Would Keystone Pipeline Unload 'Carbon Bomb' or Job Boom?" *USA Today*,
March 10, 2014. http://www.usatoday.com/story/news/nation/2014/03/01/keystonexls
-myths-debunked/5651099/.

Kolbert, Elizabeth. "Unconventional Crude: Canada's Synthetic-Fuels Boom." *The New
Yorker*, November 12, 2007, 46–51.

Lizza, Ryan. "The President and the Pipeline." *The New Yorker*, September 16, 2013,
38–51.

Nikiforuk. Andrew. "A Forest Threatened by Keystone XL." *The New York Times*, November
18, 2014. http://www.nytimes.com/2014/11/18/opinion/a-forest-threatened-by-keystone
-xl.html.

Tollefson, Jeff. "Climate Science: A Line in the Sands." *Nature* 500 (August 8, 2013): 136–137.
http://www.nature.com/news/climate-science-a-line-in-the-sands-1.135150 (no longer
available).

See also: Unburnable Carbon

WILDFIRES

Increases in wildfire frequency and severity have become a worldwide phenomenon. Natural wildfires and those started by people (usually to clear farmland) in
Malaysia, Indonesia, Borneo, and Papua New Guinea between 2000 and 2006 contributed as much carbon dioxide to the atmosphere as all other human sources
combined, according to data from a carbon-detecting NASA satellite and computer
models. The results were presented in December 2008 in *Proceedings of the National
Academy of Sciences* ("NASA Study Says" 2009).

During several years from 2000 to 2016, western North America experienced
its worst wildfire seasons on record. The fires were fueled by the worst drought in
at least 500 years and rising temperatures. Nearly 4 million acres of forest from
Alaska to Southern California had burned by the end of July. Visitors to Yosemite
National Park were advised not to overexert themselves because of the "very
unhealthy quality" of the fire-polluted air. Usually, western wildfires reach their peak
in October. In 2004, however, they started earlier than any other season on record.
By early May, 77 fires had started in the state of Washington alone, compared to 22
in 2003 (Lean 2004).

Warming Associated with Longer Fire Seasons

Fire seasons have become longer in many parts of the world, according to a NASA study that associated much of the increase with global warming ("NASA Study" 2009). "Climate strongly influences global wildfire activity, and recent wildfire surges may signal fire weather-induced pyrogeographic shifts," wrote W. Matt Jolly, U.S. Forest Service ecologist, and colleagues in *Nature Communications* (Jolly et al. 2015). A study of maximum temperatures, minimum relative humidity, the number of rain-free days, and maximum wind speeds indicated that fire seasons have lengthened on a quarter of Earth's land surface over 35 years (1979 through 2013).

Increases have been most notable in parts of the western United States and Mexico, Brazil, and East Africa, which by 2013 faced wildfire seasons that were at least a month longer than 35 years earlier. In a few areas, such as Western Africa and the Pacific coast of South America, easing of droughts probably helped shorten fire seasons ("Longer, More Frequent" 2015).

Further Reading

Jolly, W. Matt, et al. "Climate-Induced Variations in Global Wildfire Danger from 1979 to 2013. *Nature Communications* 6 (7537) (July14, 2015). http://www.nature.com /ncomms/2015/150714/ncomms8537/full/ncomms8537.html.

"Longer, More Frequent Fire Seasons." NASA Earth Observatory, July 28, 2015. http:// earthobservatory.nasa.gov/IOTD/view.php?id=86268&src=eoa-iotd.

"NASA Study Says Climate Adds Fuel to Asian Wildfire Emissions." NASA Earth Observatory, April 30, 2009. http://earthobservatory.nasa.gov/Newsroom/view.php ?id=38433&src=eoa-nnews (no longer available).

The area burned by forest fires in the western United States has nearly doubled over the past three decades ending in 2015, a trend that was related to human-related climate change by John Abatzoglou at the University of Idaho in Moscow and Park Williams at Columbia University. According to a report in *Nature*,

They found that warming temperatures made the forests drier, increasing fire risk, and expanded the area burned in the western part of the United States between 1984 and 2015 by about 4.2 million hectares. Climate change also accounted for about half the increase in both the length of the fire season and the number of days with a high risk of fire. ("Wildfires Burn" 2016, 292)

Fires in California

Reports from fires in Southern California during October 2007 were apocalyptic—more than a million people routed from their homes, many huddled in stadiums and fairgrounds, highways choked with fleeing multitudes, hundreds of homes burned to the ground, firefighters completely overwhelmed as hurricane-force Santa Ana winds grew hotter as they descended from the mountains and drove flames through brush dried by a record drought. Half a million acres burned within a week in 100°F afternoon temperatures and air nearly devoid of humidity.

As is so often the case, climatic extremes have combined with other factors to produce disaster on an epic scale. An extremely strong Santa Ana wind fanned flames into suburbs that had expanded into land prone to wildfires. Winds gusted as strong as 111 miles an hour. An inventory by University of Wisconsin researchers found that some two-thirds of new building in Southern California over the past decade was on land susceptible to wildfires, said Mike Davis, a historian at the University of California at Irvine and author of environmental and social histories of the region that have anticipated the toll of wildfires. "It gives you some parameters for understanding the current situation," Davis said. "Another way to look at it is you simply drive out the San Gorgonio Pass, where the winds blow over 50 mph over a hundred days a year and you have new houses standing next to 50-year-old chaparral. You might as well be building next to leaking gasoline cans" (Vick and Geis 2007).

Wildfires and Boreal (Northern) Forests

During early May 2016, a time of year when the last snows of winter used to be melting, a ferocious wildfire stoked by drought and record heat scorched a large area of northern Alberta province's tar sands mining region, including the city of Fort McMurray, sending 90,000 refugees fleeing for their lives. This was one vivid example of a new reality in the world's far-northern forests of pine, spruce, and larches. Some of the most devastating fires were being ignited by an increasing number of lightning-bearing thunderstorms. In Alberta, temperatures had been as much as 30°F above average before the fires. Temperatures soared to 32°C (90°F) on May 3 as the fire spread ("Heat Fuels" 2016). The intense heat in Alberta coincided with an omega block, a stationary high pressure system that inhibits the passage of storms eastward and westward, a pattern that some scientists believe becomes more common as temperatures warm. An intense El Niño pattern also contributed warmth.

"It's clear that the warming temperatures and extraordinary drought are major players here," said Thomas W. Swetnam, an emeritus scientist at the University of Arizona who studies the ecology and history of wildfires. "We probably wouldn't be seeing the scale of some of these fires if it weren't for those factors" (Gillis and Fountain 2016). The fires have been adding carbon dioxide to the atmosphere as well as soot to Arctic ice, both of which aggravate global warming. In an interview, Brian J. Stocks, who retired from the Canadian Forest Service to work as a consultant, said, "We're kind of at a crossroads. . . . We anticipate more fires, and more intense fires, in the future" (Gillis and Fountain 2016).

A report by Greenpeace International suggests that between 50 percent and 90 percent of Earth's existing boreal forests are likely to disappear if atmospheric levels of carbon dioxide and other greenhouse gases double. These forests comprise one-third of Earth's remaining tree cover, some 15 million square kilometers, across Russia, Canada, the United States, Scandinavia, China, Mongolia, Japan, and parts of the Korean peninsula. Large forests also clothe many mountain ranges outside of these zones. In total, boreal forests cover approximately 10 percent of the world's land area.

Fires Burn Farther North in Siberia

During 2012 and 2013, Russia experienced two of its most severe wildfire seasons on record accompanied by unusual summer heat in large parts of northern Siberia north of 57° north, the southern limit of the *taiga* (dense forests), where such activity has been rare. Temperatures reached 32°C (90°F) as far north as Nofrilsk, 30°F above July daily average highs, under a long-lasting "blocking" high-pressure system that brought "stable air and exceptional heat" ("Heat Intensifies" 2013). Fires in 2012 and 2013 reached as far north as 65° north. "High temperatures play an important role in promoting wildfires," said the NASA Earth Observatory.

> Warm fuels burn more readily than cooler fuels because less energy is required to raise their temperature to the point of ignition. With temperatures soaring in northern Russia, it was easier for previously active fires to continue burning and for lightning to spark new ones. . . . Weather occurs within the broader context of the climate, and there's a high level of agreement among scientists that global warming has made it more likely that heat waves and wildfires of this magnitude will occur. ("Heat Intensifies" 2013)

NASA's Earth Observatory said of Irkutsk, in Siberia, in 2014:

> With dozens of forest fires burning in Russia's Irkutsk region, authorities have declared a state of emergency. Some of the blazes likely began on farms but then spread into forests due to high winds and warm temperatures. After burning at a moderate level for a few days, the size and intensity of the fires increased significantly on May 18. The *St. Petersburg Times* reported that 77 fires had burned more than 39,000 hectares (150 square miles) in Irkutsk by May 19. Fire destroyed 22 homes in the village of Dalny and forced the evacuation of hundreds of people, according to the Russian Emergencies Ministry (EMERCOM). In addition to producing thick plumes of smoke, the fires fueled numerous pyrocumulus—tall, cauliflower-shaped clouds that billowed up above the smoke. ("Wildfires in Irkutsk" 2014)

Further Reading

"Heat Intensifies Siberian Wildfires." NASA Earth Observatory, August 2, 2013.

"Wildfires in Irkutsk." NASA Earth Observatory, May 21, 2014. http://earthobservatory
    .nasa.gov/IOTD/view.php?id=83702&src=eoa-iotd

The Greenpeace International report indicates that global warming's toll on the boreal forests had begun by the early 1990s. The report warns that decaying forests may provide an extra boost to rising carbon dioxide levels, causing warming to feed on itself. The decline of boreal forests also endangers more than 1 million indigenous people who live in them, including the Dene and Cree of Canada; the Sami (Lapplanders) of Norway, Sweden, and Finland; the Ainu of northern Japan; and the Nenets, Yakut, Udege, and Altaisk of Siberia (Jardine 1994). Some of the forests' larger animals such as the Siberian tiger already are near extinction. The Greenpeace report concludes that rapid logging of the boreal forests is intensifying pressure on animal life and accelerating the release of even more carbon dioxide and other greenhouse gases into the atmosphere.

If boreal forests continue to decline, their burning and rotting could contribute to the release of as much as 225 billion tons of extra carbon dioxide into the atmosphere, raising current levels by one-third and further accelerating the pace of warming. Trees could have difficulty colonizing the thawing tundra to the north of their current ranges because their populations simply cannot migrate quickly enough and because the treeless tundra cannot evolve quickly enough to sustain them.

Kevin Jardine described the role of insect outbreaks in the anticipated destruction of boreal forests:

Pests [that] may invade boreal forests under warming conditions include the western spruce budworm, the Douglas fir tussock moth, and the mountain pine beetle (Kurz et al. 1995, 127). Kurz et al. conclude that "biospheric feedbacks from temperate and boreal forest ecosystems will be positive feedbacks that further enhance the carbon content of the global atmosphere" (Kurz 1995, 129). By 1998, spruce budworms had devoured 50 million acres (20 million hectares) of Alaskan woodland.

In moderation, the forest usually benefits from insect outbreaks because they reduce the likelihood of catastrophic fire by helping to eliminate older stands and diseased trees. However, populations of insects such as bark beetles, the Siberian silkworm, and the spruce budworm can explode and devastate millions of acres of forest. The life cycles of the spruce budworm (*Choristoneura fumiferana*) and the spruce bark beetle (*Ips typographus*) are strongly influenced by climate, with both species likely to increase in numbers during the kind of warmer, drier weather predicted in a global warming world (Jardine 1994). . . . A Canadian government study released in 1987 showed that conifers were growing on average up to 65 percent slower than in the 1940s and 1950s, because of spruce budworm outbreaks, and possibly acid rain. (Jardine 1994)

The New York Times described the foreboding of firefighters after 19 of their number were literally overrun and killed by a quickly advancing blaze in Arizona during the summer of 2013. Was this the "new normal" in which massive, exploding fires take control, overwhelming hundreds of firefighters in a hotter, more dangerous world?

One of the deadliest wildfires in a generation vastly expanded . . . sweeping up sharp slopes through dry scrub and gnarled piñon pines a day after fickle winds and flames killed 19 firefighters. The gusty monsoon winds where the Colorado Plateau begins to drop off into the Sonoran Desert continued to bedevil about 400 firefighters who were defending 500 homes and 200 businesses in the old gold mining villages of Yarnell and Peebles Valley. (Barringer and Chang 2013)

More than 2.2 million housing units were added in the U.S. West during the 1990s, many of them in areas prone to fire, according to Roger B. Hammer, an at Oregon State University demographer (Barringer and Chang 2013). The average area burned in the U.S. West doubled from about 3.5 million acres a year during the decade of the 1990s to 7 million acres just 10 years later.

During the summer of 2012, Nebraska's Pine Ridge lost the ponderosa pines that provided their name to fires. The area has become prairie (Gaarder 2013). Doak Nickerson, district forester for the Nebraska Forest Service, has seen much of the state's evergreen cover disappear during his lifetime. In a state that is less than 2 percent forests, the once-thick cover of greenery that dominated Pine Ridge had been shrinking for 50 years because of rising temperatures and decreasing precipitation when the record fires of 2012 reduced it to scattered patches. In 2012, Nebraska experienced its most extensive drought since the Dust Bowl of the 1930s. "The size of these mega-fires is symptomatic of what is changing in front of us," Nickerson said. "Increasingly hotter, drier summers, coupled with milder winters are part of it" (Gaarder 2013). These conditions have helped fires take an increasing toll.

Wildfires and Warming: How and Why

In a tinder-dry environment, a wildfire can start in an instant and consume hundreds of thousands of acres in a few days. On June 26, 2011, all that was required to ignite one major fire in northern New Mexico's Jemez Mountains was a single aspen tree falling into a power line on a hot, dry, and windy afternoon. Within five days, the fire had consumed 42,000 hectares (1 hectare=2.47 acres). When the fire finally was extinguished weeks later, what came to be called the Las Conchas fire had burned 60,000 hectares, the largest wildfire in New Mexico's history. The burned area was so completely scorched that the forest may never recover. To many people in the area, it was a new order of fiery devastation. Warmer and drier weather was colliding with human construction—the fire never would have started without that power pole—and a century of fire suppression to breed a new degree of danger.

Across the American West, the average area burned each year has increased significantly over the past several decades, a trend that scientists attribute both to warming and drying and a century of wildfire suppression and other human activities. Ecologist Craig Allen suggested that the ecosystem had entered new territory, not only in the American West but also in many other regions. In the Jemez

Mountains, for example, ponderosa pine (*Pinus ponderosa*) forests were becoming scrubland. "We're losing forests as we've known them for a very long time," said Allen. "We're on a different trajectory, and we're not yet sure where we're going" (Nijhuis 2012, 352). A report from the United States National Research Council in 2010 estimated that each 1°C rise in temperatures will increase the area burned by wildfires in the western United States by 200 to 400 percent (Hoag 2010, 425).

Climate Change, Droughts, and Fires

A warming climate punctuated by droughts (and occasional deluges) is conducive to the growing numbers of wildfires that add even more carbon dioxide to the atmosphere. "The fires in California and here in Arizona are a clear example of what happens as the Earth warms, particularly as the West warms, and the warming caused by humans is making fire season longer and longer with each decade," said University of Arizona geoscientist Jonathan Overpeck. "It's certainly an example of what we'll see more of in the future" (Silva 2014). According to the 2014 U.S. National Climate Assessment, "Increased warming, drought and insect outbreaks, all caused by or linked to climate change have increased wildfires and impacts to people and ecosystems in the Southwest. Fire models project more wildfire and increased risks to communities across extensive areas" (Silva 2014).

As climate warms, the area of forest burned by wildfires in the United States may increase more than 50 percent by 2050, with the most impact in the Pacific Northwest and Rocky Mountains, where burned areas may increase. "Wildfires, such as those in California earlier this year, are a serious problem in the United States, and this research shows that climate change is going to make things significantly worse," said Dr. Dominick Spracklen, from the School of Earth and Environment at the University of Leeds, lead author of the research. "Our research shows that wildfires are strongly influenced by temperature. Hotter temperatures lead to dryer forests resulting in larger and more serious fires," said Spracklen ("Wildfires Set" 2009).

Conditions conducive to wildfires can be forecast. Writing in *Science* (2011), Yang Chen and colleagues described a changing system:

> We found that the Oceanic Niño Index was correlated with interannual fire activity in the eastern Amazon, whereas the Atlantic Multidecadal Oscillation index was more closely linked with fires in the southern and southwestern Amazon. Combining these two climate indices, we developed an empirical model to forecast regional fire season severity with lead times of 3 to 5 months. Our approach may contribute to the development of an early warning system for anticipating the vulnerability of Amazon forests to fires, thus enabling more effective management with benefits for climate and air quality. (Chen et al. 2011, 787)

Scientists had warned that more fires were coming before the fires of 2007. Wildfires that charred nearly three-quarters of a million acres in California during the fall of 2003 presaged increasingly severe fire danger as global warming weakens more forests through disease and drought. Warmer, windier weather and longer, drier summers could result in higher firefighting costs and greater loss of lives and property, according to researchers at the Lawrence Berkeley National Laboratory and the U.S. Forest Service. Both the number of out-of-control fires and the acreage burned are likely to increase, more than doubling losses in some regions, according to a study published in the scientific journal *Climatic Change*. Although the study examined Northern California, "the concern for Southern California would be much higher" because that region is drier for longer periods, said researcher Evan Mills of the Lawrence Berkeley lab (Thompson 2003).

According to this study, a doubling of atmospheric carbon dioxide will provoke fires that "burn more intensely and spread faster in most locations." According to models developed by Jeremy S. Fried, Margaret S. Torn, and Evan Mills, the number of "escapes" (fires that exceed initial containment efforts) will be double current frequencies, according to their models. Contained fires also burn 50 percent more land under the warmer and windier conditions anticipated with a doubling of carbon dioxide concentration in the atmosphere. The researchers stressed that their projections "represent a *minimum* expected change, or best-case forecast. In addition to the increased suppression costs and economic damages, changes in fire severity of this magnitude would have widespread impacts on vegetation distribution, forest condition, and carbon storage, and greatly increase the risk to property, natural resources, and human life" (Fried et al. 2004, 169). The researchers projected at least a 50-percent increase in out-of-control fires in the southern part of the San Francisco Bay region and a 125-percent increase in the Sierra Nevada foothills, with a more than 40-percent increase in the area burned. The state's northern coast saw no significant change under the computer model and conditions used in the study (Thompson 2003).

Fires are Hotter and Move Faster

The study's projections use conservative forecasts that do not take into account increased lightning strikes and the spread of volatile grasslands into areas now dominated by less flammable vegetation. Even potentially wetter winters simply mean more growth, providing additional fuel for summer fires, according to the study. "Fires may be hotter, move faster, and be more difficult to contain under future climate conditions," Robert Wilkinson of the University of California—Santa Barbara, School of Environmental Science and Management, said in a federal report on the impact of climate change on California. "Extreme temperatures compound the fire risk when other conditions, such as dry fuel and wind, are present" (Thompson 2003). Damage may be aggravated by the constriction of homes in brushlands that are vulnerable to fires.

In 2007, fires north of Alaska's Brooks Range singed 1,000 square kilometers, a larger area than all recorded fires on Alaska's North Slope during the previous half-century, adding 1.3 million tons of carbon dioxide to the atmosphere. The number of tundra fires is accelerating, according to Adrian Rocha of the Marine Biological Laboratory in Woods Hole, Massachusetts. Because "projected changes in climate over the next century—increased aridity, thunderstorms, and warming in the Arctic—will increase the likelihood that these thresholds will be crossed and thus result in more larger and frequent fires" (Logan 2009).

A year after such a fire, "the most severely burned tundra emitted twice as much carbon as undamaged tundra normally stores away" (Logan 2009). William Bowden, a wetland ecologist of the University of Vermont, said that fire-blackened surfaces increase the amount of water in soil. "Both factors should promote soil warming, permafrost thaw, and possible thermokarst formation," he said (Logan 2009). Thermokasts are areas of collapsed terrain where structurally important permafrost has thawed—a process that can damage the foundations of homes, roads, and pipelines. Permafrost melt will also increase the amount of greenhouse gases such as methane entering the atmosphere (Logan 2009).

Health Damage from Wildfire Smoke

Wildfires have become so widespread that two-thirds of U.S. residents who lived in areas with smoky air in 2011 reported health problems, even if they were several hundred miles downwind from the burned areas. In 2011, for example, health problems from wildfires in Texas that were driven by southerly winds were reported as far north as Nebraska, Iowa, and Missouri. Some 212 million people were afflicted by wildfire smoke in 2011, according to an analysis by the Natural Resources Defense Council (NRDC) (Koch 2013). "It affects a much wider area of the United States than people realize," said report author Kim Knowlton, NRDC senior scientist and health professor at Columbia University (Koch 2013). The smoke contains fine-particle air pollution that aggravates asthma, pneumonia, and chronic heart and lung diseases.

In Texas during 2011, medium- to high-density smoke afflicted 25 million people for a week or more. Illinois recorded no wildfires, but 12 million of its citizens were exposed to smoke. The problem of wildfire smoke exposure will only get worsen, said Knowlton, as human-induced climate change plays a role in raising temperatures and intensifying drought. "Our landscapes are becoming more of a tinderbox," she said (Koch 2013). In 2011, 8.7 million acres burned in the United States, the fourth-largest number since records began in 1985. A year later, 9.3 million acres burned.

"Heat waves are very likely to occur more frequently and last longer," said Thomas Stocker, co-chair of Working Group 1 of the Intergovernmental Panel on Climate Change (IPCC). The IPCC anticipates that the wettest areas will receive more precipitation and the driest areas will get less. Western forests in the United States, usually a relatively dry area, will be "increasingly affected by large and intense fires that occur more frequently," according to the third U.S. National Climate Assessment (Koch 2013).

The Forest Service's Budget and Wildfire Frequency

In 1991, 13 percent of the U.S. Forest Service's budget was spent on firefighting; by 2006, that figure had risen to 43 percent. By 2014, more than half of all Forest

Service funds were being spent fighting fires. Between the 1970s and the year 2000, the western U.S. firefighting season expanded to 92 days a year (Harper's 2006). Increasing heat and drought spur fires that, in turn, add extra carbon dioxide to the atmosphere, increasing future warming in a feedback loop.

Increasing fires since 2007 may be just the beginning. Wildfires in the western United States probably will increase in the coming decades, according to a study that links episodic fire outbreaks in the past five centuries with periods of warming sea surface temperatures in the North Atlantic. Warming surface waters in the North Atlantic correspond with episodes of drought and fires in the West recorded in tree rings studied by researchers, according to lead researcher Thomas Kitzberger of the University of Comahue in Argentina ("Western Wildfires" 2007).

The team analyzed nearly 34,000 individual fire scar dates from tree rings, primarily ponderosa pine and Douglas fir at 241 sites, the largest record of tree rings linked to past wildfires ever assembled. Wildfire frequencies in Washington, Oregon, California, Colorado, New Mexico, Arizona, and South Dakota were found to be affected. "This study underscores the value of building large networks of high-resolution fire history data to better understand how climate may affect fire regimes over large areas of the globe," said Kitzberger ("Western Wildfires" 2007).

Fires and Atmospheric Conditions: A "New Normal"?

Over the past 35 years, warmer temperatures and changing atmospheric circulation patterns that provoke drought in the western United States have dramatically increased both the number of large forest wildfires and the length of the season during which they occur. Specifically, according to one study published in 2006, the number of fires from 1986 to 2005 increased fourfold, and the area burned increased sixfold, compared to 1979 through 1985 (Westerling et al. 2006). The same study found that the average fire season (the period of time when major fires were taking place) had increased by 78 days, and the average length of major fires had risen from 7.5 days to 37.1 days (Running 2006, 927).

Westerling et al. wrote that increases in spring and summer temperatures by 0.9°C and mountain snow pack melting one to four weeks earlier had contributed to the increased number and duration of wildfires. Compounding all of this, according to the authors, warmer and longer growing seasons at high elevations reduce the amount of carbon removed from the atmosphere by living trees, especially during droughts (Westerling et al. 2006, 943). Adding it all up, the forests of the western United States are likely to become a net source of carbon for the atmosphere rather than a sink.

Westerling et al. (2006) compiled a comprehensive time series of large forest wildfires in the western United States for the period from 1970 to 2003 and compared those data with corresponding observations of climate, hydrology, and land surface conditions. Wildfire activity increased suddenly in the mid-1980s. Hydroclimate and fires are closely related, and climate variation has been the primary cause of the increase in fires during the period of their study, although land-use changes can also be important. Longer springs and summers that could result as the world warms will continue to lengthen the fire season and cause more and larger wildfires.

As Westerling and colleagues wrote:

Western United States forest wildfire activity is widely thought to have increased in recent decades, yet neither the extent of recent changes nor the degree to which climate may be driving regional changes in wildfire has been systematically documented. Much of the public and scientific discussion of changes in western United States wildfire has focused instead on the effects of 19th- and 20th-century land-use history. We compiled a comprehensive database of large wildfires in western United States forests since 1970 and compared it with hydroclimatic and land-surface data. Here, we show that large wildfire activity increased suddenly and markedly in the mid-1980s, with higher large-wildfire frequency, longer wildfire durations, and longer wildfire seasons. The greatest increases occurred in mid-elevation, Northern Rockies forests, where land-use histories have relatively little effect on fire risks and are strongly associated with increased spring and summer temperatures and an earlier spring snowmelt. (Westerling 2006, 940)

The mountains through which the Las Conchas fire raced received an average of close to 40 centimeters of rain a year, theoretically enough to regrow a forest. After past fires, the forests did eventually revive. Some growth remained after previous fires, but circumstances have now changed:

Over the past century, however, the policy of quickly dousing fires has allowed brush and spindly young trees to build up in many western forests, so they tend to burn hotter and less patchily than before. And over the past decade, a severe drought across the southwest has weakened trees and made them vulnerable to widespread attack by beetles, leading to a die-off of more than one million hectares of piñon pines (*Pinus edulis*). Many of the dead trees are still standing, and can serve as ladder fuels that transform relatively cool surface fires into hot, fast-moving crown fires that leap from treetop to treetop. (Nijhuis 2012, 352)

Few seed sources remained, and soil had been stripped of nutrients. Soil that remained was more prone to landslides, and a hotter and drier climate favored species that could tolerate dry heat. Scrubland became the dominant ecosystem. This pattern is being repeated throughout the U.S. West.

All around the American West, scientists are seeing signs that fire and climate change are combining to create a "new normal." Ten years after Colorado's largest recorded fire burned 56,000 hectares southwest of Denver, the forest still has not rebounded in a 20,000-hectare patch in the middle, which was devastated by an intense crown fire. Only a few thousand hectares, which the

U.S. Forest Service replanted, look anything like the ponderosa-pine stands that previously dominated the landscape. (Nijhuis 2012, 352)

"Otherwise, it's grassland and shrub land, and probably will be for centuries to come," said Peter Brown, a forest ecologist and director of the not-for-profit organization Rocky Mountain Tree-Ring Research in Fort Collins, Colorado (Nijhuis 2012, 352).

Further Reading

Barringer, Felicity, and Kenneth Chang. "Experts See New Normal as a Hotter, Drier West Faces More Huge Fires." *The New York Times*, July 1, 2013 http://www.nytimes.com/2013/07/02/us/experts-see-a-hotter-drier-west-with-more-huge-fires.htm (no longer available).

Chen, Yang, et al. "Forecasting Fire Season Severity in South America Using Sea Surface Temperature Anomalies." *Science* 334 (November 11, 2011): 787–791.

Fried, Jeremy S., Margaret S. Torn, and Evan Mills. "The Impact of Climate Change on Wildfire Severity." *Climatic Change* 64 (May, 2004): 169–191.

Gaarder, Nancy. "Pines Burned Out of Pine Ridge." *Omaha World-Herald*, January 7, 2013, A1, A2.

Gillis, Justin, and Henry Fountain. "Global Warming Cited as Wildfires Increase in Fragile Boreal Forest." *The New York Times*, May 10, 2016. http://www.nytimes.com/2016/05/11/science/global-warming-cited-as-wildfires-increase-in-fragile-boreal-forest.html.

"Harper's Index." *Harper's Magazine*, September, 2006, 17.

"Heat Fuels Fire at Fort McMurray." NASA Earth Observatory, May 7, 2016. http://earthobservatory.nasa.gov/NaturalHazards/view.php?id=87992.

Hoag, Hannah. "Degree-by-Degree Breakdown of Climate Effects Published." *Nature* 466 (July 22, 2010): 425.

Jardine, Kevin. "The Carbon Bomb: Climate Change and the Fate of the Northern Boreal Forests." Ontario, Canada: Greenpeace International, 1994. http://dieoff.org/page129.htm.

Koch, Wendy. "Wildfire Smoke Becoming a Serious Health Hazard." *USA Today*, October 25, 2013. http://www.usatoday.com/story/news/nation/2013/10/24/wildfires-smoke-climate-change-harm-health/3173165/?utm.

Kurz, Werner A., et al. "Global Climate Change: Disturbance Regimes and Biospheric Feedbacks of Temperate and Boreal Forests." Pp. 119–133 in George M. Woodwell and Fred T. MacKenzie, eds., *Biotic Feedbacks in the Global Climate System: Will the Warming Feed the Warming?* New York: Oxford University Press, 1995.

Lean, Geoffrey. "Worst U.S. Drought in 500 Years Fuels Raging California Wildfires." *The Independent* (London), July 25, 2004, 20.

Logan, Tracey. "Alaska's Biggest Tundra Fire Sparks Climate Warning." *New Scientist*, July 30, 2009. http://www.newscientist.com/article/dn17537-alaskas-biggest-tundra-fire-sparks-climate-warning.html?DCMP=OTC-rss&nsref=environment.

"NASA Study Says Climate Adds Fuel to Asian Wildfire Emissions." NASA Earth Observatory, April 30, 2009. http://earthobservatory.nasa.gov/Newsroom/view.php?id=38433&src=eoa-nnews.

Nijhuis, Michelle. "Forest Fires: Burn Out." *Nature* 489 (September 20, 2012): 352–354. http://www.nature.com/news/forest-fires-burn-out-1.11424.

Running, Steven W. "Is Global Warming Causing More, Larger Wildfires?" *Science* 313 (August 18, 2006): 927–928.

Silva, Rebecca. "California Fires Fueled by Global Warming, Says Reports." HNGN (Headlines and Global News). May 19, 2014. http://www.hngn.com/articles/31623/20140519 /california-fires-fueled-by-global-warming-says-reports.htm.

Thompson, Dan. "Experts Say California Wildfires Could Worsen with Global Warming." Associated Press, November 12, 2003 (LEXIS).

Vick, Karl, and Sonya Geis. "California Fires Continue to Rage; Evacuation May Be Largest, Officials Say." *Washington Post*, October 24, 2007; A1. http://www.washingtonpost.com /wp-dyn/content/article/2007/10/23/AR2007102300347_pf.html.

Westerling, A. L., et al. "Warming and Earlier Spring Increase Western U.S. Forest Wildfire Activity." *Science* 313 (August 18, 2006): 940–943.

"Western Wildfires Linked to Atlantic Ocean Temperatures." Environmental News Service, January 3, 2007. http://www.ens-newswire.com/ens/jan2007/2007-01-03-09.asp#anchor2 (no longer available).

"Wildfires Burn More U.S. Forest." *Nature* 538 (October 20, 2016): 292–293. See also *Proceedings of the National Academies of Science USA* (2016), http://doi.org/brsj.

"Wildfires Set to Increase 50 Percent by 2050." NASA Earth Observatory, July 28, 2009. http://earthobservatory.nasa.gov/Newsroom/view.php?id=39650&src=eoa-manews (no longer available).

See also: Australia, Heat and Drought in; Deforestation; Drought, United States; Drought, Worldwide; Drought and Deluge, Scientific Analysis of

FLORA AND FAUNA

OVERVIEW

Each and every species of plant and animal on the land and in the seas lives within an environmental context in which temperature plays an important role. Alter the temperature and the animal or plant's sense of comfort changes. Alter it even further, and the organism's survival may be in peril. In addition, the habitats of every other plant and animal exist in relationships with a given species in symbiosis or as predator or prey is also changing. Given enough time, species adapt or migrate. If change comes too quickly, however, species that cannot survive may first find their populations reduced. Extinction may follow and may cascade as species on which they depend also experience problems with reproduction and survival.

This section includes general surveys of how flora and fauna have changed and will continue to evolve in a quickly warming world. By necessity, it is selective because each of Earth's many species respond differently, and only a few can be included. Given those limits, a sketch of possible extinctions is followed by a consideration of biodiversity reduction, with special attention to migratory animals on land and sea. Specific case studies describe markedly different effects on a small group of species. For example, the gender ratios of loggerhead sea turtles change with the temperature of the sand in which they breed. And in another example, moose find themselves tortured by massive insect infestations as their habitats warm.

First-Flowering Dates, England

A study of the first-flowering date of 385 British plants found that about 200 species flowered an average of 15 days earlier in the 1990s than in previous decades. The study by Alastair Fitter of York University and his father, Richard, a distinguished naturalist, claimed to have found "the strongest biological signal yet of climate change. The Fitters tracked blooming dates for half a century, long before global warming became a widespread concern. The senior Fitter was 89 years of age in 2002 and lives in Cambridgeshire. He is one of Britain's best-known naturalists, the author of dozens of books on flowers and birds. He began his record after moving to Chinnor in Oxfordshire in 1953, when his son was five years old.

A. H. Fitter and R. S. R. Fitter reported in *Science*:

The average first-flowering date of 385 British plant species has advanced by 4.5 days during the past decade compared with the previous four decades; 16 percent of species flowered significantly earlier in the 1990s than previously, with an average advancement of 15 days in a decade. Ten species flowered significantly later in the 1990s than previously. These data reveal the strongest biological signal yet of climatic change. (Fitter and Fitter 2002, 1689)

The Fitters also raised the possibility that climate change could alter evolution as different plants move into synchronous flowering patterns, raising the chances they will breed natural hybrids (Fitter and Fitter 2002, 1691).

Spring-flowering species were the most responsive to the change, with white dead nettle flowering 55 days earlier in the 1990s than during four previous decades. From 1954 to 1990, its average first flowering date was March 18, but from 1991 to 2000 the plant bloomed on or around January 23. Ivy-leafed toadflax flowered five weeks earlier, hornbeam and hairy bittercress four weeks earlier. Lesser periwinkle flowered 25 days earlier and lesser celandine three weeks earlier. The opium poppy flowered 20 days earlier during the 1990s than its 1954 to 1990 average, according to the Fitters (Meek 2002). "Not everything is flowering earlier—otherwise you would just say that the beginning of spring has moved forward eight weeks," Fitter said (Clover 2002).

"I did an analysis of his data eight or nine years ago, up to 1990, thinking there might be some sign of climate warming, but there wasn't," the elder Fitter said. "When we came to look at the last 10 years, out of the data jumped this incredible signal that it's really happening fast, and it's no coincidence that the 1990s were the warmest decade on record" (Meek 2002).

The leafing and flowering dates do not show a one-for-one correlation with changing temperatures, however. In most temperate tree species, phenological events such as flowering and autumnal cessation of growth are not primarily controlled by temperature.

Each plant responds to a mix of temperature cues as well as photoperiod (length of day and night) and the intensity of the previous winter's chilling. Each plant species has its own ratio of these factors. "Temperature plays a modulating role and triggers the visible progress of phenology, such as leaf coloration, in many species," wrote Christian Körner and David Basler in *Science* (2010, 1461). "It is thus a misconception to linearly extrapolate a few days' advance of leafing during warm years into a proportional lengthening of the growing season in climate warming scenarios," they wrote (2010, 1461).

Further Reading

Clover, Charles. "Global Warming Is Driving Fish North." *Daily Telegraph* (London), May 31, 2002, 14.

Fitter, A. H., and R. S. R. Fitter. "Rapid Changes in Flowering Time in British Plants." *Science* 296 (May 31, 2002): 1689–1691.
Körner, Christian, and David Basler. "Phenology Under Global Warming." *Science* 327 (March 19, 2010): 1461–1462.
Meek, James. "Wildflowers Study Gives Clear Evidence of Global Warming." *The Guardian* (United Kingdom), May 31, 2002, 6.

What becomes of an overheated Earth in which the habitats of animals and plants change rapidly? David Quammen, an environmental writer, believes that Earth during the next century will become a "planet of weeds" in which human domination forces the extinction of most undomesticated living things. Quammen expects that the flora and fauna of Earth will eventually include food crops, animals raised to be eaten or petted, and a few stubborn weed species that will benefit from a harsher, hotter world. By 2050, Quammen believes that deforestation will cause half the world's wild birds and two-thirds of other wild animal species to become extinct (Quammen 1998, 61, 69). "Wildlife," he writes, "Will consist of pigeons, coyotes, rats, roaches, house sparrows, crows, and feral dogs" (Quammen 1998, 67). Human beings—"remarkably widespread, prolific, and adaptable"—are "the consummate weed" (Quammen 1998, 68).

Palms as Bioindicators of Warming

Gian-Reto Walther and colleagues tracked the expansion of *Trachycarpus fortunei*, the windmill palm (similar to the palmetto), into southern Switzerland following rising winter minimum temperatures and a lengthening growing season. *Trachycarpus fortunei* not only reproduces naturally in the foothills of the southern Alps but also has been observed spreading in seminatural habitats or seeding in gardens and parks as far north as southern costal England, Brittany in France, the Netherlands, and coastal southwestern British Columbia, all areas where warmer nights have extended the average annual growing season to well over 300 days a year (Walther et al. 2007).

The palms of Switzerland have now been observed some 300 kilometers (more than 200 miles) outside their historical range. Scientists conclude that the spread of these palms is a "significant global bioindicator across continents for present-day climate change and the projected global warming of the near future" (Walther et al. 2007).

Further Reading

Walther, Gian-Reto, et al. "Palms Tracking Climate Change." *Global Ecology and Biogeography*. New York: John Wiley & Sons, 2007. doi: 10.1111/j.1466-8238.2007 .00328.x.

Reflecting Quammen's thesis, Lewis Ziska of the U.S. Department of Agriculture says that a warming atmosphere during the 20th century doubled the per-plant production of ragweed (a plant that incites human allergies). With a carbon dioxide level of 280 parts per million (ppm), which was typical of the year 1900, ragweed plants produced an average of 5.5 grams of pollen each. With a level of 370 ppm, a level reflective of the year 2000, the same plants produce an average of 10 grams of pollen. At 600 ppm, Ziska says ragweed pollen production doubles again to approximately 20 ppm.

Albert K. Bates supports Quammen's opinion: "Sixty-five million years ago, 60 to 80 percent of the world's species disappeared in a cataclysmic mass extinction, possibly caused by an asteroid's impact with Earth. Human population, not an asteroid, will cut the remaining number of species in half again, in just the next few years" (Bates and Project Plenty 1990, 137). Another observer, Robert L. Peters, factors climate change into a similar picture of Earth's biological future:

> Habitat destruction in conjunction with climate change sets the stage for an even larger wave of extinction than previously imagined, based on consideration of human encroachment alone. Small, remnant populations of most species, surrounded by cities, roads, reservoirs, and farmland, would have little chance of reaching new habitat if climate change makes the old unsuitable. Few animals or plants would be able to cross Los Angeles on the way to the promised land. (Peters 1989, 91)

Global warming could destroy or fundamentally alter a third of the world's plant and animal habitats within a century, bringing extinction to thousands of species, according to a study by Great Britain's World Wide Fund for Nature, an affiliate of the World Wildlife Fund. According to the study, the most vulnerable plant and animal species will be in Arctic and mountain areas, where as many as 20 percent could be driven to extinction. In the north of Canada, Russia and Scandinavia, where warming was predicted to be most rapid, as much as 70 percent of habitat could be lost. The report, *Global Warming and Terrestrial Biodiversity Decline*, was written by Jay Malcolm, professor of forestry at Toronto University, and Adam Markham, the executive director of Clean Air-Cool Planet. Pests and weedy species would fare best, the report said. In addition, "If past fastest rates of migration are a good proxy for what can be attained in a warming world, then radical reductions in greenhouse gas emissions are urgently required to reduce the threat of biodiversity loss" (Clover 2000). To adapt and survive the expected rate of warming during the next century, the report's authors say, plants may need to move 10 times more quickly than they did when recolonizing previously glaciated land at the end of the last ice age. Few plant species can move at a rate of one kilometer per year, the speed that will be required in many parts of the world.

Further Reading

Bates, Albert K., and Project Plenty. *Climate in Crisis: The Greenhouse Effect and What We Can Do.* Summertown, TN: The Book Publishing Co., 1990.

Clover, Charles. "Thousands of Species 'Threatened by Warming.'" *Daily Telegraph* (London), August 31, 2000, 9.
Peters, Robert L. "Effects of Global Warming on Biological Diversity." Pp. 82–95 in Edwin Abrahamson, ed., *The Challenge of Global Warming*. Washington, DC: Island Press, 1989.
Quammen, David. "Planet of Weeds: Tallying the Losses of Earth's Animals and Plants." *Harper's*, October 1998, 57–69.

ADAPTATION, ANIMALS, LIZARDS, AND WARMING HABITATS

Rising temperatures have already played a role in the extinction of one-eighth of Mexico's lizards, according to a 2010 study described in *Science*. It concluded that "lizards have already crossed a threshold for extinctions," which may increase to 20 percent by 2080 unless the pace of warming slows (Gill 2010). The study was led by Barry Sinervo of the Department of Ecology and Evolutionary Biology at the University of California–Santa Cruz, who said, "We are actually seeing lowland species moving upward in elevation, slowly driving upland species extinct, and if the upland species can't evolve fast enough then they're going to continue to go extinct" (Gill 2010). Many lizards live at the edge of their thermal tolerance. Rising temperatures inhibit their ability to gather food and regulate their body temperatures.

Armadillos Spreading Northward

In North America, the armadillo (Spanish for "little armored thing"), a subtropical animal with a hard shell, can be used as an indicator of climate change. The first nine-banded armadillo (the only species in the United States), which had migrated from South America over an extremely long period of time, was sighted in the United States in 1849. It migrated from Mexico into Texas, where armadillos were sighted in the 1850s. A century ago, the armadillo's range was restricted mainly to Mexico, southern Texas, and parts of deserts in Arizona and New Mexico. By the late 20th century, however, they had been found in Oklahoma, Arkansas, and Georgia. The armadillo was chosen as Texas's official state small mammal. During the Great Depression, they were often called "Hoover hogs" by down-on-their-luck Americans who caught and ate the animals instead of the "chicken in every pot" that President Herbert Hoover had promised when running for president (Suhr 2006).

The armadillos' accustomed range then expanded rapidly as motorists reported hitting them (they have little street sense) as far north as southern North Carolina, Kentucky, Illinois, and Nebraska, where one was reported licking up insects on U.S. Route 50 in August 2005. "Over the last seven or eight years, we've been getting more regular reports of armadillos," said Mike Fritz, a natural heritage zoologist with the Nebraska Game and Parks Commission (Laukaitis 2005, A1).

Further Reading

Laukaitis, Algis J. "Talmage Man: No Kiddo . . . It's an Armadillo!" *Lincoln Journal Star*, August 18, 2005, A1.
Suhr, Jim. "Armadillos Making Northern March." Associated Press, Illinois State Wire, September 16, 2006 (LEXIS).

The study was summarized by its authors:

Here, we compare recent and historical surveys for 48 Mexican lizard species at 200 sites. Since 1975, 12 percent of local populations have gone extinct. We verified physiological models of extinction risk with observed local extinctions and extended projections worldwide. Since 1975, we estimate that 4 percent of local populations have gone extinct worldwide, but by 2080 local extinctions are projected to reach 39 percent worldwide, and species extinctions may reach 20 percent. Global extinction projections were validated with local extinctions observed from 1975 to 2009 for regional biotas on four other continents, suggesting that lizards have already crossed a threshold for extinctions caused by climate change. (Sinervo et al. 2010, 894)

A model developed by the team forecast locations in North and South America, Africa, Europe, and Australia where lizard populations that had been studied previously had gone locally extinct. "We did a lot of work on the ground to validate the model and show that the extinctions are the result of climate change," Sinervo said. "None of these are due to habitat loss. These sites are not disturbed in any way, and most of them are in national parks or other protected areas" ("Study Documents" 2010). As local populations continue to disappear, species extinctions will follow, he said. "Most of these species currently registering local extinctions will be completely extinct by 2080, unless we change and limit the carbon dioxide production that is driving global warming," he said ("Study Documents" 2010).

Gelada Baboon on Its Last Legs?

Human encroachment and warming temperatures in the highlands of Ethiopia are eliminating the habitat of the gelada baboon. Primate expert Chadden Hunter said, "Our research suggests that for each 2 degrees C (3.6 F) degree increase in temperature, the gelada's lower limit for grazing will rise 500 meters (547 yards). . . . Global temperatures only need to rise a few degrees, and [their habitat] will be, in effect, lifted off the tops of the mountains" (Smucker 2001). A Swiss study of the Ethiopian highlands has shown that the specific grasses on which the baboons feed have receded steadily upward over the past several hundred years. A separate study commissioned by the World Wide Fund for Nature in 2000 cited the gelada baboon as one of several mammals most at risk from the effects of global warming.

American and Ethiopian primate experts concur that temperature shifts may soon destroy the high-altitude "islands of grass" on which the gelada survives. "Since the gelada survive at the physical limits of the landscape, the likelihood of future global warming raises serious questions over the species' survival," said Jacinta Beehner, an American baboon expert with long experience in Ethiopia (Smucker 2001).

Theropithecus gelada is the last remaining species of its type. Hunter believes that the graminivorous (grass-eating) monkeys may have lost most of their close relatives to past episodes of climate change. "They are the last relic of a great dynasty, barely surviving in the Ethiopian highlands," he said (Smucker 2001). The gelada, which scamper and leap through the highlands in "herds" of 600 or more, maintain a complex social structure that offers "irreplaceable insight into human social behavior," said Hunter (Smucker 2001).

Warming may provoke the baboons' demise within a century, but the "most immediate threat to the gelada's survival comes from farming," said Tesfaye Hundessa, manager of the Ethiopian Wildlife Conservation Organization. "They could well be entirely eliminated by humans before the global warming has time to take its toll over the coming decades" (Smucker 2001).

Further Reading

Smucker, Philip. "Global Warming Sends Troops of Baboons on the Run: Rising Temperatures and Humans Encroaching on Grasslands Are Endangering the Ethiopian Primates." *Christian Science Monitor*, June 15, 2001, 7.

Lizard extinctions affect the food chain because they are prey for birds, snakes, and other animals. They, in turn, consume insects. "We could see other species collapse on the upper end of the food chain, and a release on insect populations," Sinervo said ("Study Documents" 2010).

Loggerhead Sea Turtles' Gender Ratio

Climate change threatens loggerhead turtles nesting on Florida beaches, according to the work of British researchers. Anticipated temperature increases could destroy male North American loggerhead populations and thereby bring about the extinction of the species. Researchers analyzed 26 years of loggerhead turtle nesting and climate data and compared the findings with models for future temperatures. Warmer temperatures during incubation produce a higher proportion of female loggerhead turtles; cooler conditions produce more males.

Increasing incubation temperatures 1°C eliminates male turtles in some locales, according to the research team. An increase of 3°C "would lead to extreme levels of infant mortality and declines in nesting beaches across the United States" ("Climate Change Threatens" 2007). Loggerheads already are listed as threatened under the U.S. Endangered Species Act. Loggerhead nest counts declined 22.3 percent between 1989 and 2005. In addition to the threat of rising temperatures, some loggerheads drown in fishing trawler nets before they can breed.

A Warmer World Constricts the Range of Bumblebees

Rising temperatures are constricting the range of several bumblebee species in North America and Europe, according to research published in *Science* during 2015. Warmer temperatures drive the bees out of habitats in such areas as the southern United States, and they have not been spreading northward. Scientists speculate that environments there are hostile to the bees.

"Bumblebee species across Europe and North America are declining at continental scales, and our data suggest that climate change plays a leading, or perhaps the leading, role in this trend," said Jeremy T. Kerr, a University of Ottawa conservation biologist and lead author of the study (St. Fleur 2015). Having evolved in a relatively cool climate and unable to adapt by moving into new northern habitats, many of the 67 species of bees studied by Kerr and colleagues have been caught in a trap. "Climate change is crushing [bumblebee] species in a vise," said Kerr (Carswell 2015, 126). "This is a huge loss, and it is happening very quickly," he said (Watson 2015).

Many of the bees are vital pollinators but have been subjected to diseases that may be linked to pesticides and other factors as their ranges have been reduced. Researchers found that bumblebee ranges began shrinking "even before the neonicotinoid pesticides came into play in the 1980s," said ecologist and coauthor Alana Pindar, a postdoctoral fellow at the University of Guelph in Canada (Carswell 2015, 126–127).

The study cataloged 423,000 bee sightings, comparing the period 1901–1974 with 1974–2010, when temperatures were rising rapidly (Kerr et al. 2015, 177). In summarizing the study, *The New York Times* reported that "*Bombus affinis* [one species of bumblebee] once buzzed as far south as Georgia, but now is only rarely seen in states like Illinois, Maine, and Wisconsin, while *Bombus terricola*, which thrived in North Carolina and the Mid-Atlantic, is now mostly seen in parts of Maine, New Hampshire, Ontario, and Quebec" (St. Fleur 2015).

Further Reading

Carswell, Cally. "Bumblebees Aren't Keeping Up with a Warming Planet." *Science* 349 (July 10, 2015): 126–127.

Kerr, Jeremy T., et al. "Climate Change Impacts on Bumblebees Converge across Continents." *Science* 349 (July 10, 2015): 177–180.

St. Fleur, Nicholas. "Climate Change Is Shrinking Where Bumblebees Range, Research Finds." *The New York Times*, July 9, 2015. http://www.nytimes.com/2015/07/10/science/bumblebees-global-warming-shrinking-habitats.html.
Watson, Traci. "Bumblebees Stung by Climate Change." *USA Today*, July 10, 2015, 4A.

The research, which was published in the journal *Global Change Biology*, was conducted in partnership with the Bald Head Island Conservancy and North Carolina Wildlife Resources. "We are stunned by these results and what they could mean for the species in the future," said Brendan Godley of the University of Exeter's School of Biosciences in the United Kingdom. "In particular, we are concerned that populations that are already predominantly female could become 100 percent female if temperatures increase by just one degree. This is a major issue for nesting populations further south, in Florida, for example, where males are already in short supply" ("Climate Change Threatens" 2007).

Alligators' Northward Migration

In mid-May 2006, alligators were sighted in the backwaters of the Mississippi River as far north as Memphis. "It's possible that alligators have had a northern range expansion due to the mild winters we've had in the past 10 years," wildlife agent Gary Cook said ("Gators Spotted" 2006). The Tennessee Wildlife Resources Agency received reports of alligator sightings on McKellar Lake, a backwater of the Mississippi just south of Memphis, and at T.O. Fuller State Park, north of the city. As many as five alligators may have been seen, including one said to be close to 7 feet long that was reportedly spotted on a bank beside McKellar Lake. "It was just laying out in the sun," said Kay Vescovi, a chemical plant manager at a nearby industrial park. "Nobody was concerned, just kind of shocked that they're this far north" ("Gators Spotted" 2006)

Stan Trauth, a zoology professor at Arkansas State University, said alligators spotted in the Memphis region could have been pets that were released into the wild. "Moving north is not what they would want. They would want to stay in the more moderate climate," said Trauth, who has tracked alligators released in Arkansas by wildlife agents. Trauth said that the region is along the northern edge of animals' survival range. "They're reaching their physiological tolerance in the winter in this area," he said ("Gators Spotted" 2006).

Further Reading

"Gators Spotted in Mississippi River Backwaters at Memphis." Associated Press in *Daytona Beach News-Journal*, May 13, 2006.

In Florida, which currently hosts more than 90 percent of loggerhead nests in the United States, 90 percent of hatchlings are female. By contrast, in North Carolina, where nesting sites are generally cooler, 42 percent are male. Males from more northerly locations may be migrating southward to help balance the gender ratio there. The turtles, which also nest in Oman and South Africa, reach maturity in 20 to 30 years and can weigh as much as 400 pounds. In South Africa, nesting populations have been increasing, according to Richard Penn Sawers, head of the World Wildlife Fund and Green Trust Turtle Monitoring and Community Development Project. These turtles breed almost entirely within the iSimangaliso Wetland Park, which is a designated marine protected area and World Heritage site ("Climate Change Threatens" 2007).

By 2007, loggerhead nesting reached the lowest level since Florida began keeping official records in the 1980s. The turtles were being endangered not only by rising sand temperatures that skew their sex ratio but also by seawalls and coastal development, which invade their nesting areas, and by collisions with fishing boats and becoming stranded in nets far at sea. On the Atlantic shore, nestings have declined 50 percent since 1998. Some homeowners along the coast also are installing sand-filled geotubes, 1,000-ton sandbags that block the encroaching ocean. These bags also destroy nesting areas ("Loggerhead Turtle" 2007).

Leatherback Sea Turtles

Pacific leatherback sea turtles' populations are falling rapidly as temperatures rise, human settlement devastates breeding grounds, seas slowly rise, and turtles become snared in fishing nets. They may go extinct and tourist sites that used to thrive on them will likely go with them. "We do not promote this as a turtle tourism destination anymore because we realize there are far too few turtles to please," said Álvaro Fonseca, a park ranger in Costa Rica (Rosenthal 2009). Costa Ricans who used to eat turtle eggs still poach them even after the practice became illegal.

The turtles had lived in this region for 150 million years, but they are now threatened in a wide variety of ways. They feed on coral reefs, and these have been dying as ocean waters warm. Their breeding sites are being inundated by rising seas. As in Florida, rising temperatures in the sands where they breed is skewing the gender of hatching turtles; even a small rise (to approximately 30°C) produces offspring that are mainly female. At 32°C, the gender ratio is 100 percent female. At 34°C (93°F), the eggs become sterile and produce no offspring at all. Local humans have been moving some of the eggs to hatcheries to maintain gender balance. By 2009, fewer than 3,000 leatherbacks survived in the Pacific Ocean, compared to some 90,000 just 20 years earlier (Rosenthal 2009). In Costa Rica, the "ecotourists" who visit to view the turtles and other species have provoked a building boom that is also contributing to the devastation of habitat.

"The turtles are very good storytellers about the effect of climate change on coastal habitats," said Carlos Drews, the regional marine species coordinator for the conservation group World Wildlife Fund. "The climate is changing so much faster than before, and these animals depend on so much for temperature" (Rosenthal 2009).

Moose Feel the Heat

Moose are on the move as their populations decline in traditional habitats of Montana, British Columbia, Minnesota, Manitoba, and New Hampshire. Their metabolisms are averse to heat, which makes the moose extremely uncomfortable in summer, and they are now appearing in sub-Arctic locales where no one has seen them before. Some local economies are also feeling the heat as moose populations decline. In New Hampshire, moose watching by tourists in 2012 was bringing in $115 million a year for local businesses.

Moose have historically ranged as far south as Minnesota, where Native peoples made use of them, but as temperatures warm they have been retreating northward. A warm summer punishes moose, which must expend energy to stay cool. Their low ratio of surface area to volume not only conserves body heat but also inhibits the release of that heat. Heat also makes moose more likely to suffer weight loss, parasites, and disease. Warming weather also allows ticks that afflict moose to survive winter.

Minnesota's moose population, a staple food source for many Native peoples as well as a trophy hunting quarry for non-Indians, began a significant decline in the late 1980s. Within two decades, the number of moose in northwestern Minnesota had dwindled from roughly 4,000 to 100 as the herd in the northeastern part of the state declined by more than half—from more than 8,000 to approximately 4,000 (Thompson 2015).

Red Squirrels' Reproductive Cycle

The births of red squirrels in the Yukon advanced almost three weeks within a decade because of a warming habitat, according to scientists who assert that the change is so profound that it seems to have affected the creatures' genetic makeup (Munro 2003).

"It's a phenomenal change in such a short period of time," said Stan Boutin, a biologist at the University of Alberta. Boutin has tracked hundreds of squirrels in the southwest corner of the Yukon. "We feel we've found some of the first evidence indicating climate change is leading to evolution in animal populations," said Boutin, who worked with Andrew McAdam (also from the University of Alberta), Denis Reale (from McGill University), and Dominique Berteaux (from the Universite du Quebec) (Ogle 2003).

Such changes emulate what occurs the selection process used by animal breeders who gradually breed certain traits into the animals over several generations, Boutin said. In this case, however, climate change is driving the evolutionary change among the squirrels through natural selection. "It's typical Darwinian natural selection," Boutin said. "As the conditions get warmer in spring, it favors those individuals that carry genes that allow them to breed earlier" (Munro 2003). Boutin studied four generations of squirrels over 15 years at Kluane National Park. The study, which was published in the

March 2003 edition of the *Proceedings of the Royal Society* (Réale et al. 2003) is the first to suggest that warming has triggered genetic change.

The squirrels' breeding season has been markedly affected by warmer spring temperatures and an increased food supply from spruce trees, which produce more cones as temperatures warm. Pups are now born as early as March 1. "It's good news in the sense that we are really impressed by the adaptability of these characters and their ability to keep up with the change," said Boutin. "The bad news is that these rates of change are definitely very rapid. Our human activities are affecting organisms in a variety of ways, even to the point where we are causing evolution in them" (Munro 2003).

Even as this study was released in 2003, observers in British Columbia were reporting that a mild winter was prompting some animals in that province to procreate several weeks earlier than usual. The first house sparrow hatchlings, for example, appeared in the city several weeks earlier than usual, said Stanley Park Ecology Society spokesman Robert Boelens. Boelens expected that other urban critters may be mating earlier than usual—everything from ducks to squirrels to swans to raccoons. "In fact," according to a *Vancouver Sun* report, "Squirrel love is definitely under way. . . . Several people around the Lower Mainland [near Vancouver's urban area] have reported significant squirrel nest-building activity weeks earlier than usual. The same behaviors have been reported for raccoons, skunks, rabbits, mice, crows, and rats" (Reed 2003, B-1).

Further Reading

Munro, Margaret. "Global Warming Affecting Squirrels' Genes, Study Finds: Research in Yukon: 'Phenomenal Change' Seen as Rodents Breed Earlier." *National Post* (Canada), February 12, 2003, A2.

Ogle, Andy. "Squirrels Get Squirrelier Earlier: Climate Change to Blame. Breeding Season in Yukon Advances 18 Days in Decade." *Edmonton Journal* in *Montreal Gazette*, February 12, 2003, A12.

Réale, Denis, Andrew G. McAdam, Stan Boutin, and Dominique Berteaux. "Genetic and Plastic Responses of a Northern Mammal to Climate Change." *Proceedings of the Royal Society B*, March 22, 2003. doi: 10.1098/rspb.2002.2224.

Reed, Nicholas. "Mild Winter Stirs Wildlife to Early Thoughts of Love." *Vancouver Sun*, February 12, 2003, B1.

The Denendeh Environmental Working Group in the Yukon of Canada and parts of Alaska has gathered accounts of moose and bison moving northward, along with several species of birds and insects never before seen in their homelands. Pine and spruce parasites and diseases also have increased with warming temperatures. A United Nations worldwide survey examined global warming's effects on indigenous peoples (McLean 2010).

Also in the Northern Hemisphere, Hudson Bay could be free of ice by the middle of this century. "We're starting to warm up very, very fast," said Peter Scott, scientific coordinator of the Churchill Northern Studies Center ("Hudson Bay" 2001). Scott said that climate change is bringing some species to the Arctic that have never been seen there before. "We may get southern animals. We're seeing a lot of moose now. . . . We now have moose in Churchill, where 20 years ago moose virtually didn't exist" ("Time to Act" 2001).

Several factors seem to be provoking moose migration, but climate change is a common thread. As temperatures rise in New Hampshire, for example, longer fall seasons with less snowfall encourage an explosion of winter ticks that colonize moose hides—as many as 100,000 at a time (Robbins 2013). The ticks torture the moose, who scratch themselves ceaselessly and tear off hunks of their own hair. "Some moose lose so much hair they look pale, even spectral . . . some people call them 'ghost moose,'" wrote Robbins (2013). When it rains in the spring, the moose, deprived of their warm coats, become hypothermic. Warmer weather in Minnesota also brings liver flukes and brain worms to moose.

Moose also work harder to stay cool when temperatures rise above 23°F in winter, leading to exhaustion and death. In British Columbia, pine beetles that ravage forests as temperatures warm have been depriving moose of habitat, exposing them to predators, including human hunters. Moose have become disoriented as they wander into human settlements. As Jim Robbins of *The New York Times* reported, "In Smithers, British Columbia, in April [2013], a moose—starving and severely infested with ticks—wandered into the flower section of a Safeway market. It was euthanized" (Robbins 2013).

"It's complicated because there's so many pieces of this puzzle that could be impacted by climate change," said Erika Butler, who has worked as a wildlife veterinarian with the Minnesota Department of Natural Resources (Robbins 2013).

Can Polar Bears Adapt?

Climate-change skeptics who assert that polar bears recently adapted from brown bears—and could easily transform themselves back in a few years—ignore recent explorations of their genome by scientists that indicate the species is 4 million to 5 million years old, not the roughly 600,000 as previously thought. According to Webb Miller and colleagues in the *Proceedings of the National Academy of Sciences*,

> [P]olar bears are superbly adapted to the extreme Arctic environment and have become emblematic of the threat to biodiversity from global climate change. Their divergence from the lower-latitude brown bear provides a textbook example of rapid evolution of distinct phenotypes. However, limited mitochondrial and nuclear DNA evidence conflicts in the timing of polar bear origin as well as placement of the species within versus sister to the brown bear lineage. We gathered extensive genomic sequence data from contemporary polar, brown, and American black bear samples. . . .
>
> [P]olar and brown bears [are] sister species. [in an] ancient admixture between the two species. . . . We also provide paleodemographic estimates that suggest bear evolution has tracked key climate events, and that polar bears in

particular experienced a prolonged and dramatic decline in its effective population size during the last c. 500,000 years. We demonstrate that brown bears and polar bears have had sufficiently independent evolutionary histories over the last 4–5 million years to leave imprints in the polar bear nuclear genome that likely are associated with ecological adaptation to the Arctic environment. (Miller et al. 2012)

Polar Bears as Land Animals

The loss of summer sea ice is pushing polar bears more onto land occupied by people in northern Canada and Alaska, causing inflated estimates of their numbers, said NASA scientist Claire Parkinson, who studies the bears ("Arctic Ice Melting" 2006). Charles Wohlforth, writing in *The Last Polar Bear* (Kazlowski 2008), described how the Alaskan North Slope town of Barrow was inundated by land-bound polar bears during the early fall of 2002 after sea ice retreated too far offshore to be used as a hunting platform.

The strange congregation of dozens of stranded bears . . . threw the community into fear and consternation. A 1,100-pound bear—a large male—was shot when it wouldn't leave the school. Visitors were warned not to walk outdoors. Biologists employed by the local government were constantly on the run to keep polar bears off the streets. Polar bears normally don't eat people, but they easily can and, when hungry enough, they occasionally have. Bears on land in the summer fall have been more common all along the coast since the trend of warmer weather shortened the winters and diminished the sea ice. But no one had seen an invasion like this before. Biologists counted more than one hundred bears near town. (Kazlowski 2008, 63)

During the summer of 2004, hunters found half a dozen polar bears that had drowned about 200 miles north of Barrow on Alaska's northern coast. They had tried to swim for shore after the ice had receded 400 miles. A polar bear can swim 100 miles—but not 400 (Buncombe and Carrell 2005). As many as 36 other polar bears may have died as they tried to swim to land after the Arctic ice shelf along the north shore of Alaska receded (Tidwell 2006, 57).

Without ice, polar bears can become hungry, miserable creatures, especially in unaccustomed warmth. During Iqaluit's weeks of record heat in July 2001, two tourists were hospitalized after they were mauled by a bear in a park south of Barrow. On July 20, in a similar confrontation in northern Labrador, a polar bear tried to claw its way into a tent occupied by a group of Dutch tourists. The tourists escaped injury, but the bear was shot to death. "The bears are looking for a cooler place," said Ben Kovic, Nunavut's chief wildlife manager (Johansen 2001, 18).

Until recently, polar bears had their own food sources and usually went about their business without trying to steal food from humans. Beset by late freezes and early thaws, hungry polar bears are coming into contact with people more frequently. In Churchill, Manitoba, polar bears waking from their winter's slumber have found that the ice on Hudson's Bay has melted earlier than usual. Instead of making their way onto the ice in search of seals, the bears walk along the coast until they get to Churchill, where they block motor traffic and pillage the town dump. Churchill now has a holding tank for wayward polar bears that is larger than its jail for people. TIME magazine described Churchill:

Polar bears that ordinarily emerge from their summer dens and walk north up Cape Churchill before proceeding directly onto the ice now arrive at their customary departure point and find open water. Unable to move forward, the bears turn left and continue walking right into town, arriving emaciated and hungry. To reduce unscheduled encounters between townspeople and the carnivores, natural-resource officer Wade Roberts and his deputies tranquilize the bears with a dart gun, temporarily house them in a concrete-and-steel bear "jail" and move them 10 miles north. In years with a late freeze—most years since the late 1970s—the number of bears captured in or near town sometimes doubles, to more than 100. (Linden 2000)

By 2007, several tour operators were charging $3,000 to $5,000 for two- to four-day "last chance" excursions to Churchill to witness the bears in their natural habitats. In the meantime, the amount of time that polar bears had access to ice for hunting had been reduced by a month in just two decades. Residents of Churchill remarked at wearing flip-flops and shirtsleeves in mid-October, a time when Hudson Bay was once beginning to freeze. The average weight of female bears had fallen from 600 pounds to a few pounds over 400, which inhibits reproduction. The mortality for bear cubs under 5 years of age was up 50 percent, according to Robert Buchanan, president of Polar Bears International (Clark 2007, 2D). The town dump was closed in 2005 after hungry bears made a habit of pillaging it for food scraps. The emaciated bears were even tackling garbage trucks before they reached the dump itself.

Polar bears face other problems related to warming. For example, some polar bear dens have collapsed because of lightning-sparked brush fires on the tundra, which may be increasing because of global warming, according to Stirling, who said that the Canadian government should consider fighting fires in prime bear denning areas in the north ("Brush Fires" 2002). Some pregnant bears have been digging dens only to have them collapse, Stirling said.

"The fires burn off all of the trees and bushes that are on the upper part of the banks holding the roofs together," he said ("Brush Fires" 2002). "The fires melt the permafrost in the adjacent areas, and in particular the roofs, so there's nothing to hold the roofs together. They just collapse," he said. The bears then must find another denning site where they can give birth over the winter. "The fires are definitely affecting the ability of the bears to use some of the prime areas," said Stirling. After a fire, he continued, vegetation may require 70 years to grow back to a density that is suitable for denning sites ("Brush Fires" 2002). Yet another potential risk to polar bears is the increased chance that rain in late winter will cause polar bear dens to collapse before females and cubs have departed. Scientists surveying polar bear habitat in Manitoba, Canada, have observed large snow banks used for dens that had collapsed under the weight of wet snow (Stirling and Derocher 1993, 244).

Polar Bears Increasing—or Just More Visible?

The semisovereign Canadian Inuit territory of Nunavut increased its annual hunting quotas by 29 percent in 2005 to 518 kills, an increase of 115, saying that polar bear Inuit hunters were actually seeing an increase in polar bear populations. That impression, some scientists and environmentalists say, is simply a matter of the bears' greater visibility as shrinking ice pushes them closer to Inuit communities (Krauss 2006). Those experts tick off a list of stresses on the polar bear. Global warming is melting the bears' icy migration routes, which are critical for breeding and catching seals for food around Hudson Bay and in Alaska. Poaching is threatening populations in Russia. Pollution is causing deformities and reproductive failures in Norway (Krauss 2006).

Other experts see a healthier population. They note that more than 20,000 polar bears now roam the Arctic compared to as few as 5,000 just 40 years ago and before Canada, Denmark, Norway, the Soviet Union, and the United States agreed to strong restrictions on trophy hunting in the 1970s (Krauss 2006).

Mitchell Taylor, manager of wildlife research for the Nunavut government, said warming trends had so far seriously affected only western Hudson Bay, just one of 20 areas where polar bears live. He acknowledged that overhunting could be a problem in Baffin Bay, between Canada and Greenland. "In other areas, polar bears appear to be overabundant," he added. "People have to quit thinking of polar bears as one big continuous mass of animals that are all doing the same thing" (Krauss 2006).

Some Polar Bears Engage in Cannibalism

The survival crisis facing polar bears took a fresh turn for the worse after 2004 when scientists from the United States and Canada found evidence that lack of access to food sources (mainly ringed seals) caused by shrinking ice cover might be forcing some of them to engage in cannibalism in the southern Beaufort Sea north of Alaska and western Canada. The scientists found three examples of polar bears preying on each other between January and April 2004, including the killing of a female in a den shortly after she gave birth. The bears use the sea ice not only for feeding but also for mating and giving birth. The predation study was published in an online version of the journal *Polar Biology* (Amstrup et al. 2006).

According to the study's principal author, Steven Amstrup of the U.S. Geological Survey's Alaska Science Center, polar bears sometimes kill each other for population regulation, dominance, and reproductive advantage, but killing for food heretofore had been highly unusual (Joling 2006). "During 24 years of research on polar bears in the southern Beaufort Sea region of northern Alaska and 34 years in northwestern Canada, we have not seen other incidents of polar bears stalking, killing, and eating other polar bears," the scientists said (Joling 2006). In February 2005, the Center for Biological Diversity in Joshua Tree, California, petitioned the federal government to list polar bears as threatened under the federal Endangered Species Act (Joling 2006). "It's very important new information," said Kassie Siegal, lead author of that petition. "It shows in a really graphic way how severe the problem of global warming is for polar bears" (Joling 2006).

Less Ice, Thinner Polar Bears

During 2002, according to "Polar Bears at Risk," a World Wildlife Fund study, the combination of toxic chemicals and global warming could cause extinction of roughly 22,000 surviving polar bears in the wild within 50 years. Lynn Rosentrater, report coauthor and a climate scientist in the WWF's Arctic program, said, "Since the sea ice is melting earlier in the spring, polar bears move to land earlier without having developed as much fat reserves to survive the ice free season. They are skinny bears by the end of summer, which in the worst case can affect their ability to reproduce." The same report said that lower body weight reduces female bears' ability to lactate, leading to fewer surviving cubs. Already, fewer than 44 percent of cubs now survive the ice-free season, according to the report ("Thin Polar Bears" 2002).

In western Hudson Bay where warmer temperatures during the 1990s provoked earlier ice melting in the summer and later freezing in the fall, polar bears suffered substantial weight loss. For every week that the ice broke up earlier, bears came ashore 20 to 25 pounds lighter, said zoologist Ian Stirling of the Canadian Wildlife Service in Edmonton (Kerr 2002, 1492). Stirling has documented a 15-percent decrease in the average weight and number of pups born to polar bears in western Hudson's Bay between 1981 and 1998. Such conclusions are amply supported from many other sources.

The polar bear population in the Hudson Bay area has dropped from 1,200 in 1989 to 950 in 2004 and the bears that are around are 22 percent smaller than they used to be ("Arctic Ice Melting" 2006). Canadian Wildlife Service scientists reported during December 1998 that polar bears around Hudson Bay are 90 to 220 pounds lighter than 30 years earlier, apparently because earlier ice melting has given them less time to feed on seal pups.

According to Michael Goodyear, director of the Churchill (Manitoba) Northern Studies Center, "For every week a bear has not been hunting, it is 10 kilograms (22 pounds) lighter, which can be dangerous as polar bears need to fatten up for the five months in the summer and fall that they are forced to fast" (Clavel 2002).

Weight loss affects the bears' ability to reproduce and weakens future generations that are born. Polar bears may soon reach a point where they have lost enough weight to render them infertile. Lara Hansen, chief scientist at the World Wildlife Fund, told *The Independent* (London), "If the population stops reproducing, that's the end of it" (Connor 2004).

Further Reading

"Arctic Ice Melting Rapidly, Study Says." *The New York Times*, September 14, 2006. http://www.nytimes.com/aponline/us/AP-Warming-Sea-Ice.html (no longer available).

Clavel, Guy. "Global Warming Makes Polar Bears Sweat." Agence France Presse, November 3, 2002.

Connor, Steve. "Meltdown: Arctic Wildlife Is on the Brink of Catastrophe; Polar Bears Could Be Decades from Extinction." *The Independent* (London), November 11, 2004, n.p. (LEXIS).

Kerr, Richard A. "A Warmer Arctic Means Change for All." *Science* 297 (August 30, 2002): 1490–1492.

"Thin Polar Bears Called Sign of Global Warming." Environmental News Service, May 16, 2002. http://ens-news.com/ens/may2002/2002L-05-16-07.html (no longer available).

According to Dan Joling's summary of the predation study for the Associated Press, researchers discovered the first kill in January 2004. A male bear had pounced on a den, killed a female and dragged it 245 feet away, where it ate part of the carcass. Females are about half the size of males. "In the face of the den's outer wall were deep impressions of where the predatory bear had pounded its forepaws to collapse the den roof, just as polar bears collapse the snow over ringed seal lairs," the paper said, continuing: "From the tracks, it appeared that the predatory bear broke through the roof of the den, held the female in place while inflicting multiple bites to the head and neck. When the den collapsed, two cubs were buried, and suffocated, in the snow rubble" (Joling 2006).

Three months later, the researchers found a partially eaten carcass of an adult female that had been walking with a cub. The killer, a male, had not followed the cub, indicating to the scientists that he was motivated by a need for food, not a desire to mate. (The male bear showed no interest in the cub's mother.) A few days after that, the remains of a year-old cub were found. Tracks indicated that the cub had been stalked and killed.

Further Reading

Amstrup, S. C., et al. "Recent Observations of Intraspecific Predation and Cannibalism among Polar Bears in the Southern Beaufort Sea." *Polar Biology*, 29(11) (2006): 997–1002. doi: 10.1007/s00300-006-0142-5.

"Arctic Ice Melting Rapidly, Study Says." *The New York Times*, September 14, 2006. http://www.nytimes.com/aponline/us/AP-Warming-Sea-Ice.html (no longer available).

"Brush Fires Collapsing Bear Dens." *Calgary Sun*, November 2, 2002, 18.

Buncombe, Andrew, and Severin Carrell. "Melting Planet: Species Are Dying Out Faster Than We Have Dared Recognize, Scientists Will Warn This Week," *The Independent* (London), October 2, 2005.

Clark, Jayne. "Tours Bear Witness to Earth's Sentinel Species." *USA Today*, November 2, 2007, 1D, 2D.

"Climate Change Threatens Loggerhead Turtles." Environment News Service, February 22, 2007. http://www.ens-newswire.com/ens/feb2007/2007-02-22-02.asp (no longer available).

Gill, Victoria. "Climate Change Link to Lizard Extinction." British Broadcasting Corp., May 14, 2010. http://news.bbc.co.uk/2/hi/science_and_environment/10113949.stm.

"Hudson Bay Ice-Free by 2050, Scientists Say." CBC News, May 14, 2001. http://www.cbc.ca/news/canada/hudson-bay-ice-free-by-2050-scientists-say-1.286276.

Johansen, Bruce E. "Arctic Heat Wave." *The Progressive*, October, 2001, 18–20.

Joling, Dan. "Study: Polar Bears May Turn to Cannibalism." Associated Press, June 14, 2006 (LEXIS).

Kazlowski, Steven. *The Last Polar Bear: Facing the Truth of A Warming World*. Seattle: Braided River Books, 2008.

Krauss, Clifford. "Bear Hunting Caught in Global Warming Debate." *The New York Times*, May 27, 2006. http://www.nytimes.com/2006/05/27/world/americas/27bears.html.

Linden, Eugene. "The Big Meltdown: As the Temperature Rises in the Arctic, It Sends a Chill around the Planet." *Time* 156(10), September 4, 2000. http://www.time.com/time/magazine/articles/0,3266,53418,00.html (no longer available).

"Loggerhead Turtle Nesting Down by Half Since 1998." Environment News Service, November 8, 2007. http://www.ens-newswire.com/ens/nov2007/2007-11-08-094.asp

McLean, Kirsty Galloway. *Advance Guard: Climate Change Impacts, Adaptation, Mitigation and Indigenous Peoples*. Darwin, Australia, 2010. http://archive.ias.unu.edu/resource_centre/UNU_Advance_Guard_Compendium_2010_final_web.pdf.

Miller, Webb, et al. "Polar and Brow Bear Genomes Reveal Ancient Admixture and Demographic Footprints of Past Climate Change." *Proceedings of the National Academy of Sciences* 109(36) (September 4, 2012): E2382–E2390. doi: 10.1073/pnas.1210506109.

Robbins, Jim. "Moose Die-Off Alarms Scientists." *The New York Times*, October 14, 2013. http://www.nytimes.com/2013/10/15/science/earth/something-is-killing-off-the-moose.html.

Rosenthal, Elisabeth. "Turtles Are Casualties of Warming in Costa Rica." *The New York Times*, November 14, 2009. http://www.nytimes.com/2009/11/14/science/earth/14turtles.html.

Sinervo, Barry, et al. "Erosion of Lizard Diversity by Climate Change and Altered Thermal Niches." *Science* 328 (May 14, 2010): 894–899.

Stirling, Ian, and Andrew Derocher. "Possible Impacts of Climatic Warming on Polar Bears." *Arctic* 46(3) (September 1993): 240–245.

"Study Documents Widespread Extinction of Lizard Populations Due to Climate Change." NASA Earth Observatory, May 13, 2010. http://earthobservatory.nasa.gov/Newsroom/view.php?id=43996&src=eoa-manews (no longer available).

Thompson, Douglas. "Minnesota Tribes Collaborate to Save State's Disappearing Moose Population." Indian Country Today Media Network, February 11, 2015. http://indiancountrytodaymedianetwork.com/2015/02/11/minnesota-tribes-collaborate-save-states-disappearing-moose-population-159151.

Tidwell, Mike. *The Ravaging Tide: Strange Weather, Future Katrinas, and the Coming Death of America's Coastal Cities*. New York: Free Press, 2006.

"Time to Act on Nunavut Climate Change." Canadian Broadcasting Corporation, March 16, 2001. http://north.cbc.ca/cgi-bin/templates/view.cgi?/news/2001/03/16/16nunmoose (no longer available).

See also: Amphibians and Warming; Animal Life, Antarctic; Animal Life, Arctic; Biodiversity; Coral Reefs; Extinctions; Fisheries; Jellyfish; Migrations; Phytoplankton; Salmon and Trout

AMPHIBIANS AND WARMING

By 2006, an increasing number of scientific studies pointed to warming temperatures as a crucial reason—but not the sole reason—why several species of amphibians (frogs and toads) have been going extinct or facing sharp population declines around the world. One primary reason for some extinctions is the prevalence of skin diseases such as chytridiomycosis, but these may be aggravated by climate change (Alford 2011, 461).

These include studies of the Monteverde harlequin frog (*Atelopus* sp.), which vanished along with the golden toad (*Bufo periglenes*) during the late 1980s from the mountains of Costa Rica. According to analysis by J. Alan Pounds, resident scientist at the Tropical Science Center's Monteverde Cloud Forest Preserve in Costa Rica, and his colleagues (Pounds et al. 2006), an estimated 67 percent of roughly 110 species of *Atelopus* that were once endemic to the American tropics are now extinct, having been afflicted by a pathogenic chytrid fungus (*Batrachochytrium dendrobatidis*). Pounds and colleagues concluded with more than 99 percent confidence that large-scale environmental warming is a key factor in the disappearances. They found that rising temperatures in many highland localities have encouraged *Batrachochytrium* growth and outbreaks. "With climate change promoting infectious disease and eroding biodiversity, the urgency of reducing greenhouse gas concentrations is now undeniable," they concluded (Pounds et al. 2006, 161).

"Disease is the bullet killing frogs, but climate change is pulling the trigger," said Pounds, who has taken a lead role in these studies. "Global warming is wreaking havoc on amphibians and will cause staggering losses of biodiversity if we don't do something first" (Eilperin 2006). The chytrid fungus kills frogs by growing on their skin, attacking their epidermis and teeth, and releasing a toxin. Higher temperatures allow more water vapor into the air, which forms a cloud cover that leads to cooler days and warmer nights. These conditions favor the fungus, which grows and reproduces best at temperatures between 63 and 77°F. "There's a coherent pattern of disappearances, all the way from Costa Rica to Peru," Pounds said. "Here's a case where we can show that global warming is affecting outbreaks of this disease" (Eilperin 2006).

Pounds and Karen L. Masters, both of whom have been active in studying amphibian die-offs in Costa Rica's Monte Verde Cloud Forest Preserve, have criticized monocausal interpretations that attribute frogs' and toads' problems to one factor, such as chytrid fungus, in isolation. In a review of a book—James P. Collins and Martha L. Crump's *Extinction in Our Time: Amphibian Decline* (2009)—that draws that conclusion, they wrote that concentration on the fungus in isolation ignores many other factors, one of which is climate change. It is, they wrote, like blaming an automobile accident on excessive speed while ignoring the alcohol on a driver's breath. "Amphibians belong to a chorus of canaries telling us one thing: Earth's life-support system is on trouble" (Pounds and Masters 2009, 39).

Amphibians Threatened Worldwide

In 2004, the first worldwide survey of 5,743 amphibians species (frogs, toads, and salamanders) indicated that one in every three species was in danger of extinction, many of them likely victims of an infectious fungus possibly aggravated by drought and warming (Stokstad 2004). Findings were contributed by more than 500 researchers from more than 60 countries. "The fact that one-third of amphibians are in a precipitous decline tells us that we are rapidly moving toward a potentially epidemic number of extinctions," said Achim Steiner, executive director of the World Conservation Union based in Geneva (Seabrook 2004).

"Amphibians are one of nature's best indicators of the overall health of our environment," said Whitfield Gibbons, a herpetologist at the University of Georgia's Savannah River Ecology Laboratory (Seabrook 2004; Stokstad 2004). Amphibians are more threatened and are declining more rapidly than birds or mammals; "The lack of conservation remedies for these poorly understood declines means that hundreds of amphibians species now face extinction," wrote the scientists who conducted the worldwide survey (Stuart et al. 2004).

In North and South America, the Caribbean and Australia, a major culprit appears to be the highly infectious fungal disease chytridiomycosis. New research shows that prolonged drought may cause outbreaks of the disease in some regions, although some scientists attribute the disease's spread to global warming. Other threats include loss of habitat, acid rain, pesticides and herbicides, fertilizers, consumer demand for frog legs, and a depletion of the stratospheric ozone layer that leaves the frogs' skin exposed to radiation (Seabrook 2004). Gibbons said that the loss of wetlands and other habitat to development, agriculture, and other human enterprises may be the leading cause of amphibian declines in Georgia and elsewhere in the southeastern United States. Pollution also may be playing a significant role, he said. "It's hard to find a pristine stream anymore" (Seabrook 2004).

Warming and the Decline of Oregon's Western Toad

Global warming could be playing a role in the decline of the western toad in Oregon, according to a report that was among the first to link climatic change with amphibian die-offs in North America (Pounds 2001, 639). Toad population declines may be part of an ecological chain reaction in which rapid global warming plays a role. Pounds has researched the decline of the golden toad in Costa Rica from his post with the Monteverde Cloud Forest Preserve and Tropical Science Center, and has described the situation in Oregon: "In the crystal-clear waters surrounded by snow-capped peaks in the Cascade range, the jet-black [western toad] embryos are suffering devastating mortality. They develop normally for a few days, then turn white and die by the hundreds of thousands" (Pounds 2001, 639).

Research by Pennsylvania State University's Joseph M. Kiesecker and colleagues indicates that the toads' fatal infection results from a complex sequence of events provoked by warming temperatures. Kiesecker and colleagues wrote, "Elevated sea-surface in this region [the western United States] since the mid-1970s, which have affected the climate over much of the world, could be the precursor for pathogen-mediated amphibian declines in many regions" (Kiesecker et al. 2001, 681). "Reductions in water depth due to altered precipitation patterns expose the embryos to damaging ultraviolet b (UV-B) radiation, thereby opening the door to lethal infection by a fungus, *Saprolegnia ferax*" (Pounds 2001, 639).

Depth of water influences the amount of UV-B radiation that reaches the toads' eggs; less water depth is related to the El Niño–La Niña cycle, which, Kiesecker and colleagues theorize is related to climate change. Similar patterns have been detected in cases of massive frog and toad mortality from other parts of the world. In some areas, fungus-borne diseases afflict lizards as well as frogs and toads (Pounds 2001, 639–640). Given what Kiesecker and others have found, Pounds has asserted there "is clearly a need for a rapid transition to cleaner energy sources if we are to avoid staggering losses of biodiversity" (2001, 640). Kiesecker's results support those of Pounds who, in 1999, reported that warmer and drier periods in the cloud forest atop the Continental Divide at Monte Verde were tied to El Niño events and had caused massive population reductions in more than 20 frog species, including the disappearance of the golden toad (Souder 2001).

The role of warming involves several factors implicated in the decline of amphibians. For example, Ross A. Alford and colleagues wrote in *Nature*:

> Is global warming contributing to amphibian declines and extinctions by promoting outbreaks of the chytrid fungus *Batrachochytrium dendrobatidis*? Analyzing patterns from the American tropics, Pounds et al. envisage a process in which a single warm year triggers die-offs in a particular area (for instance, 1987 in the case of Monteverde, Costa Rica). However, we show . . . that populations of two frog species in the Australian tropics experienced increasing developmental instability, which is evidence of stress, at least two years before they showed chytrid-related declines. Because the working model of Pounds et al. is incomplete, their test of the climate-linked epidemic hypothesis could be inconclusive. (Alford et al. 2007)

Further Reading

Alford, Ross A., Kay S. Bradfield, and Stephen J. Richards. "Ecology: Global Warming and Amphibian Losses." *Nature* 447 (May 31, 2007): E3, E4.

Kiesecker, Joseph M., Andrew R. Blaustein, and Lisa K. Belden. "Complex Causes of Amphibian Population Declines." *Nature* 410 (April 5, 2001): 681–684.

Pounds, J. Alan. "Climate and Amphibian Decline." *Nature* 410 (April 5, 2001): 639–640.

Souder, William. "Global Warming and a Toad Species' Decline." *Washington Post*, April 9, 2001, n.p. (LEXIS).

Amphibian Deformations and Ultraviolet Radiation

Three studies published in the July 1, 2002, edition of *Environmental Science & Technology* indicated that ultraviolet radiation could be responsible for increasing numbers of malformed frogs and other amphibians. "We really wanted to fill the gap between the findings of other laboratory research and what might happen in natural environments," said Dr. Steve Diamond, an environmental toxicologist at the U.S. Environmental Protection Agency and author of all three papers ("UV Radiation" 2002).

In one of the three studies, Diamond and his colleagues kept frog eggs in small outdoor containers while exposing them to varying degrees of UV radiation—from 25 percent to 100 percent of natural sunlight. As the eggs developed, the researchers observed hatching success, tadpole survival, and the presence of malformations. The scientists found that the frequency of malformations rose with increasing UV radiation, with half of the frogs experiencing malformations at 63.5 percent of the intensity of natural sunlight.

According to an Environment News Service report,

> In the second study, the researchers measured levels of dissolved organic carbon (DOC) in water in wetlands in Wisconsin and Minnesota. The top five to 20 centimeters of wetlands absorb as much as 99 percent of UV radiation, the researchers found. The third study involved a survey of 26 wetlands in the same region to estimate the specific level of risk of frogs living in these environments. Using a combination of computer models, historical weather records and DOC measurements, they concluded that UV radiation posed a risk to amphibians living in three of the 26 wetlands. ("UV Radiation" 2002)

Further Reading

Alford, Ross A. "Bleak Future for Amphibians." *Science* 480 (December 22 and 29, 2011): 461–462.

Alford, Ross A., Kay S. Bradfield, and Stephen J. Richards. "Ecology: Global Warming and Amphibian Losses." *Nature* 447 (May 31, 2007): E3, E4.

Collins, James P., and Martha L. Crump. *Extinction in Our Time: Amphibian Decline.* New York: Oxford University Press, 2009.

Eilperin, Juliet. "Warming Tied to Extinction of Frog Species." *Washington Post*, January 12, 2006, A1. http://www.washingtonpost.com/wp-dyn/content/article/2006/01/11/AR20 06011102121_pf.html.

Pounds, J. Alan, et al. "Widespread Amphibian Extinctions from Epidemic Disease Driven by Global Warming." *Nature* 439 (January 12, 2006): 161–167.

Pounds, J. Alan, and Karen L. Masters. "Amphibian Mystery Misread." *Nature* 462 (November 5, 2009): 38–39.

Seabrook, Charles. "Amphibian Populations Drop." *Atlanta Journal-Constitution*, October 15, 2004, 1C.

Stokstad, Erik. "Global Survey Documents Puzzling Decline of Amphibians." *Science* 306 (October 15, 2004): 391.

Stuart, Simon N., et al. "Status and Trends of Amphibian Declines and Extinctions Worldwide." *Science* 306 (December 3, 2004): 1783–1786.

"UV Radiation Linked to Deformed Amphibians." Environment News Service, June 2, 2002. http://www.ens-newswire.com/ens/jun2002/2002-06-21-09.html.

See also: Adaptation, Animals, Lizards, and Warming Habitats; Animal Life, Antarctic; Animal Life, Arctic; Biodiversity; Coral Reefs; Extinctions; Fisheries; Jellyfish; Migrations; Phytoplankton; Salmon and Trout

ANIMAL LIFE, ANTARCTIC

Animals that depend on sea ice for food have been migrating southward in the Southern Hemisphere. Tom Spears described the Antarctic peninsula's changing environment in 2001, "A gentle Antarctic winter [that] is bringing mild breezes along the seashore, weather for pleasant walks in a light jacket, snow that's mushy enough to make snowballs. This is the depth of winter in the southern hemisphere, yet part of Antarctica remains balmy, even above freezing, as shown by many of the scientific research stations that report daily weather" (Spears 2001). Scientists from the British Antarctic Survey reported in December 2004 that grass was growing on some parts of the Antarctic peninsula in places that until recently had been covered by ice year round (Rohter 2005). Around 1950, four of five winters at Palmer Station produced extensive sea ice; by the end of the 20th century, the average was two winters in five. By 2015, sea ice was becoming a rarity on some of the peninsula's shores.

Penguins under Assault

One prominent example of a species driven to exile or death by climate change is the Adélie penguin, an 18-inch bird that looks as if it is wearing a tuxedo. These birds have become scarcer near Palmer Station on the Antarctic Peninsula because they eat the krill that arrives with sea ice. Adélie penguins that have not died have been moving their range southward toward areas that were previously too cold and barren for them.

Adélie populations within two miles of Palmer Station have declined by some 50 percent in 25 years, including a 10-percent decline during two years in the late 1990s (Petit 2000, 68). By the late 1980s, penguin biologist William R. Fraser was among the first to associate retreating sea ice on the Antarctic peninsula with the decline of Adélie penguins. As ice-loving species such as the Adélies declined, ice-avoiding ones flourished, including chinstrap penguins and fur seals. "It reeked of habitat change due to a warming climate," Fraser said (Montaigne 2010, 160).

The lives of Magellanic penguins are being complicated by weather extremes that are increasing mortality for those not already dying from predators and starvation. Increasing storm intensity is one such threat. "Rainfall is killing a lot of penguins, and so is heat," said P. Dee Boersma, a University of Washington conservation biologist, in describing the penguins' largest breeding colony, approximately 200,000 breeding pairs, near Punta Tombo on Argentina's southeast coast. "And those are two new causes" (Fountain 2014). Even without heavy rain and heat, two-thirds of hatchlings die. With heavy rain and heat, mortality rises, in large part because soaking rain can cause hypothermia for a newly hatched penguin. The size of the colony declined 24 percent from 1987 to 2013 (Fountain 2014).

Ecologist Bill Fraser led a team of researchers that followed the decline of Adélie penguins, which he describes as "smart and fussy little men in evening clothes" (Montaigne 2010, 40). In 1976, 30,000 breeding pairs could be found on seven islands around Palmer Station on the Antarctic peninsula; by 2010, less than 20 percent of that number remained. Sea ice, under which the penguins feed, has declined along the peninsula, and krill populations, their main food source, has also declined 80 percent since 1976, which is in line with Adélie penguin numbers (Le Maho 2010, 1035). An increase of only 0.3°C in sea surface temperatures near the ice shelves contributed to a 10-percent drop in survival rates of king penguins (Le Maho 2010, 1035).

Less Krill, Fewer Penguins

Different types of penguins are suited to varying ecological niches shaped by climate. As coastal Antarctica has warmed, they have died or moved. Gentoo penguins, which are habituated to relatively mild weather, have been replacing Adélie penguins on the Antarctic peninsula as it warms rapidly and sea ice retreats. The Adélies feed on krill that lives on the undersides of eroding ice sheets. Warmer weather also produces more snowfall, which interferes with the Adélies' ability to breed and incubate their eggs. Farther south, where climate has become more amenable to the Adélies, their numbers are increasing.

A warming climate poses several problems for the Adélies. For example, warmer ocean waters west of the Antarctic peninsula have opened ocean surface as ice has retreated, allowing more evaporation and snowfall that inhibits the Penguins' ability to create and maintain nests. The penguins evolved in a polar desert, and are a snow-intolerant species. Ice by 2015 was present near major Adélie rookeries three months less per year on average compared to 1979. Less ice provides less krill to feed chicks, which die in larger numbers from malnutrition. All of these factors combine to destroy Adélie penguins on the west Antarctic peninsula (Montaigne 2010, 212, 216–17).

As sea ice has declined off the West Antarctic peninsula, scientists expected to see declines in ice-dwelling Adélie penguins and increases in chinstrap penguins, which forage in ice-free waters. However, Wayne Trivelpiece and his colleagues at the U.S. National Marine Fisheries Service in La Jolla, California, say there is now "overwhelming evidence" that both populations are declining in the region. The authors' data, which covers populations at two sites over 30 years, suggests that sea-ice losses resulting from climate change have reduced the availability of Antarctic krill, the prey of both birds. If temperatures rise in future, sea ice and krill will decline further, and both species of penguin are likely to follow. ("Sea Ice Penguin" 2011)

Different species of penguins feed in various ways that have affected them variously as ice retreats. Adélie penguins usually browse the undersides of sea ice and have gone hungry as this ice recedes or breaks up. Their numbers have been declining. By contrast, chinstrap and gentoo penguins, which usually forage along open shorelines, have found a greater range as ice recedes, encouraging their reproduction (Pollack 2009, 212–213).

In spring 2009, sea ice formed 54 days later than in the autumn and retreated 31 days earlier compared with 1979. Eighty-seven percent of glaciers on the peninsula are retreating as well (Montaigne 2009, 78, 82). Chinstraps, which prefer open water, are not declining on the peninsula, whereas Adélies, denizens of the retreating ice, are disappearing from the area. About 2.5 million pairs of Adélies still breed in Antarctica, but their range is changing. On Avian Island, for example, they have doubled in 35 years, but scientists expect that they will decline here as well when sea ice begins to retreat with future warming (Montaigne 2009, 82).

At Palmer Station, winter temperatures rose 11°F during the 60 years ending in 2010, one of the most rapid warming rates on Earth; average annual temperatures rose 5°F in 50 years, five times the worldwide average. Great Britain's Rothera Station on the western shore of the Antarctic peninsula may be the most rapidly warming locale on Earth. Average temperatures there rose 20°F during the last quarter of the 20th century (Bowen 2005, 33).

"It has become increasingly clear," wrote Fen Montaigne in *Fraser's Penguins* (2010, 7, 8) "that humanity is every bit as responsible for the decline of the Adélie penguins along the northwestern Antarctic peninsula as Nathaniel Palmer and his mates were for the near extirpation of Antarctica's fur seals nearly two centuries ago." Some 2.5 million breeding pairs of Adélie penguins still existed in Antarctica as a whole in 2010, even as shrinking ice drove them to near extinction near Palmer Station. They will continue to exist, but perhaps not thrive, as the range of oceanic ice shrinks over time.

Some of the penguin mortality that has been attributed to climate change may actually be caused by scientists' use of leg bands to track them. The banding of penguins also reduces some penguins' lifespan and ability to feed their young because the bands create drag when they swim, inhibiting their gathering of krill. Rory P. Wilson, a zoologist at Swansea University, Wales, said that "if we stopped marking, we wouldn't have any way of measuring how birds are being affected by the climate, or anything else" (Bhanoo 2011; Saraux et al. 2011, 203–206; Wilson 2011, 164–185). Scientists may replace tags with microchips, although these cannot be tracked by sight.

South African Penguins Face Starvation

Around 1900, at least 1.5 million African penguins waddled and swam on and along the coasts of Namibia and South Africa. In 2001, the population was down to 120,000, and by 2007 roughly 20,000 survived. With few exceptions, they occupied only a few islands near South Africa's tip, including Robben Island where Nelson Mandela (among others) were imprisoned. The penguin population has been declining rapidly as the anchovies and sardines they eat migrate southward out of warming water.

The African penguin, the continent's only native penguin, is less than two feet tall "with large eyes set in a field of red and a distinctive black stripe circumnavigating its belly. Its calls range from a delicate trill to a loud nighttime bray, which leads some to call it the 'jackass penguin'" (Wines 2007). Rob Crawford, a marine scientist with South Africa's department of marine and coastal management, said that the penguins' decline is related to the shifting of sardine schools southward and "has to be either fishing or climate or a mixture of both," he said, "but my hunch is that it's environmental" (Wines 2007).

The African penguins can find no land base farther south, so they are limited to food that is available within waters 25 miles out. They have been starving as they wait for the sardines and anchovies to return. Commercial fishing also has depleted sardine stocks. Scientists have concluded the sardines move with changes in the cold, nutrient-rich Benguela Current, which flows from Antarctica along the southwest African coast. In addition to climate change, the sardines may move in a 50-year cycle (Wines 2007).

Crabs Obliterate a Warming Ecosystem

Warming in Antarctica causes ecosystems to change. For example, Douglas Fox wrote in *Nature* (2012), "Crabs invading the Antarctic continental shelf could deal a crushing blow to a rare ecosystem." Fox described sea floor life that had existed for 30 million years: "Mostly invertebrates: sea lilies waving in the currents; brittlestars with their skinny, saw-toothed arms; and sea pigs, a type of sea cucumber that lumbers along the sea floor on water-inflated legs." In some areas, however, sea life was largely absent, having been wiped out by an invader, moving in as water temperatures rose: "a red-shelled crab, spidery, and with a leg-span as wide as a chessboard." An estimated 1.5 million crabs by 2012 had invaded the depths of the Palmer Sea, obliterating local fauna. All of these changes are taking place because local waters have risen from between 1°C and –2°C to about 1.8°C, which has been just enough to allow the rapidly reproducing crabs to become the dominant species over a decade or two.

Many of the local fauna are adapted to water at just above the freezing point and struggle to survive in anything warmer. "Many of the species here are exquisitely sensitive to increases in temperature. The brittlestars and other invertebrates have extremely slow metabolisms—an adaptation to the cold water—and only meager ability to absorb and transport oxygen," wrote Fox (2012, 172). "'So what do those guys do if it warms up and their metabolic rate speeds up?'" asked Lloyd Peck, a biologist at the British Antarctic Survey in Cambridge, who has monitored these creatures in aquarium warming experiments. "Their oxygen demand revs beyond what their gills can supply—and they slowly suffocate" (Fox 2012, 172).

"There are no hard-shell-crushing predators in Antarctica," said Craig Smith, a marine ecologist from the University of Hawaii at Manoa. "When these come in they're going to wipe out a whole bunch of endemic species" (Fox 2012, 171). The invading crabs have been wiping out a sea floor ecology that is unique on Earth. "It's a fascinating thing," said Richard Aronson, a marine biologist at Melbourne's

Florida Institute of Technology. "A little scary, because it's a very obvious footprint of climate change" (Fox 2012, 171). The crabs are wiping out an ecosystem that may have been preserved in the chilly waters for millions of years. "All of this stuff has got a very Palaeozoic flavour to it," said Aronson. "Westerly winds are strengthening and the circumpolar current is intensifying, driven by atmospheric warming and a hole in the ozone layer over Antarctica," wrote Fox. "These changes are lifting warm, dense, salty water from 4,000 meters down in the Southern Ocean up over the lip of the continental shelf" (Fox 2012, 171).

Further Reading

Bhanoo, Sindya N. "Penguins Harmed by Tracking Bands, Study Finds." *The New York Times*, January 18, 2011, D3.

Bowen, Mark. *Thin Ice: Unlocking the Secrets of Climate in the World's Highest Mountains*. New York: Henry Holt, 2005.

Fountain, Henry. "For Already Vulnerable Penguins, Study Finds Climate Change Is Another Danger." *The New York* Times, January 29, 2014. http://www.nytimes.com/2014/01/30/science/earth/climate-change-taking-toll-on-penguins-study-finds.html.

Fox, Douglas. "Polar Research: Trouble Bares Its Claws." *Nature* 492 (December 2012): 170–172.

Le Maho, Yvon. "Conservation: After the Ice." *Nature* 468 (December 23, 2010): 1034–1035.

Montaigne, Fen. "The Ice Retreat: Global; Warming and the Adélie Penguin." *The New Yorker*, December 21 and 28, 2009, 72–82.

Montaigne, Fen. *Fraser's Penguins: A Journey to the Future of Antarctica*. New York: Henry Holt, 2010.

Petit, Charles W. "Polar Meltdown: Is the Heat Wave on the Antarctic Peninsula a Harbinger of Global Climate Change?" *U.S. News and World Report*, February 28, 2000, 64–74.

Pollack, Henry. *A World Without Ice*. London: Avery/Penguin, 2009.

Rohter, Larry. "Antarctica, Warming, Looks Ever More Vulnerable." *The New York Times*, January 25, 2005. http://www.nytimes.com/2005/01/25/science/earth/25ice.html.

Saraux, Claire, Céline Le Bohec, Joël M. Durant, et al. "Reliability of Flipper-Banded Penguins as Indicators of Climate Change." *Nature* 469 (January 13, 2011): 203–206.

"Sea Ice Penguin Theory Sinks." *Nature* 472 (April 21, 2011): 263.

Spears, Tom. "Antarctica Rides Global 'Heatwave': Continent's Warm Coast Causes Concern." *Ottawa Citizen*, August 8, 2001, A-1.

Wilson, Rory P. "Animal Behaviour: The Price Tag." *Nature* 469 (January 13, 2011): 164–165.

Wines, Michael. "Dinner Disappears, and African Penguins Pay the Price." *The New York Times*, June 4, 2007. http://www.nytimes.com/2007/06/04/world/africa/04robben.html.

See also: Adaptation, Animals, Lizards, and Warming Habitats; Animal Life, Arctic; Biodiversity; Coral Reefs; Extinctions; Fisheries; Jellyfish; Migrations; Phytoplankton; Salmon and Trout

ANIMAL LIFE, ARCTIC
A Winter without Walrus

Hunting animals, including walrus, is still key to survival for Native peoples across the Arctic. Imported food is not only culturally out of character but also so expensive that few Native people can afford it. Food imported by air to the one small grocery store in Gambell, Alaska (in extreme northwest Alaska and within sight of Siberia's Chukchi peninsula), is no alternative to country food. A single chicken costs $25. Wild swings in weather and ice conditions have devastated harvests of walrus, the main subsistence staple in some Alaskan Native villages. In recent years, ice, the walrus' primary habitat, has been too scarce—except, as in the winter of 2012–2013, when weather was so cold and ice so thick that hunters could not get out often enough to provide for their families.

Melting Arctic sea ice is separating walrus young from their mothers and sometimes killing them. A Coast Guard icebreaker during two months in 2004 found nine lone walrus calves in deep waters flowing from the Bering Sea to the shallower continental shelf of the Chukchi Sea. Researchers reported in the journal *Aquatic Mammals* that the pups, living with their mothers, probably fell into the sea following collapse of an ice shelf that broke up in water where the temperature had increased 6°F in two years. "We were on a station for 24 hours, and the calves would be swimming around us crying," said Carin Ashjian, a biologist at Woods Hole Oceanographic Institution in Massachusetts and a member of the research team. "We couldn't rescue them" (Kaufman 2006). Walrus pups usually live with their mothers for two years.

Breakup of sea ice close to the shore also deprives adult walruses of platforms from which to dive for food. Remaining sea ice offshore is in waters that are too deep for diving. The researchers, who were funded by the National Science Foundation and the Office of Naval Research, said it was possible that the walrus young were separated from their mothers by a severe storm or that the mothers were killed by hunters from nearby native communities. They said that local hunting ethics frown on killing walrus mothers with young, so it was far more likely that retreating sea ice—rather than a storm—caused the young to become separated (Kaufman 2006).

Walrus Assemble Onshore

Families once put 20 to 25 walruses in their meat lockers each fall to last to the next spring. In recent years, they have been storing only a handful of slaughtered animals because of retreating ice. "If this continues, we will seriously starve," said 38-year-old Jennifer Campbell, mother of five, whose family caught only two walruses in the autumn of 2013, down from 20 or so previously (Carlton 2013). Historically, the Yupik Eskimos harvested walrus from passing ice floes in May and June using small boats. In 2013, however, the hunt lasted only a week. "The ice was so bad we couldn't get out," said Brian Aningayou, a hunter who took only four walrus but needs to feed an extended family of 40 people (Carlton 2013). He has been shooting birds, which provide little meat compared to a walrus that can be the size and weight of a small car.

With ice receding hundreds of miles offshore of Alaska and Russia during the late summer of 2007, walrus gathered by the thousands onshore in Alaska and Siberia. Joel Garlich-Miller, a walrus expert with the U.S. Fish and Wildlife Service, said that walrus began to gather onshore late in July, a month earlier than usual. A month later, their numbers had reached record levels from Barrow to Cape Lisburne, some 300 miles southwest on the Chukchi Sea.

Walrus usually feed on clams, snails, and other bottom-dwelling creatures from the ice. In recent years, the ice has receded too far from shore to allow the usual feeding pattern. A walrus can dive 600 feet deep, but water under ice shelves in late summer is now several thousands of feet deep. The walruses have been forced to swim much farther to find food using energy that cause increase calf mortality. More calves are being orphaned. Russian research observers have also reported many more walrus than usual on shore, tens of thousands in some areas along the Siberian coast. They would have stayed on the sea ice in earlier times (Joling 2007, October 4).

Walrus are prone to stampedes once they are gathered in large groups. Thousands of Pacific walrus above the Arctic Circle were killed on the Russian side of the Bering Strait, where more than 40,000 had hauled out on land at Point Schmidt, as ice retreated farther northward. A polar bear, a human hunter, or noise from an airplane flying close and low can send thousands of panicked walrus rushing into the water.

"It was a pretty sobering year, tough on walruses," said Joel Garlach-Miller, a walrus expert for the U.S. Fish and Wildlife Service. Several thousand walrus died in late summer 2007 from internal injuries suffered in stampedes. The youngest and the weakest animals, many of them calves born the previous spring were crushed. Biologist Anatoly Kochnev of Russia's Pacific Institute of Fisheries and Oceanography estimated 3,000 to 4,000 walruses out of a population of perhaps 200,000 died, which would be two or three times the usual number on shoreline haulouts (Joling 2007, December 14).

The reports support expectations of the fate of walrus once ice retreated, said wildlife biologist Tony Fischbach of the U.S. Geological Survey. "We were surprised that this was happening so soon, and we were surprised at the magnitude of the report," he said (Joling 2007, December 14). Walrus lacking summer sea ice from which they used to dive for clams and snails, may strip coastal areas of food and then starve in large numbers. Shrinking Arctic ice is taking a toll on walrus numbers, as well as those of polar bears. The toll is unknown. "The ice is melting three weeks earlier in the spring than it did 20 years ago, and it's re-forming a month later in the fall," said Carleton Ray of the University of Virginia, who has studied walruses since the 1950s (Angier 2008).

The walrus are imperiled by expanding human industry as well as eroding sea ice. In February 2008, for example, the U.S. Interior Department granted Royal Dutch Shell oil-drilling rights in the Chukchi Sea near the northwestern shore of Alaska, the first exploitation of this prime walrus hunting ground. Walrus find their food in shallow water and haul out to rest on ice or land. Females and their young use the ice year-round if it is available. Ice in the Chukchi Sea, however, has been retreating from the shore into waters too deep for walrus to feed.

"As a result," according to an account in *The New York Times*, females and calves have been forced to abandon the ice in midsummer and follow the males to land.

The voyage leaves them emaciated and easily panicked. With the slightest disturbance, the herd desperately heads back into the water, often trampling one another to death as they flee. "The ones that take the brunt of it are the calves," said Chad Jay of the walrus research program at the Alaska Science Center in Anchorage. "Our Russian colleagues have observed thousands of calves killed in episodes of beachside mayhem" (Angier 2008).

During 2009, for the third consecutive year, the range for walrus along the Chukchi Sea was mostly open water in the summer. On September 14, 2009, 131 walruses were found dead near Icy Cape, Alaska, having been crushed during stampedes. "I think there is reason to be concerned," said University of Alaska marine biologist Brendan P. Kelly (Revkin 2009). Stampedes certainly occurred before recent ice melt, but open water has increased their frequency.

Warming Affects Arctic Wildlife

Steven Kooneeliusie and other Inuit who hunt caribou, seal, and other animals say the signs of a gradual increase in temperatures are everywhere around them. "When I went hunting years ago, I used to wear a full-length caribou skin coat, but now I just wear a light parka. It is so hot these days my snowmobile often overheats," Kooneeliusie said in Pangnirtung, a small town 100 miles north of Iqaluit and astride the Arctic Circle (Ljunggren 2000).

In Alaska, Native peoples have recounted seeing ringed seal pups being abandoned by their disoriented mothers as sea ice melts sooner than ever before and impeding breeding cycles. These pups often are too young to fend for themselves, and not having been properly weaned they die in large numbers. Spotted seals have had similar problems, but bearded seals, for unknown reasons, seemed to be in relatively good condition, at least until 2016. Walrus have been forced to travel farther as sea ice melts earlier and have been losing weight. Because of violent weather and early ice melt, polar bears have been less able to find shelter. All of this puts pressure on Native Alaskan hunters, who must range far afield of their homes to find seals and walrus, and they are also forced to negotiate unfamiliar and often treacherous conditions on ice that breaks up under their feet (Voggesser 2010).

Polar Bears Facing Extinction?

The steady melting of Arctic ice threatens extinction for polar bears before the end of this century. Seymour Laxon, senior lecturer in geophysics at UCLA's Center for Polar Observation and Modeling, said that serious concern exists regarding the long-term survival of polar bears as a species. "To put it bluntly," he said, "No ice means no bears" (Elliott 2003).

Andrew Derocher, a professor of biology at the University of Alberta, supports Laxon's beliefs. "If the progress of climate change continues without any intervention, then the prognosis for polar bears would ultimately be extinction," he said ("Expert Fears" 2003). Although reluctant to give an exact date for the extinction of polar bears, Derocher said changes in the northern ecosystem are occurring so quickly that the habitat may be incapable of supporting polar bears within 100 years. "What we're looking at is major changes in terms of the distribution of polar bears," he said. "But there are so many variables to put a number of years on something like extinction of a species. I guess ultimately I'm an optimist that humans will turn the events causing global warming back a bit" ("Expert Fears" 2003).

The International Union for Conservation of Nature and Natural Resources, a network of 10,000 scientists, forecast in May 2007 that global warming and over-hunting could diminish the polar bear population by at least 30 percent in coming decades. "Given what the climate models predict for continuing warming and melting of sea ice, the whole thing leads to an extinction curve," said Peter Ewins, director of the World Wildlife Fund Canada's Arctic Conservation Program. "And it's not a question of if, it's a clear question of when" (Krauss 2006).

During the fall of 2007, after the Arctic ice cap had lost almost a quarter of its mass in one summer, warnings from scientists became more strident. "Just 10 more years of current global warming pollution trajectories will commit us to enough warming to melt the Arctic and doom the polar bear to extinction," said Kassie Siegel, director of the Center for Biological Diversity's Climate, Air, and Energy Program. "We urgently need to address global warming, not just for the sake of the polar bear but for the sake of people and wildlife around the world" ("Comment Period" 2007).

Polar bears are swimming longer distances, often more than 30 miles, searching for suitable ice, according to Derocher, who took part in a project that tracked more than 100 bears in the Beaufort Sea north of Alaska and the Yukon between 2007 and 2012. A polar bear can swim 30 miles in a day, but without a place to haul out, rest, and search for food, female bears with cubs face "a death sentence" under such conditions, Derocher said ("Bears Face" 2016). One female polar bear swam 250 miles in nine days. In 2004, a quarter of the bears observed swam 31 miles or more; by 2012, 69 percent were swimming that far. "Coupled with an earlier study," wrote Derocher and colleagues, "The yearly proportions of . . . adult females swimming in 2004–2012 were positively associated with the rate of open water gain. Results corroborate the hypothesis that long-distance swimming by polar bears is likely to occur more frequently as sea ice conditions change due to climate warming" (Pilford et al. 2016).

Polar Bears' Regional Decline

Shrinking ice from global warming provoked a 50-percent decline in polar bear populations in the southern Beaufort Sea during the decade ending in 2010 largely because of the starvation of cubs, according to the U.S. Geological Survey (USGS) and Environment Canada. The bear population, estimated at 1,600 in 2004, fell to about 900 in 2010. "These estimates suggest to me that the habitat is getting less stable for polar bears," said study lead author Jeff Bromaghin, a USGS statistician. Bromaghin said two of some 80 polar bear cubs the scientists tracked during the study survived—against a previously recorded survival rate of approximately half. "We suspect that they are dying of starvation," Bromaghin said (Borenstein 2014;

Bromaghin et al. 2015). He said that warming had caused sea ice retreat and a drastic decline in the polar bears' supply of seals.

Arctic summer sea ice, declining since the 1970s, has nearly disappeared in the southern Beaufort Sea. "We've seen over the past decade, decade-and-a-half, the rate of decline has really accelerated," said Mark Serreze, director of the National Snow and Ice Data Center in Colorado. "There is definitely a relationship here between what's happening to the bears and what's happening to the ice," said Serreze, who was not involved of the polar bear study (Borenstein 2014).

The offshore ice-based ecosystem is sustained by upwelling nutrients that feed plankton, shrimp, and other small organisms that are eaten by fish. These, in turn, are eaten by seals, which are eaten by bears. The Native people of the area also occupy a position in this cycle of life. When the ice is not present, the entire cycle collapses.

Ringed and bearded seals, the polar bears' usual prey, are among many larger animals that have adapted to life on, in, and under the ice. "Ranging year-round as far as the [North] pole," wrote Kevin Krajick in *Science*, "They never leave the ice pack, keeping breathing holes open all winter, and making lairs under snow mounds. During the spring, the snow lairs camouflage their new pups from polar bears and protect them from cold air" (Krajick 2001, 425).

Ice in many areas of the Arctic now sometimes melts as early as March, just when the seals are having their pups. Because the ice breaks up too early, the pups often have not been fully weaned, so many of them starve or mature in a weakened state. During two decades—1980 to 2002—ice broke up two weeks earlier than previously. Biologist Lois Harwood of the Canadian Department of Fisheries and Oceans said that ice in the western Arctic broke up three weeks earlier than usual in 1998, dumping many hungry pups into the water before they had been weaned. Adult seals were thinner than usual as well, despite available prey exposed by the early breakup of the ice. "They were starving in the midst of plenty," Harwood said (Krajick 2001, 425). Peary caribou also have been observed falling through unusually thin ice during their migrations.

Musk Oxen, Reindeer, Harp Seals, and Caribou

Retreating ice may imperil caribou populations in the Arctic as well as polar bears and walrus. Reporting from Iqaluit, Dwane Wilkin, working for the *Nunatsiaq News*, summarized the consequences of global warming for that area: "The good news is that sailing through the Northwest Passage will finally be a cinch. The bad news? Global warming will probably drive musk oxen, polar bears, and Peary caribou into extinction. And other species—including humans—will face declining food sources. . . ." (Wilkin 1997). These combined stresses could lead to "complete reproductive failure of the caribou in a worst-case scenario" (George 2000).

Ice and Animal Mortality

Arctic precipitation may increase as much as 25 percent with the advent of rapidly rising temperatures during a century. When precipitation falls as heavy and wet snow and ice, caribou and other grazing animals have a harder time finding food during lean winter months. Heavier snow cover may lead to smaller, thinner animals who will be forced to go farther afield for food in winter. The same animals will be plagued by increasing numbers of insects during warmer summers. Climate change is already making it tougher for browsing animals in the Arctic such as caribou and reindeer to survive the seven months of each year when they must find and eat large amounts of lichen and moss buried beneath snow and ice (Calamai 2002).

Temperatures near the freezing point are more likely to cause precipitation to fall as freezing rain rather than snow, forming a barrier several centimeters thick. Given warmer weather in the Arctic, chances of ice replacing snow are rising, according to U.S. climate expert Jaakko Putkonen, an earth-sciences professor at the University of Washington. Based on two decades of work at the Spitsbergen archipelago in the Scandinavian Arctic, Putkonen calculated that five centimeters of rain falling on snow is enough to form an impenetrable ice barrier for grazing animals (Calamai 2002).

"You only need one really big event to have a disastrous effect," said Putkonen. A single heavy rainfall on Russia's snowy Kamchatka peninsula led to the death of 5,000 reindeer, he said (Calamai 2002). During the winter of 1996–1997, an estimated 10,000 reindeer died of starvation on Russia's far northeast Chukotka peninsula after ice formed a thick crust over pastures, which prevented reindeer from grazing. Three thousand Sami in northern Sweden tend reindeer that have been suffering from climate change that often has changed snow to ice. During the 1990s, unusually rainy autumns caused ice crust on the moss that feed reindeer. According to one account, "Herd sizes . . . tumbled, bringing not just financial hardship but also the threat of cultural decay" (Roe 2001).

In addition to separating grazing animals from their food, ice also raises temperatures at ground level by sealing in latent heat, encouraging the spread of fungi and toxic molds that attack the lichen. Signs of this problem were reported among reindeer in the Scandinavian Arctic; experts say the same threat also exists for more than 3 million caribou in migratory herds that sustain scores of aboriginal communities in northern Canada. Computer modeling suggests that the area vulnerable to such rain-on-snow events could increase by 40 percent worldwide by the 2080s under current global warming projections, with significant expansion into parts of central Canada that are home to both woodland and barren ground caribou (Putkonen and Roe 2003, 1188).

"They're the same animal genetically, only those in Europe and Asia have been domesticated while here they roam wild," said Don Russell, a biologist who studies caribou in the Yukon for the Canadian Wildlife Service. "They need about three kilograms dry weight every day—about a garbage-bag full. And they can spend about half their time just digging though the snow with their feet and feeding," said Russell (Calamai 2002).

Shifting Seasons Affect Many Species

In the Arctic National Wildlife Refuge, spring has been arriving earlier. As a consequence, caribou have had difficulty migrating from wintering areas in time to take advantage of periods of maximum springtime plant growth. Spring warming since 1990 has been the earliest in nearly 40 years. By the time animals reach the plain, their principal food plant has gone to seed. Caribou herd populations could decline significantly should future climate and vegetation patterns prevent the proper nourishment of calves. High-Arctic Peary caribou and musk oxen may even become extinct.

The changing climate affects the behaviors of many animals, according to observations by local residents. In the Kitikmeot Region of Canada, for example, caribou have been altering their migration routes because of early cracks in sea ice. Changes in vegetation types and abundance also affect caribou foraging strategies. Increases in massive caribou drownings because of thinner ice have been observed ever since 1996. Caribou also have died due to exhaustion from extreme heat as they are tortured by swarms of mosquitoes. Seals come up through the unusually high number of cracks in sea ice in early spring around Hope Bay, which attracts polar bears (Thorpe 2000). Grizzly bears have been observed crossing from the mainland northward to Victoria Island for the first time in 1999. New birds seen for the first time include robins and an unidentified yellow songbird (Thorpe 2000). Near Baker Lake in Nunavut, 1,330 kilometers west of Iqaluit, caribou are not as healthy. Meat is tough, and the skin around their neck areas tears too easily. Observers see more liquid than usual in caribou's joints and more white pustules on meat. Caribou have become leaner and undernourished because of summer heat and dryness. More diseased caribou have been found with sores in their mouths and on their tongues (Government of Nunavut 2001).

Even as record snows plagued parts of the United States during the winter of 2009–2010, several thousand harp seal pups died in Canada's Gulf of St. Lawrence because of a lack of river ice. Starving pups were abandoned on beaches along Prince Edward Island, victims of the lowest ice conditions in the 41 years that the province has kept records. Off Newfoundland, another major seal hunting ground, ice by mid-March 2010 had formed only off the northern peninsula at a time when it usually extended along all of the island's northeast coast. Observers from the International Fund for Animal Welfare reported that the Gulf of St. Lawrence, the annual birthing ground of hundreds of thousands of harp seals, is "essentially devoid of both ice and seals" ("Worst Ice" 2010).

Climate-Provoked Chains of Events

Climate change can set in motion an entire chain of events that affect Inuit daily life. For example, one such sequence was presented by a hunter. Changes in the environment affect the diets of caribou, which in turn affect the traditional food security of Inuit. Ice layers formed by freezing rain seal in lichen that caribou are unable to reach. As a consequence, hungry caribou have been raiding town garbage dumps—not unlike polar bears that found themselves stranded on shore when ice receded prematurely. One town reported that hungry bears had even hitched rides on garbage trucks bound for the town dump in Churchill, Manitoba (Nickels et al. 2005, 76).

Further Reading

Angier, Natalie. "Ice Dwellers Are Finding Less Ice to Dwell On." *The New York Times*, May 20, 2008. http://www.nytimes.com/2008/05/20/science/20count.html.

"Bears Face Days-Long Swims." *Omaha World-Herald*, April 23, 2016, 3A.

Borenstein, Seth. "Study: Polar Bears Disappearing from Key Region." Associated Press, November 17, 2014. http://bigstory.ap.org/article/28aa5cf3d43f44e1aa1f37b836a0e82b/study-polar-bears-disappearing-key-region.

Bromaghin, Jeffrey F., et al. "Polar Bear Population Dynamics in the Southern Beaufort Sea during a Period of Sea Ice Decline." *Ecological Applications*, April 1, 2015. http://online library.wiley.com/doi/10.1890/14-1129.1/abstract.

Calamai, Peter. "Global Warming Threatens Reindeer." *Toronto Star*, December 23, 2002, A23.

Carlton, Jim. "A Winter without Walrus: Harvesting of Food Staple for Remote Eskimo Villages Plummets; Disaster Declared." *Wall Street Journal*, October 4, 2013, A4.

"Comment Period Extended on Polar Bear Extinction Threat." Environment News Service, October 2, 2007. http://www.ens-newswire.com/ens/oct2007/2007-10-02-091.asp (no longer available).

Elliott, Valerie. "Polar Bears Surviving on Thin Ice." *The Times* (London), October 30, 2003, n.p. (LEXIS).

"Expert Fears Warming Will Doom Bears." *Times-Colonist* (Victoria, BC), January 5, 2003, C8.

George, Jane. "Global Warming Threatens Nunavut's National Parks." Nunatsiaq News, May 19, 2000. http://www.nunatsiaq.com/archives/nunavut000531/nvt20519_18.html.

Government of Nunavut. *Inuit Qaujimajangit Hilap Alanguminganut/Inuit Knowledge of Climate Change: A Sample of Inuit Experiences of Climate Change in Nunavut. Baker Lake and Arviat, Nunavut.* January–March 2001. Government of Nunavut, Department of Sustainable Development, Environmental Protection Services.

Joling, Dan. "Walruses Abandon Ice for Alaska Shore." *Washington Post,* October 4, 2007. http://www.washingtonpost.com/wp-dyn/content/article/2007/10/04/AR2007100402299_pf.html (no longer available).

Joling, Dan. "Thousands of Pacific Walruses Die; Global Warming Blamed." Associated Press, December 14, 2007 (LEXIS).

Kaufman, Marc. "Warming Arctic Is Taking a Toll." *Washington Post*, April 15, 2006, A7. http://www.washingtonpost.com/wp-dyn/content/article/2006/04/14/AR2006041401368_pf.html.

Krajick, Kevin. "Arctic Life, on Thin Ice." *Science* 291 (January 19, 2001): 424–425.

Krauss, Clifford. "Bear Hunting Caught in Global Warming Debate." *The New York Times*, May 27, 2006. http://www.nytimes.com/2006/05/27/world/americas/27bears.html.

Ljunggren, David. "Effects of Global Warming Clear in Canada Arctic." Environmental News Network, April 20, 2000. http://www.enn.com/enn-subsciber-news-archive/2000/04/04202000/reu_arctwarm_12170.asp (no longer available).

Nickels, S., et al. *Unikkaaqatigiit: Putting the Human Face on Climate Change: Perspectives from Inuit in Canada.* Inuit Tapiriit Kanatami, Nasivvik Center for Inuit Health and Changing Environments at Université Laval and the Ajunnginiq Center at the National Aboriginal Health Organization, Ottawa, 2005. https://www.itk.ca/sites/default/files/unikkaaqatigiit01_0.pdf (no longer available).

Pilford, Nicholas W., Alysa McCall, Andrew E. Derocher, et al. "Migratory Response of Polar Bears to Sea Ice Loss: To Swim or Not to Swim." *Ecography*, April 14, 2016. http://onlinelibrary.wiley.com/doi/10.1111/ecog.02109/abstract.

Putkonen, J. K, and G. Roe. "Rain-on-Snow Events Impact Soil Temperatures and Affect Ungulate Survival." *Geophysical Research Letters* 30(4) (2003): 1188–1192.

Revkin, Andrew C. "Walruses Suffer Substantial Losses as Sea Ice Erodes." *The New York Times*, October 3, 2009. http://www.nytimes.com/2009/10/03/science/earth/03walrus.html.

Roe, Nicholas. "Show Me a Home Where the Reindeer Roam." *The Times* (London), November 10, 2001, n.p. (LEXIS).

Thorpe, N. "Contributions of Inuit Ecological Knowledge to Understanding the Impacts of Climate Change on the Bathurst Caribou Herd in the Kitikmeot Region, Nunavut." Unpublished master's thesis, School of Resource and Environmental Management. Project No. 268. Simon Fraser University, Burnaby, British Columbia, 2000.

Voggesser, Garrit. "The Tribal Path Forward: Confronting Climate Change and Conserving Nature." *The Wildlife Professional*, Winter 2010.

Wilkin, Dwayne. "A Team of Glacial Ice Experts Say Mother Nature's Thermostat Has Kept the Eastern Arctic at about the Same Temperature since 1960." *Nunatsiaq News*, May 30, 1997. http://www.nunanet.com/~nunat/week/70530.html#7 (no longer available).

"Worst Ice Year Kills Canadian Seals before Hunters Can." Environment News Service, March 26, 2010. http://www.ens-newswire.com/ens/mar2010/2010-03-26-01.html.

See also: Adaptation, Animals, Lizards, and Warming Habitats; Animal Life, Antarctic; Arctic Hunters; Biodiversity; Carbon Cycle Feedbacks; Climate Change, Abrupt Nature of; Extinctions; Fisheries; Salmon and Trout; Sea Ice, Arctic

BIODIVERSITY

Studies by many wildlife advocacy groups support scientists' projections of impending mass extinctions partly because of warming temperatures. Other human-induced factors often play a role as well, including exposure to toxins and destruction of habitat. These studies usually define the problem in terms of declining biodiversity. The studies do more than document current and potential extinctions of several species. They also advocate emphatic action to combat climate change. These groups call on U.S. lawmakers to help curtail greenhouse gas emissions, for example, by enacting higher fuel-efficiency standards for ground transportation and more energy-efficient building codes (Lazaroff 2001).

Two such reports were compiled by the National Wildlife Federation in the United States and the World Wildlife Fund (WWF), both of which contend that species from the tropics to the poles are at risk. Many species may be unable to move to new areas quickly enough to survive habitat changes that rising temperatures will bring to their historic habitats. Based on a doubling of atmospheric carbon dioxide levels that many scientists expect within a century, a WWF report asserted that one-fifth of the world's most vulnerable natural areas may be facing a "catastrophic" loss of species (Lazaroff 2001). "It is shocking to see that many of our most biologically valuable ecosystems are at special risk from global warming. If we don't do something to reverse this frightening trend, it would mean extinction for thousands of species," said Jay Malcolm, author of the WWF report and professor at the University of Toronto (Lazaroff 2001).

According to this WWF report, the areas and species most vulnerable to devastation from global warming include those in the Canadian low Arctic tundra; the central Andean dry puna of Chile, Argentina, and Bolivia; the Ural Mountains; the Daurian steppe of Mongolia and Russia; the Terai-Duar savanna of northeastern India; southwestern Australia; and the Fynbos of South Africa. Among the U.S. ecosystems at risk, areas in California, the Pacific Northwest, and the Northern Prairie may be hardest hit, the WWF said. The changes could devastate the shrub and woodland areas that stretch from Southern California to San Francisco, prairies in the northern heart of the United States, the Sierra Nevada, the Klamath–Siskiyou forest near the California–Oregon border, and the Sonoran and Baja deserts across the southwestern United States.

Banking Seeds against Extinction

In England, the Royal Botanic Gardens marked International Biological Diversity Day in 2007 by banking its billionth seed in the vaults of the Millennium Seed Bank as insurance against risks of global warming. Genes of wild plants that are related to widely cultivated crops may be used to assist them in resisting pests and tolerating drought, according to a report by scientists of the Consultative Group on International Agricultural Research. "Kew's Millennium Seed Bank must be one of the most significant conservation projects ever," said Barry Gardiner, U.K. Minister for Biodiversity, Landscape, and Rural Affairs. "It is a global insurance policy against the loss of uniquely valuable plant species through land pressures or dangerous climate change" ("Saving Earth's" 2007).

Climate change, for example, could threaten wild relatives of important food crops such as potatoes and peanuts. "Our results would indicate that the survival of many species of crop wild relatives, not just wild potato and peanuts, are likely to be seriously threatened even with the most conservative estimates regarding the magnitude of climate change. . . ."At the moment, existing collections are conserving only a fraction of the diversity of wild species that are out there" said the study's lead author, Andy Jarvis ("Saving Earth's" 2007). In the next 50 years, said Jarvis, as much as 60 percent of the 51 wild peanut species analyzed and 12 percent of the 108 wild potato species analyzed could become extinct because of warming.

"The irony here is that plant breeders will be relying on wild relatives more than ever as they work to develop domesticated crops that can adapt to changing climate conditions," said Annie Lane, the coordinator of a global project on crop wild relatives led by Bioversity International, the world's largest international research organization dedicated to the use and conservation of agricultural biodiversity. "Yet because of climate change, we could end up losing a significant amount of these critical genetic resources at precisely the time they are most needed to maintain agricultural production," said Lane ("Saving Earth's" 2007).

Further Reading

"Saving Earth's Plant Diversity from Global Warming." Environment News Service, May 22, 2007. http://www.ens-newswire.com/ens/may2007/2007-05-22-02.asp (no longer available).

Even during the 20th century, warmer weather (especially milder winters) has been associated with changes in forest species. In one study, Gian-Reto Walther documented the increasing growth of broad-leaved evergreen trees in lower areas of southern Switzerland, where the growing season had increased to as long as 11 months in some areas by roughly the year 2000 (Walther 2002, 129). These areas have remarkably precise records of frost-free days. Between 1902 and 1998, for example, the onset of winter (the first day of frost) at Lugano had changed from an average of the second week in November to the first week of December. The last day of frost had changed from an average of the fourth week in March to the fourth week of February (Walthur 2002,132).

Threats to Wildlife

Research compiled by the National Wildlife Federation (NWF) suggests that global warming will probably pose an intensifying threat to U.S. wildlife, including more trouble with invasive species and significant environmental changes that will jeopardize the quality of human life in the near future (Lazaroff 2001). "Global warming has come down to Earth for the wildlife right in our backyards," said Mark Van Putten, NWF president. "The effects are already happening and will likely worsen unless we get serious about reducing emissions of carbon dioxide and other heat-trapping gases to help slow global warming" (Lazaroff 2001). The NWF's findings appeared in *Wildlife Responses to Climate Change* (2001), a book edited by the late Stephen Schneider of Stanford University and Terry Root of the University of Michigan, which includes eight case studies by researchers that demonstrate how global warming and associated climate change is affecting North American wildlife.

Seabirds Starve as Waters Warm

The survival of seabirds on New Zealand's sub-Antarctic islands is related to supplies of krill, which are declining as waters warm. Populations of rockhopper penguins, a species with brilliant yellow eyebrows that nests on the rocks of windblown Campbell Island, have declined by more than 95 percent since the 1940s. A breed of albatross called the gray-headed mollymawk has declined by 84 percent. Muttonbirds are down by one-third and elephant seals by approximately half, according to a report in the *New Zealand Herald* (Collins 2002).

A study by New Zealand's National Institute of Water and Atmospheric Research (NIWAR) said that the collapse of the rockhopper penguin population is a result of the birds' inability to find enough food in a less-productive marine ecosystem. "The cause is reduced productivity of the ocean where they are feeding," said NIWAR scientist Paul Sagar. "They eat mainly krill but they do also eat small fish and squid" (Collins 2002).

According to Sagar, further research was required to discern why phytoplankton populations had declined. Some scientists assert that the declining populations result from water that has become too warm for a species that is adapted to the sub-Antarctic. "It could be a long-run natural cycle, with the possibility that polar water coming up from the south may not be coming as far north as it used to," Sagar said. "It could be that the position of the currents has changed so the productive areas have moved further south or north, away from where the penguins are feeding." Another study at New Zealand's Otago University indicated that muttonbird numbers have dropped by a third on one of their major breeding islands. Conservation Department scientist Peter Moore, who studied the rockhoppers on Campbell Island in 1996, said their numbers declined from 1.6 million breeding pairs in 1942 to 103,000 pairs in 1985 and have continued to fall at a similar rate ever since (Collins 2002). Sagar said there was evidence that some of the adult birds were feeding better-quality food to their chicks while their own diet declined, but Moore said the orphaned chicks often died as well. During 1990, many surviving chicks from a yellow-eyed penguin colony were fed and reared by humans after their parents died. After the chicks were released, however, 99 percent of them disappeared (Collins 2002).

Seabird population declines have spanned the world for similar reasons. In Scotland, for example, guillemots, Arctic terns, kittiwakes, and other seabird colonies on the Shetland and Orkney Islands in 2004 experienced one of their worst breeding seasons ever recorded. The Royal Society for the Protection of Birds observed exceedingly few chicks on the breeding cliff ledges. Sandeels, the small fish on which the birds feed, have migrated northward, probably because of warming waters, placing the birds' traditional food supply largely out of reach; as a result, they have had difficulty reproducing and feeding their young.

Further Reading

Collins, Simon. "Birds Starve in Warmer Seas." *New Zealand Herald*, November 14, 2002, n.p. (LEXIS).

The NWF report asserted that invasive species such as tamarisk shrubs in the U.S. Southwest may expand their ranges, reducing water and food available to native wildlife and humans. Imported fire ants in the U.S. Southeast also may expand their range, dominating native ant species and creating an enhanced health risk to humans

(Lazaroff 2001). Species such as the sachem skipper butterfly in the Pacific Northwest and the Bay checkerspot butterfly in California already are responding to climatic and weather changes. Changes in climate also may alter habitats for grizzly bears, red squirrels, and other wildlife in the Greater Yellowstone Ecosystem region of Wyoming, Montana, and Idaho by contributing to a reduction in whitebark pine trees, an important food source for the animals.

Roughly 70 percent of the plants and animals in the Fynbos of the Western Cape of South Africa are unique to that area, according to the WWF. In summer, the area can become so parched by drought and heat that it is often ravaged by fire, with many plants dependent on regular fires to stimulate seeding. Almost half of this area will become uninhabitable for its major species within a century, according to the WWF (Browne 2002). Some animal species such as the springbok of southern Africa are also extremely attached to their habitats and are expected to die out rather than move.

Lake Tanganyika's Warming Waters Choke Life

Two independent teams of scientists studying Central Africa's Lake Tanganyika, Africa's second-largest body of freshwater, have found a microcosm of crisis vis à vis global warming. The scientists have found that warming at the lake's surface has impaired the mixing of nutrients and thus seen a reduction in fish populations. These reductions have affected the local economy because fishing yields fell by a third or more over 30 years, with more declines anticipated. Until recently, Lake Tanganyika's fish had supplied 25 percent to 40 percent of the protein consumed by neighboring peoples in parts of Burundi, Tanzania, Zambia, and the Democratic Republic of Congo.

"Climate warming is diminishing productivity in Lake Tanganyika," wrote Catherine M. O'Reilly and colleagues. "In parallel with regional warming patterns since the beginning of the 20th century," they continued, "a rise in surface-water temperature has increased the stability of the water column" (O'Reilly et al. 2003, 766).

Lake Tanganyika is especially vulnerable because year-round tropical temperatures accelerate biological processes, "and new nutrient inputs from the atmosphere or rock weathering cannot keep up with the high rates of algal photosynthesis and decomposition" (Verschuren 2003, 732).

Writing in *Nature*, O'Reilly and colleagues suggested that the lake's productivity as measured by the amount of photosynthesis has diminished by 20 percent, which could easily account for the 30-percent decline in fish yields. The scientists said that climate change rather than overfishing was mainly responsible for the collapse in Tanganyika's fish stocks. With additional warming, scientists expect that fish populations in the lake will decline further (O'Reilly 2003, 766).

"The human implications of such subtle, but progressive, environmental changes are potentially dire in this densely populated region of the world, where large lakes are essential natural resources for regional economies," the scientists said. Dirk Verschuren, a freshwater biologist at Ghent University in Belgium, noted that both studies could explain why sardine catches in Lake Tanganyika have declined between 30 percent and 50 percent since the late 1970s (Verschuren 2003, 732). "Since overexploitation is at most a local problem on some fishing grounds, the principal cause of this decline has remained unknown," Verschuren said. "Taken together . . . the data in the two papers provide strong evidence that the effect of global climate change on regional temperature has had a greater impact on Lake Tanganyika than have local human activities. Their combined evidence covers all the important links in the chain of cause and effect between climate warming and the declining fishery."

Further Reading

O'Reilly, Catherine M., et al. "Climate Change Decreases Aquatic Ecosystem Productivity of Lake Tanganyika, Africa." *Nature* 424 (August 14, 2003): 766–768.
Verschuren, Dirk. "Global Change: The Heat on Lake Tanganyika." *Nature* 424 (August 14, 2003): 731–732.

"Migration routes have become increasingly impassable due to human activities," wrote Gian Reto-Walther, summarizing several scientific studies (Walther 2003, 177). At the same time, human communication and transportation facilitate invasive species worldwide. Climate change also plays favorites among flora and fauna by removing climatic constraints; witness the explosion of spruce budworms throughout the pine forests of North America.

"Many habitats will change at a rate approximately 10 times faster than the most rapid changes since the last ice age, causing extinctions," Collier said (Browne 2002). Washington state's Olympic peninsula rain forest is also on the WWF's list of endangered habitats. The rain forest cannot move as climate changes, so the peninsula is likely to undergo a drastic change, the report said, although scientists say it is difficult to determine how quickly that change will occur.

A report titled "Global Climate Change and Wildlife in North America" by the Wildlife Society, an association of nearly 9,000 wildlife managers, research scientists, biologists, and educators, based in Washington, D.C., anticipates that many animals may find their migratory paths blocked by cities, transportation corridors, or farmland. Predators and their prey may not move at the same time, impeding natural balances. Plants could suffer if the birds and insects that pollinate them head to cooler climes. Wetlands in the Midwest and central Canada are expected to dry up, causing some duck species to decline by as much as 69 percent over the next 75 years. Nesting habitat could be lost as wetlands become more suitable for row crops.

Further Reading

Browne, Anthony. "How Climate Change Is Killing Off Rare Animals." *The Observer* (London), February 10, 2002, 15.

Lazaroff, Cat. "Climate Change Threatens Global Biodiversity." Environment News Service, February 7, 2001. http://ens-news.com/ens/feb2002/2002L-02-07-06.html (no longer available).

Schneider, Stephen, and Terry Root, eds. *Wildlife Responses to Climate Change*. Washington, DC: Island Press, 2001.

Walther, Gian-Reto. "Weakening of Climatic Constraints with Global Warming and Its Consequences for Evergreen Broad-leaved Species." *Folia Geobotanica* 37 (2002): 129–139.

Walther, Gian-Reto. "Plants in a Warmer World." *Perspectives in Plant Ecology, Evolution, and Systematics* 6:3 (2003): 169–185.

See also: Adaptation, Animals, Lizards, and Warming Habitats; Animal Life, Antarctic; Animal Life, Arctic; Arctic Hunters; Carbon Cycle Feedbacks; Climate Change, Abrupt Nature of; Extinctions; Fisheries; Sea Ice, Arctic

EXTINCTIONS
Paleoclimatic Precedents

According to Albert K. Bates, "Sixty-five million years ago, 60 to 80 percent of the world's species disappeared in a cataclysmic mass extinction, possibly caused by an asteroid's impact with Earth. Human population, not an asteroid, will cut the remaining number of species in half again, in just the next few years" (Bates and Project Plenty 1990, 137).

Research continues into the nature of mass extinctions in the past. The End Permian Mass Extinction, the most severe such event known to science, occurred 228 million years ago and lasted approximately 200,000 years or less. Shu-zhong Shen and colleagues wrote in *Science* (2011, 1367) that "Associated charcoal-rich and soot-bearing layers indicate widespread wildfires on land. A massive release of thermogenic carbon dioxide . . . or methane may have caused the catastrophic extinction." They further drew an analog to anticipated conditions in a rapidly warming future world with accelerating ocean acidification and warming and declining levels of oxygen:

> Our studies indicate that both marine and terrestrial ecosystems collapsed very suddenly, and massive release of thermogenic CO_2 as well as methane, is a highly plausible explanation . . . [resulting] in increased continental aridity by rapid global warming, which caused widespread wildfires and accelerated deforestation in the world . . . [which] further enhanced the continental weathering and finally resulted in catastrophic soil erosion. . . . (Shen et al. 2011, 1372)

Pika Populations Plunge in the Rocky Mountains

As described by Usha Lee McFarling in the *Los Angeles Times*, pikas are "tennis ball-sized critters that whistle at passing hikers and scamper over loose, rocky slopes of the High Sierra and the Rocky Mountains." This "shy, flower-gathering mammal and longtime icon of the West's high peaks" may fall victim to global warming. By 2003, they already had disappeared from almost 30 percent of the areas where they were common during the early parts of the 20th century (McFarling 2003).

Pikas resemble hamsters and are biological relatives of rabbits. Over many millennia, they adapted to a degree of intense cold that is becoming scarcer in the mountains. Living above 7,000 feet, Pikas cannot withstand heat. A comprehensive survey found that sites that have lost pikas were on average drier and warmer than sites where the animals remain, according to Erik Beever, a U.S. Geological Survey biologist based in Corvallis, Oregon (McFarling 2003). Even with the 2 to 3°F of warming during the last century in its mountainous range, the pika has retreated to higher elevations. Since the 1990s, some of these animals have vanished, having effectively run out of suitable habitat.

Beever has theorized that such factors as cattle grazing and proximity to roads had some effect on the animals, but warmer and drier conditions in recent decades have been a major factor in their rapid disappearance. Earlier surveys in the United States found pikas missing from much of their previous range. Observation of pikas in the Yukon revealed that 80 percent of the animals died in some areas after unusually warm winters (McFarling 2003). Their demise has been extremely rapid, according to Beever. He was most surprised to find groups of animals disappearing over decades rather than in centuries or millennia as was seen during climatic swings of the distant past (McFarling 2003).

Further Reading

McFarling, Usha Lee. "A Tiny 'Early Warning' of Global Warming's Effect; The Population of Pikas, Rabbit-Like Mountain Dwellers, Is Falling, a Study Finds." *Los Angeles Times*, February 26, 2003, A17.

As anticipation of mass extinctions abetted by global warming have become more common, scientists have sought paleoclimatic parallels. According to some scientists, the worst mass extinction in the history of the planet could be replicated in as little as a century if global warming continues at the pace forecast by the Intergovernmental Panel on Climate Change (IPCC). Researchers at England's Bristol University have estimated that a 6°C increase in global temperatures was enough to play a role in the annihilation of as many as 95 percent of species that were alive

Sea Lion Mortality and Warming Water

Dead and starving sea lions have been washing up on California shores in record numbers, most probably because warming waters are provoking a decline in their food supply on offshore islands where they usually breed and wean their pups. Mothers have been forced to range farther for food for their starving pups. Searching for food, they swim from these islands to shore between San Francisco and San Diego. "This year [2015]," reported Jack Healy in *The New York Times*, "animal rescuers are reporting five times more sea lion rescues than usual—1,100 last month [February 2015] alone. The pups are turning up under fishing piers and in backyards, along inlets and on rocky cliffs. One was found curled up in a flower pot." The number of dead and dying sea lions has been rising for three years by 2015, paralleling record heat and drought.

Many of the starving pups also suffer from pneumonia, "their throaty barks muted to rasping coughs. Parasites have swarmed their digestive systems. Some are so tired that they cannot scamper away when rescuers approach them with nets and towels and heft them into large pet carriers" (Healy 2015). "They come ashore because if they didn't, they would drown," said Shawn Johnson, the director of veterinary science at Marine Mammal Center in Sausalito, California. "They're just bones and skin. They're really on the brink of death" (Healy 2015).

Further Reading

Healy, Jack. "Starving Sea Lions Washing Ashore by the Hundreds in California." *The New York Times*, March 13, 2015. http://www.nytimes.com/2015/03/13/us/starving -sea-lions-washing-ashore-by-the-hundreds-in-california.html.

on Earth at the end of the Permian period 251 million years ago. This is roughly the same amount of warming expected by the IPCC if levels of greenhouse gases in the atmosphere continue to rise at current rates (Reynolds 2003).

The wave of mass extinctions at the end of the Permian period probably was caused by a series of extremely large volcanic eruptions that triggered a runaway greenhouse effect that nearly extinguished life on Earth. Conditions in what geologists have termed this "postapocalyptic greenhouse" were so severe that 100 million years passed before species diversity returned to previous levels. Michael Benton, head of Earth sciences at Bristol University, commented: "The end-Permian crisis nearly marked the end of life. It's estimated that fewer than one in 10 species survived. Geologists are only now coming to appreciate the severity of this global catastrophe and to understand how and why so many species died out so quickly" (Reynolds 2003).

The Permian heat wave was felt first and most intensely in the tropical latitudes; loss of species diversity spread from there. Reduction of vegetation, soil erosion and the effects of increasing rainfall wiped out the lush, diverse habitats of the tropics, which today would lead to the loss of animals such as hippos, elephants and all of the primates, according to Benton (Reynolds 2003). He added: "The end-Permian extinction event is a good model for what might happen in the future because it was fairly nonspecific. The sequence of what happened then is different from today because then the carbon dioxide came from massive volcanic eruptions, whereas today it is coming from industrial activity. However, it doesn't matter where this gas comes from; the fact is that if it is pumped into the atmosphere in high volumes, then that gives us the greenhouse effect and leads to the warming with all the other consequences" (Reynolds 2003).

According to a theory first advanced by Anthony Hallam and Paul Wignall (1997), the volcanic eruptions 251 million years ago provoked several biotic feedbacks that accelerated global warming of about 6°C. In his book *When Life Nearly Died: The Greatest Mass Extinction of All Time* (2003), Michael Benton sketched how the warming (which was accompanied by anoxia) may have fed on itself:

The end-Permian runaway greenhouse may have been simple. Release of carbon dioxide from the eruption of the Siberian Traps [volcanoes] led to a rise in global temperatures of 6 [°C] or so. Cool polar regions became warm and frozen tundra became unfrozen. The melting might have penetrated to the frozen gas hydrate reservoirs located around the polar oceans, and massive volumes of methane may have burst to the surface of the oceans in huge bubbles. This further input of carbon into the atmosphere caused more warming, which could have melted further gas hydrate reservoirs. So the process went on, running faster and faster. The natural systems that normally reduce carbon dioxide levels could not operate, and eventually the system spiraled out of control, with the biggest crash in the history of life. (Benton 2003, 276–277)

Greg Retallack, an expert on ancient soils at the University of Oregon, has speculated that the resulting methane "belch" was of such a magnitude that it caused mass extinction of land animals via oxygen starvation. Bob Berner of Yale University has calculated that a cascade of effects on wetlands and coral reefs may have reduced oxygen levels in the atmosphere from 35 percent to only 12 percent in only 20,000 years. Marine life also may have suffocated in the oxygen-poor water ("Suffocation" 2003). One animal, the meter-long reptile *Lystrosaurus,* survived because it had evolved to live in burrows where oxygen levels are low and carbon dioxide levels are high. According to a report by the New Scientist News Service, "It had developed a barrel chest, thick ribs, enlarged lungs, a muscular diaphragm, and short internal nostrils to get the oxygen it needed ("Suffocation" 2003).

According to Chris Lavers, writing in *Why Elephants Have Big Ears* (2000), a spike of worldwide warming contributed to this mass extinction in part because all of Earth's continents at the time were combined into one land mass known as Pangaea (Lavers 2000, 231). Tropical region warming at this time has been estimated at approximately 11°F, with larger rises near the poles that tended to create a generally warm atmosphere planetwide, "a flattening of the temperature difference between the poles and the equator" (Lavers 2000, 232), a condition that Lavers suspects drastically slowed or shut down ocean mixing and thus killed many sea creatures. "Unstirred," wrote Lavers, "the oceans begin to stagnate. Deep waters gradually lost oxygen, and species began to vanish" (Lavers 2000, 233).

Warming and Anticipated Animal Extinctions

Mark C. Urban of the Department of Ecology and Evolutionary Biology, University of Connecticut, synthesized many existing published studies on extinctions anticipated in a warmer world "to estimate a global mean extinction rate and determine which factors contribute the greatest uncertainty to climate change–induced extinction risks" (Urban 2015).

"Extinction risks will accelerate with future global temperatures, threatening up to one in six species," says Urban, if temperatures rise according to current estimates of rising greenhouse gas levels. The highest risks, wrote Urban, are in South America, New Zealand, and Australia. Urban's estimates may be conservative. The number of extinctions may well be two or three times higher, said John J. Wiens, an evolutionary biologist at the University of Arizona.

Among the many studies that Urban examined, some contend that many species will adapt to the heat and the extinction rate will be lower; others anticipate that rates will be as high as 50 percent. Many studies apply only to specific areas (such as the Amazon River valley) or types of species such as birds or butterflies. South American studies found 23 percent of species to be at risk of extinction.

Urban's study "revisit[ed] every climate extinction model ever published. He threw out all the studies that examined just a single species . . . on the grounds that these might artificially inflate the result of his meta-analysis" (Urban 2015). Urban ended up with 131 studies examining plants, amphibians, fish, mammals, reptiles, and invertebrates spread out across the planet. He reanalyzed all the data in those reports.

Urban projected that at a temperature increase of 3.6°F, 5.2 percent of species would become extinct. At 7.7°F, close to the projected increase by the Intergovernmental Panel on Climate Change for the end of the current century, 16 percent would die out. The higher the temperature, he said, the more rapidly the extinction rate will accelerate. Richard Pearson, a biogeographer at University College London, said that extinction of one species also impacts many others. "What happens to other species in an ecosystem when a species goes extinct?" Pearson asked (Zimmer 2015).

Further Reading

Urban, Mark C. "Accelerating Extinction Risk from Climate Change." *Science* 348 (May 1, 2015): 571–573.

Zimmer, Carl. "Study Finds Global Warming as Threat to 1 in 6 Species." *The New York Times*, May 1, 2015. http://www.nytimes.com/2015/05/05/science/new -estimates-for-extinctions-global-warming-could-cause.html.

What caused this spike in temperatures? The prime suspect, at least in the beginning, was coal-bearing deposits in the southern reaches of Pangaea that were oxidized after they were lifted by tectonic activity. Large volumes of carbon dioxide were released through volcanic eruption. The level of greenhouse gases in the atmosphere thus increased because of the most concentrated bout of volcanic activity on Earth in the last 600 million years. "This injection of volcanic CO_2," wrote Lavers, "was probably the decisive event that ultimately tipped the biosphere into the new era of the Mesozoic" (Lavers 2000, 235).

Flora and Fauna: Coming Extinctions

We are now in the midst of one of Earth's most intense, rapid, and pervasive mass extinctions. In harm's way are many varieties and species of flora and fauna that humans do not eat or keep as pets. Earth has experienced mass extinctions before, but all of them have resulted from natural causes. The current global warming is a product of humankind's increasing dominance of Earth and the devastation of native habitats of the many animals and plants that are being driven to extinction. Compared to past mass extinctions, which were driven by natural catastrophes such as meteor strikes or large-scale volcanism, the current human-driven wave of extinctions has been occurring with frightening speed. Given the projected rises in temperature during the decades to come, the flora and fauna of our home planet have thus far only seen their initial travails.

As time has passed, estimates of impending extinctions because of climate change have led to more realistic models. A study by Craig Moritz and Rosa Agudo (2013) as well as others ("Climate-Change Ecology" 2012) suggest that, "As the climate warms, many species are predicted to shift their ranges to stay within comfortable temperature zones. However, some species will be better able to do so than others." Mark Urban (University of Connecticut at Storrs) and colleagues created a model that factors in the competition that species will face as they migrate into new habitats. This model described anticipated effects of a 4°C rise in temperature on 40 species over a century. Results "found a much higher number of extinctions than did models that do not account for species competition and species' differing dispersal abilities."

Temperatures Skew Sex Ratios of Bearded Dragons

The temperature at which eggs are laid also affects the sex ratio of Australian bearded dragons. Warming temperatures favor females in this case, too. This propensity is so common in reptiles that scientists have given it a name: *temperature override*. Some amphibians and fish also display the same tendency. With bearded dragons, even a 1°C warming can skew the sex ratio significantly.

In *Nature*, Clare E. Holleley and colleagues wrote,

Sex determination in animals is amazingly plastic. Vertebrates display contrasting strategies ranging from complete genetic control of sex (genotypic sex determination) to environmentally determined sex (for example, temperature-dependent sex determination). . . .

Here we make the first report of reptile sex reversal in the wild, in the Australian bearded dragon (*Pogona vitticeps*), and use sex-reversed animals to experimentally induce a rapid transition from genotypic to temperature-dependent sex determination. Controlled mating of normal males to sex-reversed females produces viable and fertile offspring whose phenotypic sex is determined solely by temperature (temperature-dependent sex determination). . . . The instantaneous creation of a lineage of ZZ temperature-sensitive animals reveals a novel, climate-induced pathway for the rapid transition between genetic and temperature-dependent sex determination, and adds to the concern about adaptation to rapid global climate change. (Holleley et al., 2015)

Further Reading

Holleley, Clare E., et al. "Sex Reversal Triggers the Rapid Transition from Genetic to Temperature-Dependent Sex." *Nature* 523 (July 2, 2015): 79–82. http://www.nature.com/nature/journal/v523/n7558/full/nature14574.html?message-global =remove.

Climate Change: One of Several Drivers of Mass Extinction

Human-caused climate change is often one of several factors driving destructive environmental change. The result is a high risk of mass extinctions at a speed heretofore unknown in human history and far more quickly than anyone had anticipated, according to a report from an interdisciplinary international workshop held in April 2011. "The findings are shocking," said Dr. Alex Rogers, scientific director of the International Programme on the State of the Ocean, which convened the workshop. "As we considered the cumulative effect of what humankind does to the ocean, the implications became far worse than we had individually realized" ("Mass Extinction" 2011).

The panel urged an immediate "reduction in CO_2 emissions coupled with significantly increased measures for mitigation of atmospheric CO_2 and to better manage coastal and marine carbon sinks to avoid additional emissions of greenhouse gases. It is a matter of urgency that the ocean is considered as a priority in the deliberations of the [Intergovernmental Panel on Climate Change] and [United Nations Framework Convention on Climate Change]" ("Mass Extinction" 2011).

The Monteverde Rain Forest and the Golden Toad

In Costa Rica, the area known as Monteverde began as a colony of expatriate North American Quakers in the middle of the 20th century. Over the years, Monteverde has become a major attraction for American biologists, many of whom have devoted their careers to the flora and fauna of the region. Carol Kaesuk Yoon described the area in *The New York Times*: "A quick look at the forest's inhabitants makes clear why—from the record-breaking diversity of orchids to creatures like the resplendent quetzal, a glittering green, kite-tailed bird truly worthy of its name, and the so-called singing mice that whistle and chirp like birds" (Yoon 2001). In addition to its value as a shelter for plants and animals, Monteverde has become a center for ecotourism. Nearly 50,000 tourists visit each year to walk its many trails, "stopping for a bite to eat or perhaps picking up a golden toad mug or Monteverde T-shirt along the way." Scientists say that if the cloud forest falters, much business will be lost along with the unique species (Yoon 2001).

Deforestation and a warming environment may help to explain the disappearance of *Bufo periglenes*, the golden toad of the Monteverde tropical rain forest, which was found nowhere else on Earth. No one knows how long the golden toad had lived in the cloud forest that runs along Costa Rica's Pacific coast. By 1987, however, the level of mountain clouds had risen, reducing the frequency of mists during the dry season and probably playing a role in a massive population crash that affected most of the 50 species of frogs and toads in the forest. No fewer than 20 species became locally extinct (Moss 2001).

According to reporter Carol Yoon,

A flashy orange creature last seen in the late 1980s, the golden toad has become an international symbol of the world's disappearing amphibians. Seeing how even pristine and protected forests such as those at Monteverde can lose their crucial mists and clouds, researchers say it becomes less mysterious how a water-loving creature like the golden toad could vanish even from a forest where every tree still stands. (Yoon 2001)

Further Reading

Moss, Stephen. "Casualties." *The Guardian* (United Kingdom), April 26, 2001, 18.

Yoon, Carol Kaesuk. "Something Missing in Fragile Cloud Forest: The Clouds." *The New York Times*, November 20, 2001, F5.

The rate at which carbon dioxide is now being absorbed by the oceans vastly exceeds that of the last globally significant extinction of ocean species some 55 million years ago; then as many as 50 percent of some groups of deep-sea animals were wiped out. The time available for action is shrinking, the panel warned. "The longer the delay in reducing emissions the higher the annual reduction rate will have to be and the greater the financial cost. Delays will mean increased environmental damage with greater socioeconomic impacts and costs of mitigation and adaptation measures" ("Mass Extinction" 2011).

Robert L. Peters, describes how human-provoked climate change combines with humanity's subjugation of the planet to sketch Earth's biological future:

Habitat destruction in conjunction with climate change sets the stage for an even larger wave of extinction than previously imagined, based on consideration of human encroachment alone. Small, remnant populations of most species, surrounded by cities, roads, reservoirs, and farmland, would have little chance of reaching new habitat if climate change makes the old unsuitable. Few animals or plants would be able to cross Los Angeles on the way to the promised land. (Peters 1989, 91)

A Methane "Burp" and Noah's Ark

Gregory Ryskin, a geologist at Northwestern University, has asserted that a small-scale methane "burp" may help to explain the biblical flood allegedly navigated by Noah's Ark. The biblical flood, according to Ryskin, may be attributable to an eruption from Europe's Black Sea. Some geological evidence suggests an event of this type 7,000 to 8,000 years ago. According to an account by Tom Clarke on Nature Online (2003), "Ryskin contends that methane from bacterial decay or from frozen methane hydrates . . . began to be released. Under the enormous pressure from water above, the gas dissolved . . . [and] was trapped there as its concentration grew." A single disturbance, according to this account, "a small meteorite impact or even a fast-moving mammal, could then have brought the gas-saturated water closer to the surface. Here it would have bubbled out of solution under the reduced pressure. Thereafter the process would have been unstoppable: a huge overturning of the water layers would have released a vast belch of methane" (Clarke 2003; see also Ryskin 2003, 737).

Further Reading

Clarke, Tom. "Boiling Seas Linked to Mass Extinction; Methane Belches May Have Catastrophic Consequences." Nature Online, August 22, 2003. http://info.nature .com/cgi-bin24/DM/y/eLod0BfHSK0Ch0DYy0AL (no longer available).
Ryskin, G. "Methane-driven Oceanic Eruptions and Mass Extinctions." *Geology* 31 (2003): 737–740.

Mass Extinctions within a Century?

Global warming could play a role in the destruction or fundamental alteration of a third of Earth's plant and animal habitats within a century, bringing extinction to thousands of species, according to a study by Great Britain's World Wide Fund for Nature, an affiliate of the World Wildlife Fund. The study said that the most vulnerable plant and animal species will be in Arctic and mountain areas, where as many as 20 percent could be driven to extinction. In the north of Canada, Russia, and Scandinavia, where warming was predicted to be the most rapid, as much as 70 percent of habitat could be lost, according to this study.

The 2000 report, *Global Warming and Terrestrial Biodiversity Decline*, was written by Jay Malcolm, professor of forestry at Toronto University, and Adam Markham, the executive director of Clean Air-Cool Planet. Pests and weedy species would fare best, the report said. The report concluded, "If past fastest rates of migration are a good proxy for what can be attained in a warming world, then radical reductions in greenhouse gas emissions are urgently required to reduce the threat of biodiversity loss" (Malcom and Markham 2000, 5; Clover 2000). According to the authors, to adapt and survive the expected rate of warming during the next century, plants may need to move 10 times more quickly than they did when recolonizing previously glaciated land at the end of the last ice age. Few plant species, however, can move at a rate of one kilometer per year, the speed that will be required for their survival in many parts of the world.

Following the first study of its kind, researchers investigating a range of habitats—including northern Britain, the wet tropics of northeastern Australia, and the Mexican desert—said early in 2004 that given "midrange" climate change scenarios for 2050, they anticipate that 15 percent to 37 percent of the species in their sample of regions (covering 20 percent of Earth's surface) will be "committed to extinction" (Thomas et al. 2004, 145). The severity of extinctions is expected to vary with the severity of warming. The study used projections from United Nations authorities that world average temperatures will rise 2.5 to 10.4°F by 2100. "We're not talking about the occasional extinction—we're talking about 1.25 million species. It's a massive number," the study authors wrote (Gugliotta 2004).

This study was also described in *Nature* and marked the first time that scientists have produced a global analysis with concrete estimates of the effect of climate change on many various animal and plant habitats. Thomas led a 19-member international team that surveyed habitat decline for 1,103 plant and animal species in Europe; Queensland, Australia; Mexico's Chihuahua desert; the Brazilian Amazon River valley; and the Cape Floristic Region on South Africa's southern tip (Gugliotta 2004).

According to the researchers, climate change since about 1985 already has produced many shifts in the distribution and abundance of plants and animals. Climate change has thus already become a major driver of biodiversity change. The survey team used one of ecology's few ironclad laws—the species–area relationship—which was first postulated by Charles Darwin in his *Origin of Species* (1859). The law holds that a smaller habitable area will host a smaller number of viable species.

The researchers then projected habitat changes based on various warming scenarios. With a temperature rise of 0.8 to 1.7°C, they anticipated an 18 percent extinction rate; at 1.8 to 2.0°C, they projected 24 percent; and at more than 2.0°C, 35 percent (Pounds and Puschendorf 2004, 108).

The study emphasizes that examining possible extinctions solely in light of global warming probably understates their potential scope because extinctions in the real world also may be caused by such factors as landscape modification, species invasions, and pollution. By projecting effects of rising temperatures alone, the researchers realized they might be ignoring the effects of precipitation changes on habitat. Declines of amphibians in Costa Rica have been traced to reduced cloudiness in mountainous areas that probably are related to warming, for example.

The authors of the study considered a range of possibilities based on the ability of each species to move to a more congenial habitat to escape warming. If all species were able to move or disperse, the study said, only 15 percent would be irrevocably headed for extinction by 2050. If no species were able to move, the extinction rate could rise as high as 37 percent (Gugliotta 2004). The scientists concluded: "These estimates show the importance of rapid implementation of technologies to decrease greenhouse gas emissions and strategies for carbon sequestration" (Thomas et al. 2004, 145).

The survey's findings were disputed by skeptics such as William O'Keefe, president of the George C. Marshall Institute, who said that the research "ignored species' ability to adapt to higher temperatures" and also assumed that technologies will not arise to reduce emissions (Gugliotta 2004). Some animals and plants are adapting to warming—to a point—as they move upward in elevation or toward the poles in direction until something blocks their way. In the European Alps, for example, some plant species have been migrating upward by one to four meters each decade (Grabherr et al. 1994, 448). Across Europe, the growing season in controlled mixed-species gardens lengthened by 10.8 days a year from 1959 to 1993 (Menzel and Fabian 1999, 659). In Europe and North America, many migratory birds now arrive earlier in the spring and depart later in the autumn. Butterflies, beetles, dragonflies, and other species are now found farther north where conditions previously were too cold for them to survive.

Further Reading

Bates, Albert K., and Project Plenty. *Climate in Crisis: The Greenhouse Effect and What We Can Do.* Summertown, TN: The Book Publishing Co., 1990.

Benton, Michael J. *When Life Nearly Died: The Greatest Mass Extinction of All Time.* London: Thames & Hudson, 2003. See Chapter 6, "What Caused the Biggest Catastrophe of all Time?"

"Climate-Change Ecology: Extinctions Underestimated." *Nature* 481 (January 12, 2012): 116–117.

Clover, Charles. "Thousands of Species 'Threatened by Warming.'" *Daily Telegraph* (London), August 31, 2000, 9.

Grabherr, G., M. Gottfried, and H. Pauli. "Climate Effects on Mountain Plants." *Nature* 339 (1994): 448–451.

Gugliotta, Guy. "Warming May Threaten 37 Per Cent of Species by 2050." *Washington Post*, January 8, 2004, A1. http://www.washingtonpost.com/wp-dyn/articles/A63153-2004Jan7.html (no longer available).

Hallam, Anthony, and Paul Wignall. *Mass Extinctions and Their Aftermath*. Oxford, UK: Oxford University Press, 1997.

Lavers, Chris. *Why Elephants Have Big Ears*. New York: St. Martin's Press, 2000.

Malcolm, Jay R., and Adam Markham. *Global Warming and Terrestrial Biodiversity Decline: A Modeling Approach*. Report Prepared for the World Wildlife Fund, July 2000. http://www.wwf.se/source.php/1117008/figures.pdf.

"Mass Extinction of Ocean Species Soon to Be 'Inevitable.'" Environment News Service, June 21, 2011. http://www.ens-newswire.com/ens/jun2011/2011-06-21-01.html (no longer available).

Menzel, A., and P. Fabian. "Growing Season Extended in Europe." *Nature* 397 (1999): 659–662.

Moritz, Craig, and Rosa Agudo. "The Future of Species under Climate Change: Resilience or Decline?" *Science* 341 (August 2, 2013): 504–508.

Peters, Robert L. "Effects of Global Warming on Biological Diversity." Pp. 82–95 in Edwin Abrahamson, ed., *The Challenge of Global Warming*. Washington, DC: Island Press, 1989.

Pounds, J. Alan, and Robert Puschendorf. "Clouded Futures." *Nature* 427 (January 8, 2004): 107–108.

Reynolds, James. "Earth Is Heading for Mass Extinction in Just a Century." *The Scotsman*, June 18, 2003, 6.

Shen, Shu-zhong, et al. "Calibrating the End-Permian Mass Extinction." *Science* 334 (December 9, 2011): 1367–1372.

"Suffocation Suspected for Greatest Mass Extinction." NewScientist.com News Service, September 9, 2003. https://www.newscientist.com/article/dn4138-suffocation-suspected-for-greatest-mass-extinction/.

Thomas, Chris D., et al. "Extinction Risk from Climate Change." *Nature* 427 (January 8, 2004): 145–148.

See also: Adaptation, Animals, Lizards, and Warming Habitats; Animal Life, Antarctic; Animal Life, Arctic; Carbon Cycle Feedbacks; Climate Change, Abrupt Nature of; Fisheries; Ocean Acidity, North America; Ocean Acidity, Worldwide; Pliocene Paleoclimate; Salmon and Trout; Sea Ice, Arctic

MIGRATIONS

A report by the United Nations Environment Program found that climate change is inflicting severe impacts on migratory species from whales and dolphins to birds and turtles, and it is likely to be increasingly disruptive. Migratory species are in many ways more vulnerable than other species as they use multiple habitats and sites and many different resources during their migrations ("Climate Change Dislocates" 2006). The National Audubon Society and the American Bird Conservancy's WatchList 2007, said that 178 bird species in the United States, a quarter of its bird species, are threatened with extinction not only by global warming but also by invasive species and urban sprawl, an increase of 10 percent from 2002.

Ireland: Changing Flora and Fauna

In Ireland, by 2001 swallows were arriving earlier and the growing season for trees was steadily increasing, according to scientists at Trinity College of Dublin and National University of Ireland at Maynooth who have been examining natural responses to temperature changes. Alison Donnelly of the Climate Change Research Center, School of Botany, Trinity College, found that the beginning of the growing season, defined by the unfolding of leaves, has been occurring earlier in the spring. They also noted that the end of the growing season, the time when the leaves fall, occurred later in autumn, both of which point to an influence of rising global temperatures.

According to Donnelly, records of the dates that swallows arrive on the east coast of Ireland only extend back to the 1980s but still provided some evidence of the swallows' changing patterns of migration (Fahy 2001). Ireland's Environmental Protection Agency issued a report in late November 2002 titled "Climate Change Indicators for Ireland," which said that temperatures in Ireland have been rising 0.25°C per decade, with increases accelerating during the 1990s. The number of frosty days also had declined (Hogan 2002).

Hummingbird hawk moths, usually resident in southern Europe, have started appearing in Ireland, according to broadcaster and naturalist Eanna Ni Lamhna. Often mistaken for hummingbirds, the hawk moths hover as they swig nectar from flowers (Meagher 2004). Bird-watchers have also noticed an increase in the number of exotic birds. The little egret, a smaller version of the heron, has taken to nesting along the coast of Cork and Waterford. These birds are usually found in lands bordering the Mediterranean Sea but first started nesting in Ireland in 1997; 50 pairs are now thought to reside in the country. Other migrant species from southern climes that have recently appeared in Ireland include the pied flycatcher, the bearded tit, and the Mediterranean gull (Meagher 2004).

In the oceans surrounding Ireland, jellyfish have thrived in warmer sea temperatures, while cockles have expired from the warming waters in some enclosed bays. Karin Dubsky, the environmentalist who runs Coastwatch Ireland, said that warming ocean temperatures may cause native oysters to be outbred by Pacific oysters being farmed along the coast. "We are worried that Pacific oysters might thrive here as a result of the increase in sea temperatures. They could outbreed the Irish oyster in the same way that the gray squirrel knocked out the red squirrel" (Meagher 2004).

Further Reading

Fahy, Declan. "Nature Charts Its Own Change: Irish Researchers Are Finding Signs of Climate Change in Trees and Bird Species." *Irish Times*, September 13, 2001, n.p. (LEXIS).

Hogan, Treacy. "Still Raining in Costa del Ireland." *Belfast Telegraph*, November 26, 2002, n.p. (LEXIS).

Meagher, John. "Look What the Changing Climate Dragged in . . ." *Independent* (Dublin), July 9, 2004, n.p. (LEXIS).

The U.N. Environment Program report *Migratory Species and Climate Change: Impacts of a Changing Environment on Wild Animals* (2006) documented a wide range of current climate effects. European bee-eaters, *Merops apiaster*, birds once rare in Germany, are now breeding regularly across the country. The rosy-breasted trumpeter finch, *Rhodopechys githaginea*, is one of many bird species once previously confined to arid North Africa and the Middle East but now increasingly found in large numbers in southern Spain.

The arrival of hundreds of Bewick swans, *Cygnus columbianus*, flying in their distinctive "V" skeins, used to herald the arrival of the British winter. Ornithologists report that their numbers are now down to double figures. Warmer weather on the continent and the absence of the northeast winds that aid their migration are the likely reasons for the swans' disappearance from traditional British wintering sites. Changing wind patterns, in fact, are making it more difficult for many birds to migrate over the Caribbean Sea, where spring storms are becoming more numerous and of greater intensity ("Climate Change Dislocates" 2006).

Decline of the Cuckoo in Great Britain

The cuckoo, England's herald of spring, is a bird well known for the male's distinctive call and for laying its eggs in nests assembled by other species. The has been disappearing from the British countryside over the last three decades, their numbers declining by 30 percent in urban areas during that time, while they have declined by as much as 60 percent in woodland areas. The cuckoo migrates to all parts of the country from winter feeding grounds in sub-Saharan Africa. David Marley of the Woodland Trust said that the cuckoo's preferred woodland habitat was particularly vulnerable to global warming, which could be affecting its breeding season and food supplies (Smith 2002).

"The cuckoo is one of the most amazing birds you can come across in Britain, but it is declining by a staggering amount and we are getting reports from across the country that people just aren't hearing it any more," Marley said. Other theories for the 30-year slump in cuckoo populations include habitat loss and the spread of intensive farming practices. At their wintering places in Africa, cuckoos also may be suffering from indiscriminate use of agricultural chemicals as well as widespread drought (Smith 2002).

The British Trust for Ornithology believes that cuckoos in Britain number between 12,000 and 24,000 pairs, down from 17,500 to 35,000 in 1970. The Woodland Trust has launched a study that will record cuckoo sightings and build data to help document the changing ecology that may help explain why cuckoos are in sharp decline. Reports will be added to data collected since 1726 to help to monitor the impact of climate change. Volunteers will watch for cuckoos, as well as other wildlife, including ladybirds, bumblebees, and swallows (Smith 2002).

Although some bird species have been declining in England as weather has warmed, others are now flourishing. For example, the number of wild parrots was rising rapidly in 2004, with 100,000 expected by the end of the decade. The parrots, which are large and aggressive, have been competing with domestic species such as starlings, jackdaws, and small owls for food and territory (Prigg 2004).

Further Reading

Prigg, Mark. "Despite All the Heavy Rain, That Was the Hottest June for 28 Years." *Evening Standard* (London), July 1, 2004, A9.
Smith, Lewis. "Falling Numbers Silence Cuckoo's Call of Spring." *Times* (London), March 6, 2002, n.p. (LEXIS).

Although fears abound that rapid climate change will outstrip many species' abilities to adapt, one 47-year study of the great tit (*Parus major*) in the United Kingdom indicates that these wild birds not only adapted as the weather warmed but also flourished, adjusting their eating habits and reproductive behavior. The average egg-laying date of the great tits advanced 14 days during the study period (1961–2007), for example (Charmantier et al. 2008, 800). All species have their limits, of course, and one question left unanswered by this study is just what degree of warming will continue to favor the great tit because climate change has thus far been modest compared to what models project for the future.

Widespread Effects of Changing Weather on Birds

More than 200 species of birds in the United States are threatened by rising temperatures according to *The State of the Birds: 2010 Report on Climate Change* by the U.S. Department of the Interior's Fish and Wildlife Service, which compiled the study with aid from the U.S. Geological Survey, academics, and several environmental and wildlife groups. Many of the species such as the akikiki, a tiny olive-and-white colored bird that uses its long tongue to harvest insects from crevices in tree bark on the Hawaiian island of Kauai, will become more vulnerable as rising temperatures allow mosquitoes to colonize higher elevations, spreading avian malaria. Birds in and near ocean environments (including those on islands) are most vulnerable, according to the report (Fang 2010, 332).

Baltimore without Orioles

Baltimore orioles (*Icterus galbula*) once were so numerous in eastern North America that naturalist painter John J. Audubon wrote about the delight of hearing "the melody resulting from thousands of musical voices that come from some neighboring tree" (Pianin 2002). The bird, a Maryland icon whose namesake was adopted by a succession of Baltimore baseball teams, was officially designated the state bird in 1947. Local legend maintains that George Calvert, the first governor of Baltimore, liked the oriole's bright-orange plumage so much that he adopted its colors for his coat of arms.

Maryland's Baltimore orioles, which have been declining because of habitat loss for many years, could vanish altogether late in the 21st century due to changes in migration patterns strongly influenced by a warming climate. A study by the National Wildlife Federation and the American Bird Conservancy "suggests that the effects of global warming may be robbing Maryland and a half-dozen other states of an important piece of their heritage by hastening the departure of their state birds" (Pianin 2002).

According to the report, Earth's rising temperature "is already shifting songbird ranges, altering migration behavior and perhaps diminishing some species' ability to survive" (Pianin 2002). Iowa and Washington state may lose the American goldfinch, and New Hampshire's purple finch could become a historical relic. California could lose California quails, Massachusetts's black-capped chickadee could vanish, and Georgia could lose its brown thrasher (Pianin 2002).

The life cycles of the oriole and other birds are tied closely to weather patterns that are changing with general warming. Seasonal changes in weather patterns tell the birds when they should begin their long flights southward in the fall and northward in the spring. Temperature and precipitation also influence the timing and availability of flowers, seeds, and other food sources for the birds when they reach their destinations (Pianin 2002).

Global warming is not the only danger to the oriole and other well-known birds. Their decline results also from diminishing breeding habitat and forests in North America (where orioles spend their summers) and in Central and South America where they winter. "Climate change on top of fragmented habitat is the straw that breaks the camel's back," said Patricia Glick, an expert on climate change with the National Wildlife Federation (Pianin 2002).

Further Reading

Pianin, Eric. "A Baltimore without Orioles? Study Says Global Warming May Rob Maryland, Other States of Their Official Birds." *Washington Post*, March 4, 2002, A3.

The Fish and Wildlife Service report also said, "This autumn [2006] several large monarch butterflies, *Danaus plexippus*, which migrate in millions every year from the USA and Canada to Mexico, have been blown across the Atlantic to England 5,000 kilometers away" ("Climate Change Dislocates" 2006). Elsewhere, according to the report, desertification is increasing the size of the Sahara desert, adversely affecting the ability of Afro-European migrants to successfully cross this ecological barrier ("Climate Change Dislocates" 2006).

Worldwide, birds are suffering the escalating effects of climate change, according to a 2006 report by the World Wildlife Fund (WWF). The study found declines in some populations of 90 percent, "as well as total and unprecedented reproductive failure in others." The WWF study estimated that bird extinction rates could be as high as 38 percent in Europe and 72 percent in northeastern Australia if global warming exceeds 2°C above preindustrial levels. Currently, warming is 1.0°C above those levels, with another degree of increase "in the pipeline" ("Climate Change Pushing" 2006). The report, *Bird Species and Climate Change: The Global Status Report* (Wormworth n.d.) reviewed more than 200 scientific articles on birds on every continent to assemble a global picture of climate change's impacts.

"Robust scientific evidence shows that climate change is now affecting birds' behavior," said Karl Mallon, scientific director at Climate Risk of Sydney, Australia, an author of the report. "We are seeing migratory birds failing to migrate, and climate change pushing increasing numbers of birds out of synchrony with key elements of their ecosystems," Mallon said ("Climate Change Pushing" 2006).

"Birds have long been used as indicators of environmental change, and with this report we see they are the quintessential 'canaries in the coal mine' when it comes to climate change," said Hans Verolme, director of WWF's Global Climate Change Program. "This report finds certain bird groups, such as seabirds and migratory birds, to be early, very sensitive, responders to current levels of climate change," said Verolme. "Large-scale bird extinctions may occur sooner than we thought" ("Climate Change Pushing" 2006).

In Australia, the Mallee emu-wren, *Stipiturus mallee,* is rapidly losing population, its habitat so fragmented that a single bushfire could wipe out the species. Enduring drought in the southern and western parts of the species' range have destroyed vegetation that forms the basis of this bird's diet. In South Australia, this species has been reduced to about 100 birds confined over 100 square kilometers ("Warmer Climate" 2008).

More Birds and Butterflies Affected by Warming

In 2010, Michael Kearney of Australia's University of Melbourne and colleagues published a study describing how, between 1941 and 2005, Australia's common brown butterfly had emerged from their pupae 1.5 days earlier per decade near Melbourne. The researchers then linked this emergence to a rise in temperatures in the area as well as butterfly development (Zimmer 2011).

Across the Pacific Ocean, roughly 50,000 tufted puffins spend their summers on Triangle Island 30 miles off the British Columbia coast. These birds have been falling prey to periods of starvation because small changes on ocean temperature have driven away their food supply. Researchers say the adult birds bring back far fewer sand lance fish to their young in warm years. The sand lance, the puffin's favored food, have become less abundant in the waters around Triangle Island as water temperatures have risen. Scientists believe the lack of food, coupled with self-preservation instincts, had led the adults to abandon their chicks.

Climate Change and Seabirds

The vulnerability of seabirds to climate change is illustrated by an unprecedented breeding crash of several of the United Kingdom's North Sea species. The direct cause for the breeding failure of common guillemots, Arctic skuas, great skuas, kittiwakes, Arctic terns, and other seabirds at Shetland and Orkney colonies was a shortage of their prey, a small fish called the sandeel. Warming ocean waters and major shifts in species that underpin the ocean food web are believed to have been behind the major decline in sandeels. Nearly 7,000 pairs of great skuas in the Shetlands produced only a few chicks. A World Wildlife Fund report said that some starving adult birds ate their own young ("Climate Change Pushing" 2006).

Dutch researchers have reported that temperature increases have decoupled pied flycatchers migration dates from available food sources. The species spends its winters in West Africa and returns to woodlands in Holland each spring. Average temperatures in the spring territory have increased several degrees over the last 20 years. According to the researchers, higher spring temperatures mean that the birds' food source—insects—reach peak abundance earlier, providing birds that hatch at the same time with more food. The researchers found that over the two decades, the pied flycatcher's mean laying date has advanced by a little more than a week as selection has favored earlier laying. The dates that the birds migrate has not changed by more than a week, the researchers believe, because the birds' decision to leave their winter habitat is related to the amount of daylight, which is not affected by air temperature. The birds are arriving at the same time but laying eggs earlier. The researchers suggest that this phenomenon may be contributing to the decline of other long-distance migrants (Fountain 2001; Both et al. 2006, 81).

Further Reading

Both, Christiaan, et al. "Climate Change and Population Declines in a Long-Distance Migratory Bird." *Nature* 441 (May 4, 2006): 81–83.

"Climate Change Pushing Bird Species to Oblivion." Environment News Service, November 14, 2006. http://www.ens-newswire.com/ens/nov2006/2006-11-14-01.asp (no longer available).

Fountain, Henry. "Observatory: Early Birds and Worms." *The New York Times*, May 22, 2001, F4.

"The difference between 1998 and 1999 was one of the most striking things I have ever seen in my career in ecology," said Doug Bertram, a marine bird specialist at the Canadian Wildlife Service, as he recalled how dead chicks littered the tufted puffin colony in 1998 (Munro 2003). "We show that the extreme variation in reproductive performance exhibited by tufted puffins (*fratercula cirrhata*) was related to changes in sea surface temperatures both within and among seasons," the authors of a scientific study of the birds said. Such changes in ocean temperatures "could precipitate changes in a variety of oceanic processes to affect marine species worldwide" (Gjerdrum et al. 2003, 9377).

Dismayed researchers watched helplessly as chicks dropped dead of starvation during several warm summers in the 1990s. In other years such as 1999 when water was 1.5°C cooler, the chicks thrived (Munro 2003). By 2003, puffin populations had returned to their former levels as water temperatures cooled to near-average levels. The scientists worried, however, that the respite may be temporary if global warming causes ocean temperatures to rise again in coming years.

To the southwest, Jerram Brown has charted the breeding seasons of Mexican jays in the Chiricahua Mountains of southern Arizona for 31 years. By 1998, Brown found that the jays were laying their eggs an average of 10 days earlier than in 1971. Camille Parmesan has analyzed records tracking the distribution patterns of 57 nonmigratory butterfly species across Europe. She found that, during the last century, two-thirds of the species have shifted their ranges northward, some of them by as much as 240 kilometers (Wuethrich 2000, 795). "We ruled out all other obvious factors, such as habitat change," said Parmesan. "The only factor that correlated was climate" (Wuethrich 2000, 795).

Parmesan et al.'s tracking of butterfly ranges provided evidence of poleward shifts in the ranges of entire species. In a sample of 35 nonmigratory European butterflies, 63 percent have ranges that shifted to the north by 35 to 240 kilometers during the 20th century, whereas only 3 percent of ranges have shifted to the south (Parmesan et al. 1999, 579). The study team's evaluation of its data ends with a warning about other species:

> Given the relatively slight warming in this century compared [with anticipated temperature] increases of 2.1 to 4.6 [°C] for the next century, our data indicate that future climate warming could become a major force in shifting species distributions. But it remains to be seen how many species will be able to extend their northern range margins substantially across the highly fragmented landscapes of northern Europe. This could prove difficult for all but the most efficient colonizers. (Parmesan et al. 1999, 583)

Butterflies in Britain: Warming and Habitat Destruction

By 2000, warmer temperatures were bringing butterflies to England two weeks to a month earlier than during the 1970s. Scientists studying 35 of the estimated 60 species of British butterflies say that some such as the red admiral can now be seen a month earlier. Others such as "the peacock and the orange tip, are appearing between 15 and 25 days earlier than two decades ago," according to a report in the *Times* of London (Nuttall 2000). Roy and Tim

Sparks, both of the Center for Ecology and Hydrology at Monks Wood, Cambridgeshire, analyzed data from 1976 to 1998 provided by the Butterfly Monitoring Scheme, whose members check more than 100 sites each week from April to September. According to a news report in *The Guardian* of the United Kingdom, "One [red admiral] was monitored after crossing the [English] Channel on New Year's Day" (Vincent and Brown 2000).

Some species of British butterflies that were expected to flourish in warmer temperatures have instead declined because of severe habitat destruction. Warm summers and mild winters since the 1970s should have attracted populations of butterflies that usually steer clear of the British Isles. According to a paper by M. S. Warren and colleagues in *Nature*, however, three-quarters of the butterfly species that might have expanded northward with warmer European weather have declined (Warren et al. 2001). The findings come from an analysis of 1.6 million butterfly sightings by 10,000 amateur naturalists between 1995 and 1999.

The study's coordinator, Chris Thomas of York University, said, "Most species of butterflies that reach the northern edge of their geographic ranges in Britain have declined over the past 30 years, even though the climate has warmed. This is surprising because climate warming is expected to increase the range of habitats these species can inhabit" (Derbyshire 2001). "Our computer models show that climatically suitable areas are available for colonization, but most species have failed to exploit them either because they no longer contain suitable breeding sites or because breeding habitats are out of reach" (Derbyshire 2001; Warren et al. 2001, 65).

Threats of a warming climate to butterflies in Britain also were described in a 2002 study compiled by British biologists and ecologists that was published by the Royal Society. According to this study, at least 30 of Britain's butterfly species face extinction or an alarming drop in numbers because they are failing to cope with the effects of global warming. These include some rare species such as the large heath and purple emperor, both of whose numbers are expected to decline by as much as three-quarters. Many face eventual extinction as populations fall below replacement levels (Mason et al. 2002).

"There's no silver lining in this data," said Richard Fox, a coauthor of the study and spokesperson for the Butterfly Conservation Society. "What we will see here is a retreat and, potentially, a mass extinction in the slightly longer term, of many of our familiar species" (Mason et al. 2002). According to a report describing this work in *The Independent*, "Over the past few years, [red admirals, orange tips, and small tortoiseshells] have been seen earlier in the spring and surviving for several weeks longer each autumn, suggesting that butterflies would generally benefit from climate change. However, the researchers, led by Jane Hill, a biologist at York University, noticed that many more butterflies had failed to spread during the 1990s, even though average temperatures were beginning to rise" (Mason et al. 2002). A few species such as the ringlet and the marbled white have been prospering, moving northward and farther uphill as the summers became warmer during the 1990s, contrary to declining populations for many other species.

The butterfly study projected that as warming accelerates in Britain, species living in northern England and Scotland, including the Western Isles (Outer Hebrides), would lose two-thirds of their habitats. In southern England, butterfly habitat would decline by one-quarter. For some species, the future is particularly bleak. For example, the black hairstreak, an exceedingly rare species that has been confined mainly to the East Midlands, will lose at least half of its usual habitat (Mason et al. 2002). "We may get some butterflies from the south colonizing us, but what this paper shows is that our butterflies are going to become much, much rarer," Fox said (Mason et al. 2002).

Further Reading

Derbyshire, David. "Global Warming Fails to Boost Butterfly Visitors." *Daily Telegraph* (London), November 1, 2001, 13.

Mason, John, Jack A. Bailey, and Ardea London. "Doomsday for Butterflies as Britain Warms Up; Dozens of Native Species at Risk of Extinction as Habitats Come under Threat." *The Independent* (London), September 29, 2002, 12.

Nuttall, Nick. "Climate Change Lures Butterflies Here Early." *Times* (London), May 24, 2000, n.p.

Vincent, John, and Paul Brown. "Swoop to Conquer: Global Warming Brings Butterflies to Britain Earlier." *The Guardian* (U.K.), May 24, 2000, 9.

Warren, M. S., et al. "Rapid Responses of British Butterflies to Opposing Forces of Climate and Habitat Change." *Nature* 414 (November 1, 2001): 65–69.

Tropical Fish and Warm-Climate Birds

As cold-water fish abandon waters near Britain, species that usually live in southerly waters have taken their place. Sightings of warm-water fish have been plentiful since around 1990 in British coastal waters and have received considerable publicity in local newspapers. Warm-water species have been turning up regularly off the coasts of Devon and Cornwall for more than a decade; some scientists believe they are clear indicators of global warming. In the late 1980s, southern species such as sunfish and torpedo rays began to appear; by the late 1990s, such visitors were no longer regarded as oddities.

A series of small cold-water marine animals such as copepods have also been replaced by their warm-water cousins (McCarthy 2002). "As fish are very dependent on the temperature of the water, it is sensible to link these changes with changes in water temperature. They would be consistent with predictions of climate change," said Douglas Herdson of the National Marine Aquarium (McCarthy 2002).

Number and Frequency of Sea Creatures' Migrations

Sightings of sea creatures in migration are not unprecedented. What has changed is their number and frequency. With these migrations have come many birds that feed on them. Bob Swann, secretary of the Seabird Group, an international conservation and research organization, said, "There is much evidence of species previously more associated with more southerly regions of the Atlantic appearing around the British Isles. Climate change can influence oceanic currents and the availability of food—a prime reason for the presence of these birds. Certainly a lot of our breeding seabirds are currently doing very well because they are finding plenty to eat" (Unwin 2001).

A team of British marine biologists analyzed records dating back 40 years and in 2002 announced a "strong link" between the northward migration of fish and rising sea temperatures. These scientists related the arrival of tropical and semitropical fish off the coast of Cornwall, the southernmost tip of the British mainland, to increases in the average temperature of the North Atlantic Ocean (Connor 2002, 12).

Surveying records back to 1960, the scientists found that "more exotic species of fish are being caught or washed ashore now than ever before and that the sightings can be directly linked to a corresponding rise in sea temperatures." The link is said to be a "significant correlation" and could explain why Cornwall in particular has seen so many exotic species of marine wildlife from warmer regions of the world, according to Tony Stebbing, formerly a biologist with the Plymouth Marine Laboratory, whose work was funded by the Natural Environment Research Council (Connor 2002, 12).

Mantis Shrimp off England's South Coast

During May 2004, mantis shrimp (warm-water creatures usually found in tropical waters) were caught in trawler nets in Weymouth Bay, Dorset, on England's south coast. Fisherman took the then-unidentified orange-colored shrimp to the nearby SeaLife Center, where experts determined their species. The mantis or "toe splitter" shrimp averages three inches long. The shrimp can strike at up to 100 miles per hour with hammerlike claws, packing force as powerful as a .22-caliber bullet. Two years earlier, another colony of mantis shrimps was found at the north end of Cardigan Bay, Wales.

In midsummer 2004, a shoal of two-foot-long gray triggerfish was observed off the British coast. The triggerfish, whose usual habitat is the tropical Atlantic and the Mediterranean, were discovered two miles off the Isle of Purbeck, Dorset. During June 2002, a group of bright purple jellyfish known as "by-the-wind sailor fish" were found washed up at Kimmeridge Bay. Their usual habitat is the deep waters of the Mediterranean. Another visitor from southern seas was the four-inch weaver fish, which buries itself in sand close to shore and releases stinging venom from its dorsal spines if stepped on (Savill 2004, 7).

Further Reading

Savill, Richard. "Tropical Fish Hooked on Channel Holidays." *The Telegraph* (London, U.K.), August 12, 2004. http://www.telegraph.co.uk/news/uknews/1469221 /Tropical-fish-hooked-on-Channel-holidays.html.

According to an account in *The Independent* (London), "This summer's biggest seabird sensation was a red-billed tropic bird (*Phaethon aethereus*), which flew around a yacht about 20 miles south of the Isles of Scilly. This bird had not previously been recorded in northern European waters." The nearest colonies are on the Cape Verde Islands and islets off West Africa. Breeding also occurs on Ascension Island, St. Helena, the West Indies and on the Red Sea, Persian Gulf and Arabian Sea islands. Cape Verde and other islands off Africa are the source of the series of summer sightings of rare Fea's petrels (*Pterodroma feae*) off Scilly and the coasts of Devon and Cumbria (Unwin 2001).

Whales, Dolphins, and Jellyfish in British Waters

Whales and dolphins are being observed frequently in British waters as the area has warmed. During one ferry's round-trip between Portsmouth and Bilbao in 2001, wildlife enthusiasts logged sightings of 71 fin whales, four Cuvier's beaked whales, 37 pilot whales, 25 common dolphins, 352 striped dolphins, 120 unidentified dolphins, and seven unidentified large whales. In addition, two sperm whales were spotted in the English Channel from a ferry sailing between Portsmouth and Bilbao (in northern Spain). Newquay's Blue Reef Aquarium took custody of two loggerhead turtles that probably drifted to Britain after damaging their flippers in the open ocean (Unwin 2001).

According to Stella Turk of the Cornwall Wildlife Trust, a rise in sea temperatures may be a cause of northward migration of tropical sea creatures and birds. "It appears the sea is becoming generally warmer and it could be that such creatures are staying close to Britain throughout the year. An unusually wide range has already been reported this summer and it's still only mid-July—there could be many more surprises over the coming weeks" (Unwin 2001).

In addition to fish and birds, British observers in 2002 sighted hundreds of rootmouthed jellyfish, as well as *Rhizostoma* octopus. Large, disc-shaped sunfish (*Mola mola*) also have have been spotted more frequently in recent summers. Turk said that sightings in May 2002 began earlier than before. Another surprise was a report of a six-foot bluefin tuna, *Thunnus thynnus*. On July 14, 2002, about 40 basking sharks were spotted a mile off Perranporth, North Cornwall. A pilot whale was observed at the same time. A spokesman for Falmouth Coast Guard said, "The sharks are huge—they go up to 30 foot [long]. When the water warms up we do get basking sharks here, but it is unusual to get so many" (Unwin 2002). The flying gurnard, unknown in British waters before 1980, also has been sighted more frequently. The gurnards, with their elongated pectoral fins that enable them to move quickly through the water, were first caught in the nets of Cornish fishermen. The first sharp-nosed shark was netted in 1984 (Connor 2002, 12).

More Tropical Migrants

Global warming has brought an unexpected benefit to British West Country fishermen who have been struggling to make a living as their traditional catches dwindle. A slight rise in sea temperature has meant that valuable shellfish that were once unable to thrive north of the Channel Islands can now be farmed for export. Dislike molluscs in the genus Haliotis (known as *sea ear* and *abalone* elsewhere and as *ormers* in Britain) have also been found; their size—as much as 20 centimeters (7 inches) in length—fetches high prices in Japanese markets (Brown and Sutton 2002). Abalone is best known by tourists as a source of iridescent mother-of-pearl jewelry that is used as an inlay. In Japan, the gonads of the abalone eaten raw are regarded as a particular delicacy, while in California the abalone flesh is consumed as steaks. The people of Guernsey, England, have traditionally eaten ormer stew (Brown and Sutton 2002).

One notable warm-water visitor to English waters was a single slipper lobster (*Scyllarus arctus*), which was caught near the southwest tip of England and later displayed at the Plymouth National Marine Aquarium. The five-inch-long lobster is usually found near the coasts of the Mediterranean Sea. Only about a dozen have been recorded in United Kingdom waters during the last 250 years. The most recent specimen was the fifth to be caught in British waters since 1999. The increasing number of slipper lobsters in British waters is one of many indications that warm-water marine species have been moving northward because ocean temperatures are rising (McCarthy 2002).

Other warm-water fish have been found in English waters from two fish families, the breams and the jacks. A Guinea amberjack was first recorded off Guernsey. In 2001, there were five sightings of the almoco jack in the West Country (McCarthy 2002). A tropical zebra sea bream, which usually resides off the West African coast, was caught for the first time during 2002 in British waters. The 14-inch fish was accidentally netted by a commercial fisherman near Portland, Dorset, far from its native waters off Senegal and Mauritania. A seahorse also was sighted in the Thames River—a rare but not unprecedented event. Another seahorse was sighted in 1976.

In November 2001, Britain's first reported barracuda was caught some 40 miles from the site of the slipper lobster catch. England is not alone in observing tropical fish; two years after the English barracuda catch, another barracuda was caught on Seattle's waterfront. As they have near England, cold-water cod (of the Pacific variety) are becoming increasingly rare off of Washington's coastline (Stiffler and McClure 2003). During the summer of 2004, a giant squid, a species that usually ranges no farther north than Mexico's coastal waters, was caught near Maple Bay in southern British Columbia ("Jumbo Squid" 2004).

Charles Clover of London's *Daily Telegraph* reported that red mullet, which were restricted to waters south of the English Channel before 1990, were being caught in commercial quantities on both coasts of Scotland by 2002. "The largest geographical movement recorded over the past decade, the warmest on record," wrote Clover, "has been made by species of zooplankton, tiny shrimp-like creatures which form the base of the marine food chain. Warm-water species of copepod, as these crustaceans are known, have moved 600 miles northwards up the Bay of Biscay over the past decade, bringing warm-water fish species with them" (Clover 2002). At the same time, the cold-water copepod *Calanus finmarchicus*, the main food of the cod and the sand eel on which cod also feed, has moved north from the North

Sea. Martin Edwards of the Sir Alastair Hardy Laboratory in Plymouth said the North Sea was in a "transitional state" with the consequences for fish stocks, already endangered as a result of overfishing, hard to predict (Clover 2002).

Warm-Water Species off Ireland

Warm-water fish have been sighted in Irish waters as well. Among these have been various species of sharks, poisonous puffer fish, loggerhead turtles, and triggerfish. According to a report by Lynne Kelleher in London's *Sunday Mirror*, coastal waters that previously reached a summer maximum of 15°C now commonly reach 20°C, which draws warm-water species. According to Kelleher, "Fish never seen before are appearing in greater numbers and some are beginning to breed as they become acclimatized. A great white shark was spotted off the coast of Cornwall in recent years and exotic fish such as moray eels, mako sharks and anchovies are swimming off Irish shores" (Kelleher 2002).

Kevin Flannery, a marine biologist with the Department of Marine and Natural Resources, said the fish are arriving because of warmer water temperatures. "There is a definite correlation between the rise in temperatures and the arrival of new species of fish in Irish waters." Flannery continued:

The temperatures used to go below [5] degrees in the winter and go up to 14 or 15 degrees in the summer. Now they are between [5 and 7] in the winter and can go between 17 and 20 degrees in the summer. A moray eel, which is found in Australia and Canada, was caught by a vessel from Waterford off the coast of Cork last year. We have found three puffer fish. One was found off Fenit in Kerry last winter. A number of loggerhead turtles have been found washed up thousands of miles from where they came. (Kelleher 2002)

Large numbers of anchovies, which are usually found off the coast of Portugal, have been found swimming in Irish waters from Shannon to Kinsale. Puffer fish, which are poisonous, also have been sighted, probably for the first time (Kelleher 2002). Tropical triggerfish, which have been found in the largest numbers, are small in size but become aggressive to divers and swimmers if they are disturbed during their breeding season. Different types of bream and dory fish also have been sighted in Irish waters.

As Flannery noted,

The number of tropical species has increased dramatically. There are between 10 to 15 rare species found in recent years. One [fishing] vessel found about 80 triggerfish last year. They can do a lot of damage here to crab, lobster, and crayfish that they feed on. They have teeth like a rat and can kill a lobster. Fishermen are finding them inside the lobster pots. Their natural habitat is the tropical waters of Spain and Africa. They won't survive in waters less than 14 degrees. In 1995, the waters reached temperatures of about 20 degrees that was one of the highest temperatures. The triggerfish are staying around. They are not just coming in and out to feed. We are finding a number of pregnant triggerfish. This suggests they are acclimatizing. They wouldn't breed unless they were staying around. (Kelleher 2002)

According to Flannery, as tropical fish migrate to Ireland, traditional cold-water species such as cod have been leaving the island's coastal waters. "The rise in temperature could also mean the demise of cod, which need cold temperatures. If temperatures go over 17 degrees they will die. There has also been the demise of the Arctic char in our lakes. It is a combination of the pollution and the temperature" (Kelleher 2002).

Further Reading

Brown, Paul, and Tony Sutton. "Global Warming Brings New Cash Crop to West Country as Rising Water Temperatures Allow Valuable Shellfish to Thrive." *The Guardian* (United Kingdom), December 10, 2002, 8.

Charmantier, Anne, et al. "Adaptive Phenotypic Plasticity in Response to Climate Change in a Wild Bird Population." *Science* 320 (May 9, 2008): 800–803.

"Climate Change Dislocates Migratory Animals, Birds." Environment News Service, November 17, 2006. http://www.ens-newswire.com (no longer available).

"Climate Change Pushing Bird Species to Oblivion." Environment News Service, November 14, 2006. http://www.ens-newswire.com/ens/nov2006/2006-11-14-01.asp (no longer available).

Clover, Charles. "Global Warming 'Is Driving Fish North.'" *Daily Telegraph* (London), May 31, 2002, 14.

Connor, Steve. "Strangers in the Seas; Exotic Marine Species Are Turning Up Unexpectedly in the Cold Waters of the North Atlantic." *The Independent* (London), August 5, 2002, 12–13.

Fang, Janet. "Wildlife Service Plans for a Warmer World." *Nature* 464 (March 18, 2010): 332–333.

Gjerdrum, Carina, et al. "Tufted Puffin Reproduction Reveals Ocean Climate Variability." *Proceedings of the National Academy of Sciences* 100(16) (August 5, 2003): 9377–9382.

"Jumbo Squid Has a Message for Us: Changing Global Patterns Are Going to Bring Different Species into Our Waters." *Times-Colonist* (Victoria, BC), October 8, 2004, A10.

Kelleher, Lynne. "Look Who's Here: Tropical Fish Warming to Waters around Ireland." *Sunday Mirror* (London), October 20, 2002, 15.

McCarthy, Michael. "Climate Change Provides Exotic Sea Life with a Warm Welcome to Britain." *The Independent* (London), January 24, 2002, 13.

Munro, Margaret. "Puffin Colony Threatened by Warming: A Few Degrees Can Be Devastating. Thousands of Triangle Island Chicks Die When Heat Drives Off Their Favoured Fish." *Montreal Gazette*, July 15, 2003, A12.

Parmesan, Camille, et al. "Poleward Shifts in Geographical Ranges of Butterfly Species Associated with Regional Warming." *Nature* 399 (June 10, 1999): 579–583.

Stiffler, Linda, and Robert McClure. "Effects Could Be Profound." *Seattle Post-Intelligencer*, November 13, 2003, A8.

U.N. Environment Program. *Migratory Species and Climate Change: Impacts of a Changing Environment on Wild Animals.* Bonn, Germany: UNEP / CMS Convention on Migratory Species and DEFRA, 2006. http://www.cms.int/sites/default/files/document/ScC14_Inf_09_Migratory_Species%26Climate_Change_E_0.pdf.

Unwin, Brian. "Tropical Birds and Exotic Sea Creatures Warm to Britain's Welcoming Waters." *The Independent* (London), August 20, 2001, 7.

U.S. Department of the Interior. *The State of the Birds: 2010 Report on Climate Change.* Washington, DC: Fish and Wildlife Service.

"Warmer Climate Hurting Birds, New IUCN Red List Shows." Environment News Service, May 19, 2008. http://www.ens-newswire.com/ens/may2008/2008-05-20-03.asp (no longer available).

Wormworth, Janice. *Bird Species and Climate Change: The Global Status Report.* Fairlight, NSW, Australia: Climate Risk Pty Limited, n.d. https://www.wwf.or.jp/activities/lib/pdf_climate/environment/birdsFullReport.pdf.

Wuethrich, Bernice. "How Climate Change Alters Rhythms of the Wild." *Science* 287 (February 4, 2000): 793–795.

Zimmer, Carl. "Multitude of Species Face Climate Threat." *The New York Times,* April 4, 2011. http://www.nytimes.com/2011/04/05/science/earth/05climate.html.

See also: Adaptation, Animals, Lizards, and Warming Habitats; Animal Life, Antarctic; Animal Life, Arctic; Climate Change, Abrupt Nature of; Fisheries; Ocean Acidity, North America; Ocean Acidity, Worldwide; Salmon and Trout; Sea Ice, Arctic

OCEANS

OVERVIEW

Because humankind is a narcissistic species, when we consider the implications of global warming, we usually focus on the third of the planet that is dry land. As the land warms, however, so do the other two-thirds of Earth—and with profound implications for the species that inhabit it, including a holocaust that has already begun for coral reefs and other animals with calcium shells.

On a practical level, rising seas provoked by melting ice and thermal expansion of seawater will become the most notable challenge related to global warming and will likely range from inconvenience to disaster for many millions of people around the world. Human beings have an affinity for the open sea, so many major population centers have been built within a mere meter or two of mean sea level. From Mumbai (Bombay) to London to New York City, many millions of people will find warming seawater lapping at their heels during the coming years. Although sea levels have been rising slowly for a century or more, the pace will increase in coming years.

Melt all of Greenland's ice and the worldwide sea level rises an average of 20 feet. Melting the West Antarctic Ice Sheet adds 16 feet, the East Antarctic Ice Sheet 164 feet, and all of the planet's mountain glaciers one to two feet. Total sea level rise from melting ice could add as much as 200 feet maximum (Pilkey et al. 2016, 17).

The last time that worldwide temperatures were 3°C higher than today was during periods of the Pliocene Epoch, 2 million to 3 million years ago. The seas then were some 50 to 115 feet higher than today. The level of carbon dioxide in the atmosphere peaked at approximately 425 parts per million (ppm). The level as of 2015 had breached 400 ppm and has been increasing 2 ppm to 3 ppm per year. Thus, by the end of this century if not before we probably will have enough warming "in the pipeline" to raise sea levels as much as 80 feet within 150 to 200 years. One billion people today live within 25 meters (80 feet) of sea level. Because of thermal inertia, a century or two will be required for our Pliocene-level carbon dioxide level to melt enough ice to raise sea levels several dozen feet.

Scientists at the Lawrence Livermore National Laboratory and the U.S. National Oceanic and Atmospheric Administration (NOAA) measured the amount of heat that has been absorbed by the oceans since 1865 and learned that approximately half occurred within the past two decades ("Livermore Scientists" 2016). As Peter

J. Gleckler and colleagues reported in *Nature Climate Change* (2016), "Our model-based analysis suggests that nearly half of the industrial-era increases . . . have occurred in recent decades, with over a third of the accumulated heat occurring below 700 meters and steadily rising. . . . In recent decades the ocean has continued to warm substantially, and with time the warming signal is reaching deeper into the ocean." The oceans absorb more than 90 percent of the excess heat increase associated with global warming.

An Increasingly Acidic Ocean

The most important change in the oceans, however, has nothing to do with temperature. Rising carbon dioxide levels in the ocean make water more acidic and imperil any living thing in a calcium shell, including the phytoplankton that form the basis of the marine food chain and provide a large proportion of atmospheric oxygen. The shells of plankton are being damaged at current carbon dioxide levels, which will rise as long as the oceans continue to absorb carbon dioxide generated by human activities on land. Some shellfish on the U.S. West Coast have already gone sterile because their habitats have become too acidic.

By the early years of the 21st century, carbon dioxide levels were rising in the oceans more rapidly than any time since the age of the dinosaurs, according Ken Caldeira and Michael E. Wickett, who wrote in *Nature*, "We find that oceanic absorption of CO_2 from fossil fuels may result in larger pH changes over the next several centuries than any inferred in the geological record of the last 300 million years, with the possible exception of those resulting from rare, extreme events such as bolide impacts or catastrophic methane hydrate degassing" (Caldeira and Wickett 2003). A *bolide* is a large extraterrestrial body such as an asteroid that is usually at least a half mile in diameter, sometimes much larger, that impacts Earth at a speed roughly equal to that of a bullet in flight. In methane hydrate degassing, solid methane deposits on ocean floors are rapidly converted to their gaseous form in the atmosphere by warming temperatures.

By 2007, scientists had measured oceanic acidity at more than 30 percent above preindustrial levels, which imperils shelled animal life throughout much of the world's oceans. This danger is most notable in the colder waters of the Arctic and Antarctic (Southern) Oceans, which hold more carbon dioxide than warmer oceans (Holland 2001). As the oceans absorb more carbon dioxide and become more acidic, their capacity to hold more CO_2 in the future will be strongly reduced (Archer 2009, 112).

Moreover, ocean acidification can compound the effects of global warming in the atmosphere. Increasing seawater acidification caused by anthropogenic carbon dioxide reduces the bioavailability of iron in seawater, in turn reducing the oceans' ability to store more carbon dioxide and leaving more of it in the atmosphere to absorb heat (Sunda 2010). Writing in *Science*, Dalin Shi and colleagues explained:

> Acidification of media containing various Fe [iron] compounds decreases the
> iron uptake rate of diatoms and coccolithophores to an extent predicted by

the changes in iron chemistry. A slower iron uptake by a model diatom with decreasing pH is also seen in experiments with Atlantic surface water. The iron requirement of model phytoplankton remains unchanged with increasing CO_2. The ongoing acidification of seawater is likely to increase the iron stress of phytoplankton populations in some areas of the ocean. (Shi et al. 2010, 676)

Beginning in 2005, ocean acidification forced the near collapse of a West Coast shellfish industry that harvests $117 million annually and supports more than 3,000 jobs. Richard W. Spinrad (chief scientist of the U.S. National Oceanic and Atmospheric Administration) and Ian Boyd (chief scientific adviser to the British government's Department of Environment, Food, and Rural Affairs) wrote in *The New York Times* (2015), "Ocean currents pushed acidified water into coastal areas, making it difficult for baby oysters to use their limited energy to build protective shells. In effect, the crop was nearly destroyed" (Spinrad and Boyd 2015). By 2015, wrote Spinrad and Boyd,

The first nationwide study showing the vulnerability of the $1 billion U.S. shellfish industry to ocean acidification revealed a considerable list of at-risk areas. In addition to the Pacific Northwest, these areas include Long Island Sound, Narragansett Bay, Chesapeake Bay, the Gulf of Mexico, and areas off Maine and Massachusetts. Already at risk are Alaska's fisheries, which account for nearly 60 percent of the United States commercial fish catch and support more than 100,000 jobs. (Spinrad and Boyd 2015)

Toxic Algae Tied to Climate Change

Blooms of toxic algae called *microcystis* have become a regular feature of late summer in western Lake Erie and, in 2015, along two-thirds of the thousand-mile-long Ohio River. The blooms forced the shutdown of the Toledo, Ohio, regional water supply for 400,000 people for several days in early August 2014. The blooms have several provocations: mining; industrial pollution; phosphate and nitrate fertilizer runoffs from agricultural areas, cattle feedlots, and sewers; and a changing climate. Microcystin toxins cause diarrhea, liver damage, and vomiting in humans and may kill other animals that drink affected water.

In addition to an overabundance of phosphates and nitrates, the algae explode in stagnant water during August and September. Michael Wines reported in *The New York Times* (2015) that some scientists suspect changing weather patterns in the region: "Since the 1970s, average temperatures and rainfall have been slowly increasing in the Ohio River basin. So has the frequency of heavy rains—the steady downpours that wash fertilizer off farmlands and overwhelm sewage systems in big cities along the river's course."

Richard Stumpf, an oceanographer with NOAA who forecasts algae blooms, said, "These climate phenomena are consistent with our understanding of how these things work. We do expect them to be more common when you have wet springs followed by long, warm summers" (Wines 2015).

Further Reading

Wines, Michael. "Toxic Algae Outbreak Overwhelms a Polluted Ohio River." *The New York Times*, October 1, 2015. http://www.nytimes.com/2015/10/01/us/toxic-algae-outbreak-overwhelms-a-polluted-ohio-river.html.

Acidity is just one way in which human beings are killing plankton. Another is sunscreen, of which 6,600 tons flow from human bodies into the world's oceans each year. Nanoparticles from titanium dioxide, one of sunscreen's main ingredients, mixes with water and sun to form hydrogen peroxide, which kills phytoplankton, according to a report from the Spanish Research Council ("Lotion" 2015).

Effects of Rising Oceans

The oceans are reaching their limits as *sinks* (absorbers) of carbon dioxide and methane. Ken Caldeira and Philip B. Duffy assert in *Science* that "uptake" or removal of human-induced carbon dioxide by the oceans is less than many earlier investigators have assumed and that carbon uptake will diminish further as temperatures warm. In addition, Caldeira and Duffy contend that absorption of carbon into the oceans of the Southern Hemisphere (the focus of their study) has been diminishing since fossil fuel effluvia became a factor in the composition of the atmosphere around 1880. "Ventilation of the deep Southern Ocean was much more vigorous in the period from about 1350 to 1880 than in the recent past" (Caldeira and Duffy 2000, 620).

Warming (and rising) seas are affecting everyone who lives on or near an ocean, with coastal erosion, saltwater intrusion from higher tide levels, higher storm surges, and inundated coastal estuaries and groundwater supplies. More than 20 percent of the world's population lives within 30 kilometers of coastal areas. That population is increasing twice as quickly as aggregate population (Mathews-Amos and Berntson 1999). According to one estimate, a one-meter rise in sea level could displace 300 million people around the world (Edgerton and the Natural Resources Defense Council 1991, 70). Thirty of the world's largest cities lie near coasts and are vulnerable to a one-meter rise in ocean level (Augenbraun et al. n.d.). Most of Australia's population lives in coastal cities. In the United States, by 1990 more than 100 million people lived within 50 miles of a coastline.

Marine animals also will be affected by a warming habitat. According to a World Wildlife Fund report, "Scientific evidence strongly suggests that global climate change already is affecting a broad spectrum of marine species and ecosystems, from tropical coral reefs to polar ice-edge communities" (Mathews-Amos and Berntson 1999). The level of the world's seas and oceans have been rising slowly for much of the 20th century, a millimeter or two a year, enough to produce noticeable erosion on 70 percent of the world's sandy beaches, including 90 percent of sandy beaches in the United States (Edgerton 1991, 18). In some areas, such as the U.S. Gulf Coast between New Orleans and Houston, the withdrawal of underground water and oil is causing some areas to sink as the oceans rise.

In 2015, a team of scientists from several nations described ocean extinctions during the last 23 million years (Finnegan et al. 2015). Marine mammals have suffered a much higher rate of extinction than others, most notably those with low birthrates and limited ranges.

> Mapping the geographic distribution of these genera identifies coastal biogeographic provinces where fauna with high intrinsic risk are strongly affected by human activity or climate change. . . . Such regions are disproportionately in the tropics, raising the possibility that these ecosystems may be particularly vulnerable to future extinctions. Intrinsic risk provides a pre-human baseline for considering current threats to marine biodiversity. (Finnegan et al. 2015)

Expected Sea Level Rise

A large number of the people who could be displaced by modest sea level rise are residents of broad river deltas such as the Mississippi near New Orleans, the Nile in Egypt, and the Ganges, which joins with several other rivers in Bangladesh. Half of Bangladesh itself is less than 5 meters above sea level. A one-meter sea level rise in Bangladesh would displace some 10 percent to 20 percent of the human population and cost at least 6 percent of the nation's gross national product (GNP). An ocean-level rise of three meters would affect 27 percent of the people there and cost roughly 15 percent of GNP (Edgerton 1991, 74).

According to one observer, "Global change within the next century will be particularly hard-felt in densely populated delta areas like Bangladesh, where a substantial area is barely above sea level and vulnerable to frequent natural calamities like droughts, tropical cyclones, and tornadoes" (Mahtab 1992, 28). Mahtab projects that Bangladesh's population (115 million around 1990 and 145 million in 2000) will be 232 million in 2030. A sea level rise of one meter would put 16 percent of the country under water (Mahtab 1992, 35).

Many years from now, the oceans also could become the staging area for mass injections of greenhouse gases into the atmosphere via the so-called methane burp (or clathrate gun), which involves warming and eventual gasification of methane hydrates from the ocean floor. Scientists are studying episodes far in the past when such injections played a major role in quickly accelerating greenhouse warming.

Thus the oceans, which may be out of sight for most of humanity, should not be out of mind for anyone who is concerned with a sustainable future.

Further Reading

Archer, David. *The Long Thaw: How Humans Are Changing the Next 100,000 Years of Earth's Climate*. Princeton, NJ: Princeton University Press, 2009.

Augenbraun, Harvey, Elaine Matthews, and David Sarma. "The Greenhouse Effect, Greenhouse Gases, and Global Warming." No date. http://icp.giss.nasa.gov/research/methane/greenhouse.html (no longer available).

Caldeira, Ken, and Philip B. Duffy. "The Role of the Southern Ocean in the Uptake and Storage of Anthropogenic Carbon Dioxide." *Science* 287 (January 28, 2000): 620–622.

Caldeira, Ken, and Michael E. Wickett. "Oceanography: Anthropogenic Carbon and Ocean pH." *Nature* 425 (September 25, 2003): 365.

Edgerton, Lynne T., and the Natural Resources Defense Council. *The Rising Tide: Global Warming and World Sea Levels.* Washington, DC: Island Press, 1991.

Finnegan, Seth, et al. "Paleontological Baselines for Evaluating Extinction Risk in the Modern Oceans." *Science* 348 (May 1, 2015): 567–570.

Gleckler, Peter J., et al. "Industrial-Era Global Ocean Heat Uptake Doubles in Recent Decades." *Nature Climate Change*, January 18, 2016. doi: 10.1038/nclimate2915.

Holland, Jennifer S. "The Acid Threat: As CO_2 Rises, Shelled Animals May Perish." *National Geographic*, November 2001, 110–111.

"Livermore Scientists Find Global Ocean Warming Has Doubled in Recent Decades." Lawrence Livermore National Laboratory, January 18, 2016. https://www.llnl.gov/news/livermore-scientists-find-global-ocean-warming-has-doubled-recent-decades.

"Lotion in the Ocean." *National Geographic*, June 2015, 10.

Mahtab, Fasih Uddin. "The Delta Regions and Global Warming: Impact and Response Strategies for Bangladesh." Pp. 28–43 in Jurgan Schmandt and Judith Clarkson, eds., *The Regions and Global Warming: Impacts and Response Strategies.* New York: Oxford University Press, 1992.

Mathews-Amos, Amy, and Ewann A. Berntson. "Turning Up the Heat: How Global Warming Threatens Life in the Sea." World Wildlife Fund and Marine Conservation Biology Institute, 1999. http://www.worldwildlife.org/news/pubs/wwf_ocean.htm (no longer available).

Pilkey, Orrin H., Linda Pilkey-Jarvis, and Keith C. Pilkey. *Retreat from a Rising Sea: Hard Choices in an Age of Climate Change.* New York: Columbia University Press, 2016.

Shi, Dalin, Yan Xu, Brian M. Hopkinson, and François M. M. Morel. "Effect of Ocean Acidification on Iron Availability to Marine Phytoplankton." *Science* 327 (February 5, 2010): 676–679.

Spinrad, Richard W., and Ian Boyd. "Our Deadened, Carbon-Soaked Seas." *The New York Times*, October 15, 2015. http://www.nytimes.com/2015/10/16/opinion/our-deadened-carbon-soaked-seas.html.

Sunda, William G. "Iron and the Carbon Pump." *Science* 327 (February 5, 2010): 654–655.

CORAL REEFS

When an aquatic environment nurtures coral reefs, they are among the most productive and diverse ecosystems on Earth. Coral reefs are sometimes called the "rain forests of the sea." These reefs cover much less than 1 percent of the world's ocean floors but host more than a third of the marine species currently described by science, with many species remaining undocumented. Some of these organisms may provide new sources of anticancer compounds and other medicines. Coral reefs also protect shorelines from erosion by acting as breakwaters that can repair themselves when they are healthy.

El Niño of 2016 Intensifies Coral Devastation

Record high temperatures from 2014 to 2017 on both land and in the world's oceans intensified coral damage, threatening a third or more of reefs with eradication. Kim Cobb, a marine scientist at the Georgia Institute of Technology, was stunned by damage off of Kiritimati Island near the center of the Pacific Ocean. "The entire reef is covered with a red-brown fuzz," Cobb said. "It is otherworldly. It is algae that has grown over dead coral. It was devastating" (Innis 2016). Around Kiritimati, within a broad swath at the center of the El Niño oscillation, water temperatures rose from the usual 78°F to 88°F.

Corals (coccolithophores) support fish that feed more than a billion people, including many millions in island nations such as the Philippines and Indonesia. The largest bleaching was occurring in Australia's Great Barrier Reef, where two-thirds of northern reefs sustained significant bleaching (Innis 2016). Damage to reefs worldwide exceeded that of the El Niños of 1998 and 2010. "We are currently experiencing the longest global coral bleaching event ever observed," said C. Mark Eakin, the coral reef watch coordinator at the National Oceanic and Atmospheric Administration (NOAA) office in Maryland. "We are going to lose a lot of the world's reefs during this event" (Innis 2016). Reefs that took several centuries to grow were being destroyed in a matter of weeks.

Viability of Reefs Endangered

A worldwide survey of 845 zooxanthellate reef-building coral species using the criteria of the Red List of Endangered Species published by the International Union for Conservation of Nature in 2008 indicated that the "viability of the world's major coral reefs is endangered both by direct human disturbance and by disease and bleaching events brought on by climate change." Of the 704 species that could be assigned conservation status, 328 were listed in "elevated risk of extinction" categories. According to this study,

> [the] proportion of corals threatened with extinction has increased dramatically in recent decades and exceeds that of most terrestrial groups. The Caribbean has the largest proportion of corals in high extinction risk categories, whereas the Coral Triangle (western Pacific) has the highest proportion of species in all categories of elevated extinction risk. Our results emphasize the widespread plight of coral reefs and the urgent need to enact conservation measures. (Carpenter et al. 2008, 560)

The World Atlas of Coral Reefs

The threats to coral reefs are similar around the world. More than 80 percent of Indonesia's coral reefs, for example, have been primarily threatened by blast-fishing practices and bleaching, according to the *World Atlas of Coral Reefs* published by United Nations Environment Program (UNEP). Indonesia, along with the Philippines, Malaysia, and Papua New Guinea each have between 500 and 600 species of coral per country and are home to the world's most diverse range of corals ("Over 80 Percent" 2001).

According to Klaus Toepfer, a staff member at UNEP,

> Our new atlas clearly shows that coral reefs are under assault. They are rapidly being degraded by human activities. They are overfished, bombed, and poisoned. They are smothered by sediment, and choked by algae growing on nutrient rich sewage and fertilizer runoff. They are damaged by irresponsible tourism and are being severely stressed by the warming of the world's oceans. Each of these pressures is bad enough in itself, but together, the cocktail is proving lethal. ("Over 80 Percent" 2001)

The *World Atlas of Coral Reefs* provides detailed descriptions of the coral reefs in every country that has such reefs. As Mark Spalding, lead author of the atlas, wrote, "in every country the same problems are present" (Spalding 2001).

After traveling in the Indian Ocean, Mark Spalding, lead author of the *World Atlas of Coral Reefs* (Spalding 2001) wrote the following in *The Guardian* (U.K.):

> Over the next six weeks we watched the corals of the Seychelles die. Corals are to reefs what trees are to forests. They build the structure around which other communities exist. As the corals died they remained *in situ* and the reefs became, to us, graveyards. Fine algae grows over a dead coral within days, and so the reefs took on a brownish hue, cob-webbed. In fact, the fish still teemed and in many ways it still appeared to be business as usual, but as we traveled—over 1,500 kilometers across the Seychelles—the scale of this disaster began to sink in. Everywhere we went was the same, and virtually all the coral was dying or already dead. . . . What I witnessed in the Seychelles was repeated in the Maldives and the Chagos Archipelago. In these Indian Ocean islands alone, 80 to 90 percent of all the coral died. (Spalding 2001)

Further Reading

"Over 80 Percent of Indonesia's Coral Reefs under Threat." *Jakarta Post*, September 13, 2001, n.p. (LEXIS).

Spalding, Mark. "Coral Grief: Rising Temperatures, Pollution, Tourism, and Fishing Have All Helped to Kill Vast Stretches of Reef in the Indian Ocean. Yet, with Simple Management, Says Mark Spalding, the Marine Life Can Recover." *The Guardian* (U.K.), September 12, 2001, 8.

In late 2008, the Global Coral Reef Monitoring Network estimated that one-fifth of the world's coral reefs had died. The same group said that current trends may kill many of the remaining reefs within 40 years. "If nothing is done to substantially cut emissions, we could effectively lose coral reefs as we know them, with major coral extinctions," said Clive Wilkinson, the organization's coordinator ("Fifth of World's" 2008).

Writing in *Science*, Roberts et al. (2002) sketched the scope of possible extinctions faced by the world's coral reefs:

Analyses of the geographic ranges of 3,235 species of reef fish, corals, snails, and lobsters revealed that between 7.2 percent and 53.6 percent of each taxon have highly restricted ranges, rendering them vulnerable to extinction. . . . The 10 richest centers of endemism cover 15.8 percent of the world's coral reefs (0.012 percent of the oceans) but include between 44.8 and 54.2 percent of the restricted-range species. Many occur in regions where reefs are being severely affected by people, potentially leading to numerous extinctions. (Roberts et al. 2002, 1280)

A study involving more than 5,000 scientists and divers who monitored coral reefs adjacent to 55 countries for five years found only one reef out of more than 1,100 that was judged to be in near-pristine condition. "Coral reefs have suffered more damage over the last 20 years than they have in the last one thousand," said Gregor Hodgson, author of the report and a University of California–Los Angeles visiting professor who heads Reef Check, a monitoring program based at UCLA's Institute of the Environment. "It is the rate of decline and the global extent of the damage that is so alarming, with species reasonably abundant 30 years ago now on the verge of extinction" (Hirsch 2002). Hodgson said the Reef Check project is designed to be an early warning system and that he hopes other scientists will examine the issues raised in his report. "Although it is a big study, we have looked at only a tiny fraction of the world's reefs," he said (Hirsch 2002).

Massive Bleaching in the Indian Ocean

Some of the most severe bleaching of corals (coccolithophores) has taken place in the tropical Indian Ocean, where water temperatures during the 1998 El Niño reportedly rose to between 3 and 5°C above long-term averages in some areas. These patterns were repeated during the severe El Niño of 2015–2016. According to Clive Wilkinson et al., writing in the Swedish environmental journal *Ambio*,

Massive mortality occurred on the reefs of Sri Lanka, [the] Maldives, India, Kenya, Tanzania, and [the] Seychelles, with moralities of up to 90 percent in many shallow areas. . . . Coral death during 1998 was unprecedented in severity. . . . Coral reefs of the Indian Ocean may prove to be an important signal of the potential effects of global climate change, and we should heed that warning. (Wilkinson et al. 1999, 188)

According to another observer, "In some parts of the Indian Ocean . . . reefs in the Maldives, Sri Lanka, Kenya, and Tanzania were devastated with shallow reefs looking like graveyards" (Perry 1998).

In the Asia–Pacific region, the worst-hit coral reefs during the same El Niño episode were near Japan, Taiwan, the Philippines, Vietnam, Thailand, Singapore, Indonesia, and the islands of Palau. A statement by scientists released simultaneously in the United States and Australia relied heavily on new satellite data from the United States' National Oceanic and Atmospheric Administration (NOAA) that indicated a degree of warming never anticipated, especially in the tropics. During May 1998, marine scientists said Australia's Great Barrier Reef (the world's largest living reef stretching 1,300 miles in length) experienced its worst case of coral bleaching in recorded history. Australian marine scientists said bleaching had hit more than 60 percent of its 3,000 coral reefs, and aerial surveys showed that 88 percent of inshore reefs were bleached, with 25 percent severely bleached (Perry 1998).

By 2000, some coral reefs in the Indian and Pacific oceans had recovered somewhat from the rapid bleachings of 1998 as water temperatures cooled under La Niña conditions. Scientific studies indicated that some young corals in these areas had survived bleaching and were helping to begin rebuilding some of the reefs. Terry Done, a senior research scientist at the Australian Institute of Marine Science, said that reefs may be "more resilient than we had thought." According to a report in *Science*, Done said that many corals may "not be able to mature and recover from the repeated bleaching forecast to accompany global warming" (Normile 2000, 941). Done is well acquainted with some of the Indian Ocean corals that were characterized as looking like graveyards after the 1998 bleaching (Normile 2000, 942). He noted that most coral reefs require at least a decade to recover from bleaching. "This recovery won't do the reefs much good," Done said. "They'll no sooner get one or two years old before they'll be wiped out again" (Normile 2000, 942).

Caribbean Coral Catastrophe

Four-fifths of the corals on Caribbean reefs have died at various times since 1990. Human provocations, which include but are not limited to warming seas, are responsible for most of the destruction. The scope of the loss has astonished even scientists who have been studying the global decline of the corals. The scope of destruction has been unmatched for several thousand years, according to a study in the journal *Science* (McCarthy 2003; Gardner et al. 2003). The team leader, Isabelle Cote, a French-Canadian specialist in tropical marine ecology, said that causes of coral declines include industrial, agricultural, and other human pollution; overfishing; diseases; stronger storms; and higher sea temperatures (McCarthy 2003).

The work was carried out by researchers at the University of East Anglia, United Kingdom, and its associated Tyndall Center for Climate Change Research, using data from 263 Caribbean sites, including Mexico, Barbados, Cuba, Panama, the Florida Keys, and Venezuela. "We report a massive region-wide decline of corals across the entire Caribbean basin," the researchers wrote in *Science* (Gardner et al. 2003). The reefs of the Caribbean, seriously weakened by human depredation, may now be unable to withstand future warming. "The ability of Caribbean coral reefs to cope with future local and global environmental change may be irretrievably compromised," the team reported (McCarthy 2003).

The study involved hard corals (coccolithophores), the tiny animals that slowly build coral reefs from the calcium carbonate they excrete. The scientists found that in 1977, the start of the survey period, a typical Caribbean reef was 50 percent covered in live corals, which is considered a healthy reef. By 2002, however, a typical reef was 10 percent covered, which is regarded as potentially fatal (McCarthy 2003). "The end result surprised us, as well as all the people who gave us data," said Cote. "The rate of decline we found exceeds by far the well-publicized rates of loss for tropical forests" (McCarthy 2003; Gardner et al. 2003).

"Coral reefs in a broad swath of the Caribbean face a substantial risk of severe bleaching and die-offs through October [2009] the National Oceanic and Atmospheric Administration said on Wednesday in its latest Coral Reef Watch report." Similar conditions may develop in the southern Gulf of Mexico and central Pacific, the agency said. However, the report said that "the widest area of high risk was in the southern Caribbean, from Nicaragua's east coast across the south coasts of Haiti and the Dominican Republic and from Puerto Rico south along the Lesser Antilles. Rising ocean temperatures are contributing to the risk, the report said, noting that the National Climatic Data Center reported that in June the world's ocean surface temperature was the warmest on record" (Revkin 2009). That temperature record was exceeded in 2014 and again during the 2015–2016 El Niño.

Quoting from this report, Andrew C. Revkin reported on *The New York Times* Web page Dot.earth: "Scientists are concerned that bleaching may reach the same levels or exceed those recorded in 2005, the worst coral bleaching and disease year in Caribbean history. In parts of the eastern Caribbean, as much as 90 percent of corals bleached and over half of those died during that event. . . . The forecast said there was substantial risk of bleaching in parts of the Pacific Ocean, as well, and noted that this did not include the extra heat anticipated from a developing El Niño warming of the tropical Pacific" (Revkin 2009).

In 2010, many coral reefs in the Indian Ocean and near Southeast Asia's coasts were suffering from extensive bleaching because of rising water temperatures, the worst since a spike in coral mortality caused by a sharp El Niño in 1998. Although some of these areas later cooled and may recover, other areas warmed, including the western Pacific and the Caribbean, where NOAA issued an alert for dangerously warm water. Both sides of the Thai peninsula suffered from 80 percent to 100 percent mortality, for example. Many corals that bleached in 2010 were beginning to recover from the wave of bleachings in 1998. Large-scale bleaching was once confined to El Niño years but is now occurring during other weather patterns as well (Normile 2010).

Coral Bleaching in the Red Sea

Rising summertime sea surface temperatures have been slowing the growth rate of healthy corals in the Red Sea. According to a report by Neal E. Cantin and colleagues in *Science* (2010, 322–325):

Sea surface temperature (SST) across much of the tropics has increased by 0.4 to 1°C since the mid-1970s. A parallel increase in the frequency and extent of coral bleaching and mortality has fueled concern that climate change poses a major threat to the survival of coral reef ecosystems worldwide. Here we show that steadily rising SSTs, not ocean acidification, are already driving dramatic changes in the growth of an important reef-building coral in the central Red Sea. Three-dimensional computed tomography analyses of the massive coral *Diploastrea heliopora* reveal that skeletal growth of apparently healthy colonies has declined by 30 percent since 1998. The same corals responded to a short-lived warm event in 1941–1942, but recovered within three years as the ocean cooled. Combining our data with climate model simulations by the Intergovernmental Panel on Climate Change, we predict that should the current warming trend continue, this coral could cease growing altogether by 2070. (Cantin et al. 2010, 322)

The Condition of Coral Reefs Near Fiji

British botanist David Bellamy has warned that many of Fiji's spectacular coral reefs are being ruined as a result of bleaching caused at least partially by global warming. Bellamy visited Fiji during 2001 to inaugurate a project by a British charity to survey reefs in the Mamanuca Islands off the west coast of the main island of Viti Levu. Coral Cay Conservation has been invited by members of the tourism industry in Fiji to undertake a three-month pilot project. "Hoteliers contacted us saying they wanted to find out what had gone wrong, because without reefs there's nothing to attract high-spending divers," said Bellamy (Miles 2001). In addition to bleaching, reefs near 13 Mamanuca resorts have been threatened by overfishing, which results in an imbalance in the ecosystem, Bellamy said. Andrea Dehm, manager of Ovalau Watersports on Viti Levu, said the bleaching had caused diving in the area to become increasingly disappointing. "Our divers used to comment on the beautiful colors, but now it's all white," she said. "People don't want to see that and some operators are closing" (Miles 2001).

Coral Reefs "on the Edge of Disaster"

As global temperatures rose rapidly in 2015 and 2016, with an especially strong El Niño, corals worldwide suffered from major bleaching, as they had twice before on the instrumental record (1998 and 2010). As the *Washington Post* reported, "According to NOAA, 95 percent of all U.S. coral reefs are expected to see ocean temperatures that can lead to bleaching sometime this year [2015]. Of those areas, . . . 60 percent are expected to be hit with severe thermal stress and we're going to see a lot of corals dying." The 2015–2016 bleaching was the longest and most severe. Many of the corals are centuries old; the few that might recover require decades.

Descriptions of the world oceans' coral holocaust (occurring and impending) have been widespread in the scientific literature, peaking, with waves of bleaching, in and shortly after severe El Niño years such as 1997–1998 and 2015–2016. The journal *Science*, for example, devoted a cover story to the subject on August 15, 2003, in which T. P. Hughes and colleagues concluded, "The diversity, frequency, and scale of human impacts on coral reefs are increasing to the extent that reefs are threatened globally" (Hughes et al. 2003, 929). Prominent among these impacts is anthropogenic warming of the atmosphere and oceans that may exceed limits under which corals have flourished for half a million years. Some types of corals are more vulnerable to warming than others, however, so "reefs will change rather than disappear entirely" (Hughes et al. 2003, 929). Rising ocean temperatures, however, will certainly reduce biological diversity among corals.

The limits of acidity tolerance for many corals is now expected around 550 parts per million carbon dioxide, a level that will be reached at current rates of accretion in the middle of the 21st century. As CO_2 rises above that level, the oceans will encounter a "reef gap" or a point "when these magnificent, diverse structures disappear from the world" (Zalasiewicz and Williams 2016, 189). The last reef gap occurred 55 million years ago during the Paleocene–Eocene Thermal Maximum. The corals did eventually recover after several million years. Life will continue during such a gap because coral reefs "will be replaced by 'slime-rock' systems dominated by algal and microbial mats and jellyfish" (Zalasiewicz and Williams 2016, 191).

Clive Wilkinson, representing the Australian Institute of Marine Science, said during 2000 that 27 percent of reefs worldwide, crucial nurseries for fish and plants, had died or were in serious trouble following record sea temperatures in 1998. Another 25 percent will die during the next 25 years if existing trends continued, he said. Wilkinson, the coordinator of the Global Coral Reef Monitoring Network, cited slight cause for optimism, however, thanks to the new priority given to coral conservation by some governments. "Until recently, governments were in denial. That's no longer the case but we have a long way to go to reverse the trend," Wilkins said. However, the "climate-change models suggest things are going to get a lot worse for corals in most regions" (Nuttall 2000).

Although individual governments may be able to curtail destructive fishing practices and pollution, the gradual and pervasive warming of the oceans that kills corals is another matter. Worldwide, coral bleaching accelerated in 2002 at its worst pace since 1998, paralleling a rise in global temperatures. The year 2002 became the second worst year for bleaching after the major El Niño year of 1998. More than 430 cases of coral bleaching were documented that year, notably from the Great Barrier Reef in Australia as well as from reefs in countries such as the Philippines, Indonesia, Malaysia, Japan, Palau, Maldives, Tanzania, Seychelles, Belize, Ecuador, and the Florida coast of the United States ("New Wave" 2002).

Reports from the World Fish Center and the International Coral Reef Action Network have all documented the increasing toll on the world's corals. Coral bleaching has become endemic in the warmer regions around the world. A survey of reefs in the Seychelles in January 1999, for example, showed that 80 percent of the coral was dead, with more than 95 percent mortality in some areas (Leggett 2001). Reef-Base (available at http://www.reefbase.org/main.aspx), includes more than 3,800 records going back to 1963 that include information on the severity of bleaching.

"Zombie Ecosystems" Past and Present

Roger Bradbury, a researcher in resource management at Australian National University, calls today's corals "zombie ecosystems, neither dead nor truly alive in any functional sense, and on a trajectory to collapse within a human generation. There will be remnants here and there, but the global coral reef ecosystem with its storehouse of biodiversity and fisheries supporting millions of the world's poor—will cease to be" (Bradbury 2012). The causes of the demise of corals are many, with climate change (water warming to beyond many corals' tolerance levels) only one; overfishing, acidification, and pollution are also key causes. One thing that corals find especially difficult to survive (or adapt to) is acidity, a consequence of rising carbon dioxide levels in the oceans ("Acidic Waters" 2013). Even after several generations, overly acid ocean water can still impede the ability of corals to reproduce.

Lauren T. Toth and colleagues wrote in *Science* (2012) that coral reefs have collapsed before without any assistance from human pollution, greenhouse gas emissions, and overfishing. Approximately 4,000 years ago, increased variability of the El Niño–Southern Oscillation (ENSO) provoked a 2,500-year collapse:

> Cores of coral reef frameworks along an upwelling gradient in Panamá show that reef ecosystems in the tropical eastern Pacific collapsed for 2,500 years, representing as much as 40 percent of their history, beginning about 4,000 years ago. The principal cause of this millennial-scale hiatus in reef growth was increased variability of the [ENSO] and its coupling with the Inter-tropical Convergence Zone. The hiatus was a Pacific-wide phenomenon with an underlying climatology similar to probable scenarios for the next century. Global climate change is probably driving eastern Pacific reefs toward another regional collapse. (Toth et al. 2012)

By 2010, more than 75 percent of coral reefs worldwide were imperiled by warming water and pollution. These threats were expected to put 90 percent of reefs at risk by 2020 and all of them by 2050, according to *Reefs at Risk: Revisited* (Burke et al. 2011). NOAA administrator Dr. Jane Lubchenco said, "This report serves as a wake-up call for policy-makers, business leaders, ocean managers, and others about the urgent need for greater protection for coral reefs. . . . As the report makes clear, local and global threats, including climate change, are already having significant impacts on coral reefs, putting the future of these beautiful and valuable ecosystems at risk" ("Warming, Polluted" 2011). Nine countries—Haiti, Grenada, the Philippines, Comoros, Vanuatu, Tanzania, Kiribati, Fiji, and Indonesia—are most vulnerable to the effects of coral reef degradation. The report points out that these countries have high ratings for exposure to reef threat and reef dependence, combined with low ratings for adaptive capacity ("Warming, Polluted" 2011).

Warmer sea surface temperatures killed 90 percent of the coral reefs near the surface of the Indian Ocean in only one year, and the remaining 10 percent could die in the next 20 years, according to work by Charles Sheppard of England's University of Warwick. During 1998, according to Sheppard, rapid warming devastated shallow-water corals down to 130 feet below the surface. Some of these corals began to recover in subsequent years, but the risk continues, according to Sheppard. "It's like a forest. If you kill off 90 percent, there might be just enough left to sustain some life around it, such as squirrels and so on. But if you have the same impact again and again there's no clear line as to when it's alive or not as a forest" (Arthur 2003).

In research published by *Nature*, Sheppard used a computer model for global warming developed by Britain's Hadley Center for Climate Prediction and Research to anticipate how sea temperatures will affect coral survival (Sheppard 2003). "Most corals don't mature until [they are] five years old and, five years since the 1998 event, most sites have recovered only marginally," Sheppard said (Arthur 2003). According to Sheppard's analysis, a proportion of any given coral dies with the annual peak in sea temperature. "The warming trend is only a fraction of a degree each year, which is swamped by the annual change," Sheppard said. "But south of the Equator the probability of the temperature rising enough to kill all the corals means they could be wiped out as soon as 2020" (Arthur 2003).

The central barrier reef near Belize suffered a mass coral die-off following record high water temperatures in 1998. According to Richard Aronson, a marine biologist at the Dauphin Island Sea Laboratory in Alabama, this die-off represents the first complete collapse of a coral reef in the Caribbean Sea from bleaching (Connor 2000). Working with colleagues at the Smithsonian Institution in Washington, D.C., Aronson said in the journal *Nature* that lettuce coral was the most abundant species living in the area until 1998. Later surveys showed that "virtually all living colonies had been bleached white at almost every depth investigated" (Conner 2000). Core samples taken from the corals indicate that such a mass bleaching is unprecedented for at least the last 3,000 years. According to the research team, "Complete bleaching was also evident in almost all . . . species down to the lagoon floor at 21 meters" (Aronson et al. 2000).

The World Wildlife Fund sketched the worldwide nature of coral bleaching caused by rising ocean temperatures:

> According to NOAA, coral bleaching was reported throughout the Indian Ocean and Caribbean in 1998, and throughout the Pacific, including Mexico, Panama, Galapagos, Papua New Guinea, American Samoa, and Australia's Great Barrier Reef starting in 1997. . . . The severity and extent of coral bleaching in 1997–98 was widely acknowledged among coral reef scientists as unprecedented in recorded history. (Mathews-Amos and Berntson 1999)

The International Society for Reef Studies concluded that 1997 and 1998 had witnessed the most geographically widespread coral bleaching in the recorded histories of at least 32 countries and island nations. Reports of bleaching came from sites in every major tropical ocean in the world, some for the first time in recorded history. The 1997–1998 bleaching episode was exceptionally severe because a large number of corals died. According to one study, parts of Australia's Great Barrier Reef have been so severely affected that many of the usually robust corals, including one dated as being more than 700 years old, were badly damaged or had died during 1997 and 1998 (Mathews-Amos and Berntson 1999).

A Grim Future for the Reefs?

Terry Hughes at James Cook University in Townsville, Australia, and colleagues hold out some hope that even though climate change will alter species composition of many coral reefs, it will not completely wipe them out. Hughes and his team collected samples at 132 sites in Australia's Great Barrier Reef. "Of the 12 coral taxa sampled," they reported, "11 showed significant differences in abundance across the reef, regardless of how susceptible they were to thermal stress and bleaching. These differences in abundance did not follow changes in latitude or temperature. This flexibility may enable coral reefs to continue functioning as the environment alters with climate change" ("Can Coral Cope" 2012).

Although some isolated successes have been reported in the natural ability of corals to cope, the problems of overfishing, pollution, increasing temperatures, and ocean acidification are increasing all around the world. "They are growing broadly in line with global economic growth, so they can double in size every couple of decades," wrote Bradbury. "They have extreme inertia—there is no real prospect of changing their trajectories in less than 20 to 50 years. In short, these forces are unstoppable and irreversible" (Bradbury 2012). This is an ominous prospect for people who number in the hundreds of millions, mainly in tropical countries with many islands and long coastlines (such as Indonesia and the Philippines) and whose diets consist largely of fish that depend on the health of coral reefs to survive.

Bradbury sketches a grim future for the reefs and the ocean ecosystem generally:

> What we will be left with is an algal-dominated hard ocean bottom, as the remains of the limestone reefs slowly break up, with lots of microbial life soaking up the sun's energy by photosynthesis, few fish but lots of jellyfish grazing on the microbes. It will be slimy and and look a lot like the ecosystems of the Precambrian era, which ended more than 500 million years ago and well before fish evolved. (Bradbury 2012)

"Mortal Damage" by 2050 to 2100?

Coral reefs are extremely intolerant of small rises in water temperature that destroy the ecosystem of the reef, leaving only a lifeless, bleached exoskeleton. By the early 1990s, an increasing number of Earth's coral reefs were turning white, a sign that they had died from excessive heat. Most corals die at levels above 30 to 32°C. Bleaching results when symbiotic zooxanthellae algae are expelled from coral reefs. The algae provide reefs with most of their color, carbon, and ability to deposit limestone. Once bleached, the coral does not receive adequate nutrients or oxygen. The reefs may recover if temperatures fall, but a large part of any given reef will die if thermal stress continues.

Pollution, disease, ultraviolet radiation, too much shade, and changes in the ocean's salinity also may contribute to coral bleaching. Warming seas are one among several human-induced changes that threaten coral reefs. Other problems include overfishing, coastal development, nutrient runoff from agriculture and sewage, and sedimentation from logging in watersheds and on streams that feed into coastal waters. By one estimate, 58 percent of the world's coral reefs were being threatened by human activity by the late 1990s (Bryant et al. 1998).

The scope of devastation to corals from climate change and other human impacts rivals the losses endured by the flora and fauna of the world's great rain forests. According to several estimates, half of the world's coral reefs may be lost by 2025 unless urgent action is taken to save them from the ravages of pollution, dynamite fishing, and warming waters. Many of the coral reefs that are falling prey to human-induced destruction are among the largest living structures on Earth. Many are more than 100 million years old. Coral that has lost its ability to sustain plant and animal life turns white as if it has been doused in bleach. Afterward, the dead coral is often enveloped in a choking shroud of gray algae. Because coral polyps and their calcium carbonate skeletons "are the foundation of the entire ecosystem, fish, mollusks, and countless other species, unable to survive in this colorless graveyard, rapidly disappear, too" (Lynas 2004, 107).

Aside from the obvious ravages of fishermen who blast the reefs and pour cyanide on them, the reefs are also threatened by ocean temperature spikes that many marine biologists attribute to global warming and short-term climate phenomena such as El Niño. Most corals live extremely close to the upper limit of their temperature tolerances. Temperature rises of only a few degrees over a sustained period kill the symbiotic microorganisms living within the coral tissue that provide energy for the colony (Hirsch 2002).

Atmospheric carbon dioxide concentration at 500 parts per million and global air temperatures 2°C above levels of the year 2000 are expected between 2050 and 2100 to mortally damage many of the world's coral reefs. These carbon dioxide levels and temperatures "significantly exceed those of at least the past 420,000 years during which most extant marine organisms evolved." Such conditions will acidify the oceans and "compromise carbonate accretion, with corals becoming increasingly rare on reef systems. The result will be less diverse reef communities and carbonate reef structures that fail to be maintained," O. Hoegh-Guldberg and colleagues concluded in a literature survey on threats to corals in *Science* (Hoegh-Guldberg 2007, 1737).

Climate change is one of several human-provoked stresses on corals, including declining water quality and overexploitation of key species that are "driving reefs increasingly toward the tipping point for functional collapse." This array of threats may be the final insult to these ecosystems. A level of carbon dioxide in the atmosphere at 380 parts per million, more than 80 parts per million above any level measured in ice cores—that is, perhaps the highest level in 20 million years—is going to make the demise of many corals (along with "increasingly serious consequences for reef-associated fisheries, tourism, coastal protection, and people," an outcome that will be extremely difficult to avoid without what Hoegh-Guldberg and colleagues call "decisive action on global emissions" (Hoegh-Guldberg 2007, 1737).

Can Corals Recover?

Charles Schmidt wrote in *Science* (2013, 1517) that many corals show signs of resilience in the face of multiple insults:

Around the globe, a growing corps of scientists is searching for resilient reefs and then trying to identify what enables them to resist or bounce back from severe environmental stress. They've found tantalizing hints that heat-resistance genes, the proximity of other reefs, and even the presence of plant-eating sea life that scrub corals free of weedy algae can play a role. But they're also discovering that the factors that promote resilience can vary greatly from reef to reef.

According to another account, corals can recover some of their biodiversity, reversing worldwide declines as evidenced by more frequent, larger, and longer-lasting bleaching events observed in recent decades—"but only if anthropogenic threats can be greatly reduced or eliminated" (Polidoro and Carpenter 2013, 34). Writing in *Science* (2013, 69), James P. Gilmour and colleagues found:

Coral reef recovery from major disturbance is hypothesized to depend on the arrival of propagules from nearby undisturbed reefs. Therefore, reefs isolated by distance or current patterns are thought to be highly vulnerable to catastrophic disturbance. We found that on an isolated reef system in north Western Australia, coral cover increased from 9 percent to 44 percent within 12 years of a coral bleaching event, despite a 94 percent reduction in larval supply for 6 years after the bleaching. The initial increase in coral cover was the result of high rates of growth and survival of remnant colonies, followed by a rapid increase in juvenile recruitment as colonies matured. We show that isolated reefs can recover from major disturbance, and that the benefits of their isolation from chronic anthropogenic pressures can outweigh the costs of limited connectivity.

Some Corals Evolve Heat Tolerance

Rising ocean temperatures may kill many corals, but some are adapting, according to Stanford University scientists. "Corals are certainly threatened by environmental change, but this research, published in the journal *Marine Ecology Progress Series* in March 2009 has really sparked the notion that corals may be tougher than we thought," said Stephen Palumbi, a professor of biology and a senior fellow at Stanford's Woods Institute for the Environment. "The most exciting thing was discovering live, healthy corals on reefs already as hot as the ocean is likely to get 100 years from now," said Palumbi, director of Stanford's Hopkins Marine Station. "How do they do that?" ("Scientists Find" 2009; Oliver and Palumbi 2009, 93). This study found that some corals resist bleaching by hosting heat-resistant algae; some even allow vulnerable algae to die and replace them with tougher ones. In American Samoa, for example, the tropical coral reef marine reserve Ofu Island remains healthy even as the water gradually warms. Other studies describe similar trends amongst corals in Fiji, the Philippines, and Palmyra Atoll (Oliver and Palumbi 2009, 93).

"These findings show that, given enough time, many corals can match hotter environments by hosting heat-resistant symbionts," said Tom Oliver, a postdoctoral researcher at Stanford. "While hopeful, the work also suggests that modern environments are changing so rapidly that corals may not be able to keep up. It comes down to a calculation of the rates of environmental change versus the rates of adaptation" ("Scientists Find" 2009). Corals worldwide also are adapting to increases in acidity, Oliver said, and that corals across the tropics with the ability to switch symbionts will do so to survive.

"What I hope is that we will learn some really deep and interesting things about the cellular and genetic mechanisms that allow this symbiosis to function, and about the mechanisms that come into play when the symbiosis is breaking down under stress," said Stanford genetics professor John Pringle. "The longer-range hope is that having that understanding will contribute to efforts in coral conservation" ("Scientists Find" 2009).

Corals can adapt to a certain amount of gradual warming that precedes bleaching. However, a 2016 report in *Science* found that massive, sudden warming of corals' aquatic habitats overwhelms these defenses, killing the corals on a massive scale. Tracy D. Ainsworth and colleagues, having studied Australia's Great Barrier Reef (GBR), found that the 2015–2016 El Niño, which produced record high temperatures in the oceans as well as atmosphere, disabled the corals' natural bleaching mechanisms. Furthermore, "Near-future increases in local temperature of as little as 0.5°C result in this protective mechanism being lost, which may increase the rate of degradation of the GBR" (Ainsworth et al. 2016, 338).

Is Bleaching a Coral Reef Defense Mechanism?

Andrew C. Baker wrote in *Nature* that bleaching may be a sign of a coral reef's defenses coping with environmental change, including a warming habitat. "Coral bleaching can promote rapid response to environmental change by facilitating compensatory change in algal symbiont communities," he wrote (Baker 2001, 765). Bleaching helps the corals adjust to new conditions: "Reef corals are flexible associations that can switch or shuffle symbiont communities in response to environmental change" (Baker 2001, 765). There are costs—some symbionts that are unsuited to new conditions do die. As described by Baker, "Bleaching is an ecological gamble in that it sacrifices short-term benefits for long-term advantage" (Baker 2001, 766). Coral symbionts are, according to Baker, individually fragile, but the entire reef's flexibility gives it remarkable endurance: "[B]leaching may ultimately help reef corals to survive the recurrent and increasingly severe warming events projected by current climate models" (Baker 2001, 766).

A lively debate has developed in scientific journals over how quickly some corals may adapt to thermal stress (Baker et al. 2004; Rowan 2004). Ove Hoegh-Guldberg and colleagues take issue with Andrew C. Baker's assertions that coral "bleaching favors new host-symbiont combinations that guard populations of corals against rising sea temperature" (Hoegh-Guldberg et al. 2001, 602). They disagreed with Baker, who replied that "far fewer corals in the far-eastern Pacific Ocean died after the 1997–1998 El Niño event (0 to 26 percent) than after the 1982–83 El Niño event (52 to 97 percent) even though the magnitude and duration of sea surface temperature anomaly in the region in 1997–98 exceeded those of 1982–83" (Hoegh-Guldberg et al. 2001, 602).

Possible Rehabilitation of Coral Reefs

Once protected from various human insults, some coral reefs may recover quickly. Coral reefs have been present for 100 million years, and they can be surprisingly resilient. "Ecologists have been surprised at how quickly reef fish populations have rebounded at successful marine parks," said Gregor Hodgson, a University of California–Los Angeles visiting professor who heads Reef Check, a monitoring program based at UCLA's Institute of the Environment. "Two years is enough to replenish many species. Some types of corals also grow very quickly—like corn—but others take decades to reach a large size" (Hirsch 2002).

Proposals have been made to establish networks of marine reserves where fishing is prohibited. By minimizing the stresses of overfishing, the corals may be able to cope with other stresses such as global warming, said researcher Callum Roberts (Radford 2002). Roberts's study asserts that protection of 10 coral reef "hot spots" around the world could save a large number of marine species. The 10 reefs account for 0.017 percent of the oceans' surface area but are home to 34 percent of all species with limited ranges (Radford 2002; Roberts 2002).

Roberts and his colleagues looked at 18 areas with the greatest concentrations of species found nowhere else and selected the 10 most vulnerable. They are in the Philippines, the Gulf of Guinea, the Sunda islands in Indonesia, the southern Mascarene islands in the Indian Ocean, eastern South Africa, the northern Indian Ocean, southern Japan, Taiwan and southern China, the Cape Verde Islands, the western Caribbean, and the Red Sea and Gulf of Aden.

Eight of the 10 reefs that Roberts and colleagues would like to protect are already being dramatically altered by human activity through fishing, logging, and farming. The felling of forests, for example, means that soils are easily eroded, often depositing mud that can choke reefs. Farming releases nutrients that encourage seaweeds to grow where corals once flourished (Radford 2002). "One of the arguments is that there is nothing we can do, it is all going to go to hell, and that coral reefs are doomed. The other argument is that we should work very hard to try and do something about protecting them," Roberts said. "The question then is how? Where are we going to focus our efforts, given that we don't have the resources to do all that we would like? We cannot save all coral reefs everywhere" (Radford 2002).

Some corals not only can adapt to warmer water but also may be able to pass this tolerance on to succeeding generations. According to a report in *Nature,*

Researchers have suggested that corals physiologically acclimatize to higher temperatures rather than inherit heat tolerance. To test this idea, Line Bay at the Australian Institute of Marine Science in Townsville, Australia, Mikhail Matz at the University of Texas at Austin and their team bred corals (*Acropora millepora*) from two locations in Australia separated by 5° of latitude. Offspring produced by parents from the warmer area had an up to 10 times greater chance of survival. ("Corals Inherit" 2015)

The Brave New World of "Smart Coral"

We know we are in the Anthropocene when biologists develop "designer corals" that may be able to survive warmer, more acidic, and polluted ocean waters. They are propagating reefs that thrive in hot water so that they could become the new global norm as the rest fall victim to humankind's appetite for fossil fuels. Scientists now range the world seeking candidate corals such as the *Acropora hyacinthus* near Ofu Island in American Samoa, which can withstand water temperatures of 35°C in shallow pools under direct sunlight. These corals can be interbred with more fragile species that are being killed by warming waters. A fifth of corals have been lost since 1950 worldwide, and another 35 percent are in critical condition (Mascarelli 2014). In areas such as the Caribbean Sea, coral mortality has reached 80 percent, provoking a sense of nearly messianic urgency among developers of designer reefs.

According to an account in *Nature* (Mascarelli 2014, 444), coral designers intend "to launch a program of 'human-assisted evolution,' creating resistant corals in controlled nurseries" that have been hard hit by changing conditions created by humans: warming, acidification, disease, overfishing, and pollution. "It's a brave new world of working with corals in this way," said Ruth Gates, a marine biologist at the University of Hawaii at Manoa who, along with coral geneticist Madeleine van Oppen at the Australian Institute of Marine Science in Townsville, is helping to pioneer the field.

"Smart corals" that can resist humankind's perils are being "outplanted" among vulnerable species and coaxed to take over. Corals that can survive pollution are being sought as well: "Sometimes we find reefs that are doing very, very well in places that you would least expect to find them," said Gates. She pointed to a reef off the coast of Taiwan that lies below the wastewater outfall pipe of a nuclear power plant and experiences temperature fluctuations of between 6 and 8°C per day. "By all of our understanding, we would expect those corals to all be dead. But they're not, they're flourishing" (Mascarelli 2014).

In this brave new world, scientists give evolution a boost to assist survival in a world of hot water, rising ocean acidity, and increasing pollution, "handpick[ing] the hardiest, fastest-growing, and most heat-resistant corals for their smart reef." The wisdom of reordering nature or the necessity to amend conditions that kill weaker natural corals do not seem to be at issue. Paul Allen, cofounder of Microsoft, has given prizes for such work. "Ultimately," wrote Amanda Mascarelli in *Nature*,

> Gates and van Oppen hope to create a seed bank of gametes and fertilized embryos from extreme settings in which corals persist despite the odds—including the shallow reefs skirting Coconut Island, Hawaii, where both temperature and pH fluctuate drastically, reaching upper limits similar to those

expected in the open ocean by 2050. In another decade, before the Anthropocene, this might have been called "playing God." Now, it's simply called biology with a genetic-modification chaser. (Mascarelli 2014)

"Smart" or "super" coral may seem like an unnatural imposition on nature, but Ruth Gates, a pioneer researcher in the field who directs the Hawaii Institute of Marine Biology, says that humanity has been reshaping plants and animals for centuries, breeding food crops, food animals, and other forms of life. "In the food supply, in our pets—everywhere you turn, selectively bred stuff appears," she said. "For some reason, in the framework of conservation—or an ecosystem that would be preserved by conservation—it seems like a radical idea. But it's not like we've invented something new" (Kolbert 2016, 23).

Large-scale diffusion of "super" coral would involve problems of expense and scale that have not been confronted, much less solved. Test projects conducted thus far involve an infinitesimally small portion of the coral reefs that face destruction as oceans warm and become more acidic. Terry Hughes, director of the Australia Research Council's Center of Excellence for Coral Reef Studies, studied all ongoing restoration projects (some 250 as of 2015), and their cost (about a quarter of a billion dollars). The total area covered by all of them was about the size of two football fields. "When you consider just the Great Barrier Reef, which is a tiny fraction of the world's reefs, it has the area of Finland," he said. "So going from a test tube or an aquarium to millions of football fields is hugely expensive, obviously" (Kolbert 2016, 26).

Some Corals Resist Ocean Acidification

Some varieties of coral (notably *Porites cylindrica* in Australia's Heron Island Lagoon and the Great Barrier Reef) resist ocean acidification, according to a study by researchers at the University of Queensland and the University of Western Australia (UWA) and published in the United States in *Proceedings of the National Academy of Sciences* in 2015. This type of coral contains a calcifying fluid at a constant pH level, even as the acid level changes in water around it. "The regulatory mechanism allows the coral to grow at a relatively constant rate, suggesting it may be more resilient to the effects of ocean acidification than previously thought," said lead author Lucy Georgiou of UWA's ARC Center of Excellence for Coral Reef Studies ("Self-Regulating" 2015).

"This newly discovered phenomenon of pH homeostasis during calcification indicates that coral living in highly dynamic environments exert strong physiological controls on the carbonate chemistry of their calcifying fluid, implying a high degree of resilience to ocean acidification within the investigated ranges," the scientists wrote (Georgiou et al. 2015). This study supports the conclusions of earlier work that was also conducted in Australia. Research published in *Nature Climate Change* in 2012 "identified a powerful internal mechanism that could enable some corals and their symbiotic algae to counter the adverse impact of a more acidic ocean" ("Corals 'Could Survive'" 2015).

"The good news is that most corals appear to have this internal ability to buffer rising acidity of seawater and still form good, solid skeletons," said Malcolm McCulloch of the University of Western Australia, the paper's lead author. "Marine organisms that form calcium carbonate skeletons generally produce it in one of two forms, known as aragonite and calcite. . . . The aragonite calcifiers—such as the well-known corals *Porites* and *Acropora*—have molecular 'pumps' that enable them to regulate their internal acid balance, which buffers them from the external changes in seawater pH." "But the picture for coral reefs as a whole isn't quite so straightforward, as the 'glue' that holds coral reefs together—coralline algae—appear to be vulnerable to rising acidity," McCulloch said ("Corals 'Could Survive'" 2015).

"This species-dependent pH-buffering capacity enables aragonitic corals to raise the saturation state of their calcifying medium, thereby increasing calcification rates at little additional energy cost," McCulloch and colleagues wrote (2012). They continued: "Up-regulation of pH, however, is not ubiquitous among calcifying organisms; those lacking this ability are likely to undergo severe declines in calcification as CO_2 levels increase."

Further Reading

"Acidic Waters Do Not Toughen Corals." *Nature* 498 (June 20, 2013): 274–275.

Ainsworth, Tracy D., et al. "Climate Change Disables Coral Bleaching Protection on the Great Barrier Reef." *Science* 532 (April 15, 2016): 338–342.

Aronson, Richard B., et al. "Coral Bleach-Out in Belize." *Nature* 405 (May 4, 2000): 36.

Arthur, Charles. "Temperature Rise Kills 90 Per Cent of Ocean's Surface Coral." *The Independent* (London), September 18, 2003, n.p. (LEXIS).

Baker, Andrew C. "Reef Corals Bleach to Survive Change." *Nature* 411 (June 14, 2001): 765–766.

Baker, Andrew C., et al. "Corals' Adaptive Response to Climate Change." *Nature* 430 (August 12, 2004): 741.

Bradbury, Roger. "A World without Coral Reefs." *The New York Times*, July 13, 2012. http://www.nytimes.com/2012/07/14/opinion/a-world-without-coral-reefs.html.

Bryant, D., et al. *Reefs at Risk: A Map-Based Indicator of Threats to the World's Coral Reefs.* Washington, DC: World Resources Institute, 1998.

Burke, Lauretta, et al. *Reefs at Risk: Revisited.* Washington, DC: World Resources Institute, 2011. http://www.wri.org/sites/default/files/pdf/reefs_at_risk_revisited.pdf.

"Can Coral Cope with Climate Change?" *Nature* 484 (April 19, 2012): 290.

Cantin, Neal E., et al. "Ocean Warming Slows Coral Growth in the Central Red Sea." *Science* 329 (July 16, 2010): 322–325.

Carpenter, Kent E., et al. "One-Third of Reef-Building Corals Face Elevated Extinction Risk from Climate Change and Local Impacts." *Science* 321 (July 25, 2008): 560–563.

Connor, Steve. "Global Warming Is Blamed for First Collapse of a Caribbean Coral Reef." *The Independent* (London), May 4, 2000, 12.

"Corals 'Could Survive a More Acidic Ocean.'" University News. University of Western Australia, April 2, 2015. http://www.news.uwa.edu.au/201204024487/international/corals-could-survive-more-acidic-ocean.

"Corals Inherit Love for Heat." *Nature* 523 (July 2, 2015): 8.

"Fifth of World's Corals Dead." Discovery Channel.com. December 10, 2008. http://dsc.discovery.com/news/2008/12/10/coral-reef-death.html.

Gardner, T. A., et al. "Long-Term Region-Wide Declines in Caribbean Corals." *Science* 301 (July 18, 2003): online edition. http://science.sciencemag.org/content/301/5635/958.

Georgiou, Lucy, et al. "PH Homeostasis during Coral Calcification in a Free Ocean CO_2 Enrichment (FOCE) Experiment, Heron Island Reef Flat, Great Barrier Reef." *Proceedings of the National Academy of Sciences*, October 6, 2015. http://www.pnas.org/content/early/2015/10/01/1505586112.full.pdf?sid=26b9f3e1-4ff7-4247-821f-e806352a8708.

Gilmour, James P., et al. "Recovery of an Isolated Coral Reef System Following Severe Disturbance." *Science* 340 (April 5, 2013): 69–71.

Hirsch, Jerry. "Damage to Coral Reefs Mounts, Study Says." *Los Angeles Times*, August 26, 2002, 14. http://articles.latimes.com/2002/aug/26/science/sci-reef26.

Hoegh-Guldberg, Ove, et al. "Is Coral Bleaching Really Adaptive?" *Nature* 415 (February 7, 2001): 601–602.

Hoegh-Guldberg, O., et al. "Coral Reefs under Rapid Climate Change and Ocean Acidification." *Science* 318 (December 14, 2007): 1737–1742.

Hughes, T. P., et al. "Climate Change, Human Impacts, and the Resilience of Coral Reefs." *Science* 301 (August 15, 2003): 929–933. http://science.sciencemag.org/content/301/5635/929.

Innis, Michelle. "Climate-Related Death of Coral around World Alarms Scientists." *The New York Times*, April 9, 2016. http://www.nytimes.com/2016/04/10/world/asia/climate-related-death-of-coral-around-world-alarms-scientists.html.

Kolbert, Elizabeth. "Unnatural Selection: What Will It Take to Save the World's Reefs and Forests?" *The New Yorker*, April 18, 2016, 22–28.

Leggett, Jeremy. *The Carbon War: Global Warming and the End of the Oil Era.* New York: Routledge, 2001.

Lynas, Mark. *High Tide: The Truth about Our Climate Crisis.* New York: Picador/St. Martins, 2004.

Mascarelli, Amanda. "Climate-Change Adaptation: Designer Reefs." *Nature* 508 (April 24, 2014): 444–446. http://www.nature.com/news/climate-change-adaptation-designer-reefs-1.15073.

Mathews-Amos, Amy, and Ewann A. Berntson. "Turning Up the Heat: How Global Warming Threatens Life in the Sea." World Wildlife Fund and Marine Conservation Biology Institute, 1999. http://www.worldwildlife.org/news/pubs/wwf_ocean.htm.

McCarthy, Michael. "'Rainforests of the Sea' Ravaged; Overfishing and Pollution Kill 80 Per Cent of Coral on Caribbean Reefs." *The Independent* (London), July 18, 2003, 3.

McCulloch, Malcolm, et al. "Coral Resilience to Ocean Acidification and Global Warming through pH Up-regulation." *Nature Climate Change* 2(2012): 623–627. http://www.nature.com/nclimate/journal/v2/n8/full/nclimate1473.html.

Miles, Paul. "Fiji's Coral Reefs Are Being Ruined by Bleaching." *Daily Telegraph* (London), June 2, 2001, 4.

"New Wave of Bleaching Hits Coral Reefs Worldwide." Environment News Service, October 29, 2002. http://ens-news.com/ens/oct2002/2002-10-29-19.asp#anchor1 (no longer available).

Normile, Dennis. "Some Coral Bouncing Back from El Niño." *Science* 288 (May 12, 2000): 941–942.

Normile, Dennis. "Hard Summer for Corals Kindles Fears for Survival of Reefs." *Science* 329 (August 27, 2010): 1001.

Nuttall, Nick. "Coral Reefs 'On the Edge of Disaster.'" London *Times*, October 25, 2000, n.p. (LEXIS).

Oliver, Thomas A., and Stephen R. Palumbi. "Distributions of Stress-Resistant Coral Symbionts Match Environmental Patterns at Local but Not Regional Scales." *Marine Ecology Progress Series*, March 2009, 93–103. doi: 10.3354/meps07871.

Perry, Michael. "Global Warming Devastates World's Coral Reefs." Reuters, November 26, 1998. http://www.gsreport.com/articles/art000023.html (no longer available).

Polidoro, Beth, and Kent Carpenter. "Dynamics of Coral Reef Recovery." *Science* 340 (April 5, 2013): 34–35.

Radford, Tim. "Ten Key Coral Reefs Shelter Much of Sea Life: American Association Scientists Identify Vulnerable Marine 'Hot Spots' with the Richest Biodiversity on Earth." *The Guardian* (London), February 15, 2002, 12.

Revkin, Andrew C. "Caribbean Reefs Face Severe Summer Threat." *The New York Times*, July 22, 2009. http://dotearth.blogs.nytimes.com/2009/07/22/carribean-reefs-face-severe-summer-threat/?hpw.

Roberts, Callum M., et al. "Marine Biodiversity Hotspots and Conservation Priorities for Tropical Reefs." *Science* 295 (February 15, 2002): 1280–1284.

Rowan, Rob. "Thermal Adaptation in Reef Coral Symbionts." *Nature* 430 (August 12, 2004): 742.

Schmidt, Charles. "As Threats to Corals Grow, Hints of Resilience Emerge." *Science* 340 (March 29, 2013): 1517–1519.

"Scientists Find Heat-Tolerant Coral Reefs That May Resist Climate Change." NASA Earth Observatory, May 19, 2009. http://earthobservatory.nasa.gov/Newsroom/view.php?id=38819&src=eoa-manews (no longer available).

"Self-Regulating Coral Protect Themselves against Ocean Acidification." University News. University of Western Australia. October 6, 2015. http://www.news.uwa.edu.au/201510068035/research/self-regulating-coral-protect-themselves-against-ocean-acidification.

Sheppard, Charles R. C. "Predicted Recurrences of Mass Coral Mortality in the Indian Ocean." *Nature* 425 (September 18, 2003): 294–297.

Toth, Lauren T., et al. "ENSO Drove 2500-Year Collapse of Eastern Pacific Coral Reefs." *Science* 337 (July 6, 2012): 81–84.

"Warming, Polluted Oceans Imperil 75 Percent of All Coral Reefs." Environment News Service, February 24, 2011. http://www.ens-newswire.com/ens/feb2011/2011-02-24-02.html.

Wilkinson, Clive, et al. "Ecological and Socioeconomic Impacts of 1998 Coral Mortality in the Indian Ocean: An ENSO Impact and a Warning of Future Change?" *Ambio* 28: 2 (March 1999): 188–196.

Zalasiewicz, Jan, and Mark Williams. *Ocean Worlds: The Story of Seas on Earth and Other Planets.* New York: Oxford University Press, 2016.

See also: El Niño and La Niña; Fisheries; Jellyfish; Ocean Acidity, North America; Ocean Acidity, Worldwide

FISHERIES
Death by Carbon Dioxide

Given a high-emissions scenario in which atmospheric carbon dioxide levels reach 650 parts per million by the end of the 21st century, toxic levels of CO_2 could afflict half the fish in world oceans. This condition is known as *hypercapnia*, an "excessive amount of CO_2 in the blood that typically results in acidosis" ("Ocean Acidification" 2014). Ben I. McNeil and Tristan P. Sasse wrote in *Nature*:

> Our predicted amplification of the annual CO_2 cycle displays distinct global patterns that may expose major fisheries in the Southern, Pacific and North Atlantic oceans to hypercapnia [a toxic level of CO_2 in the blood] many decades earlier than is expected from average atmospheric CO_2 concentrations. We suggest that these ocean "CO_2 hot spots" evolve as a combination of the strong seasonal dynamics of CO_2 concentration and the long-term effective storage of anthropogenic CO_2 in the oceans that lowers the buffer capacity in these regions, causing a nonlinear amplification of CO_2 concentration over the annual cycle. The onset of ocean hypercapnia . . . is forecast for atmospheric CO_2 concentrations that exceed 650 parts per million, with hypercapnia expected in up to half the surface ocean by 2100, assuming a high-emissions scenario. (McNeil and Sasse 2016, 383)

According to the Intergovernmental Panel on Climate Change, global warming may cause some fisheries to collapse and others to expand. As Robert Gough has commented,

> Climate changes can exacerbate the effects of overfishing at a time of inherent instability in world fisheries. In addition, overfishing creates an increased imbalance in the age composition of a stock, and may reduce the resiliency of the population. Further, changes in ocean currents may result in changes in fish population location and abundance and the loss of certain fish populations. (Gough 1999, 47)

New England Cod Fishery Collapses

By 2009, cod, haddock, and winter flounder were becoming scarcer in New England's coastal waters, forcing fishermen to range farther from shore. A United States federal government study said that warmer water (half a degree Fahrenheit is enough to make a difference) was a major cause. The study appeared in the October 30, 2009, issue of the *Marine Ecology Progress* series, with lead author Janet Nye, a fisheries biologist with the National Oceanic and Atmospheric Administration (NOAA) joining two other scientists in analyzing trends in water temperature between North Carolina and Maine's border with Canada between 1968 and 2007. In addition to warming waters, overfishing was also playing a role.

The Gulf of Maine's sea surface temperatures rose faster than anywhere else on Earth from 2003 to 2014. "Over this same 10-year period, fishery managers set quotas that they felt were informed by high-quality science, and the stock just kept going down, down, down," said Andrew Pershing, chief scientist for the Gulf of Maine Research Institute in Portland, Maine (Lavelle 2015). Michael Wines and Jess Boggood wrote in *The New York Times* late in 2014 that cod, once abundant from Massachusetts shores north and east to Canada's Bay of Fundy, have been leaving that area and moving north and eastward in search of their accustomed thermal habitat. Off Massachusetts and the Bay of Fundy, cod have become so scarce that catching the fish has often been banned or severely restricted:

> Cod was once king. It paid for fishermen's boats, fed their families, and put their children through college. In one halcyon year in the mid-1980s, the codfish catch reached 25,000 tons. Today, the cod population has collapsed. Last month [November 2014], regulators effectively banned fishing for six months while they pondered what to do, and next year [2015], fishermen will be allowed to catch just a quarter of what they could before the ban. . . . The Gulf of Maine's waters are warming—faster than almost any ocean waters on Earth, scientists say—and fish are voting with their fins for cooler places to live. That is upending an ecosystem and the fishing industry that depends on it. (Wines and Bidgood 2014)

By 2015, the cod fishery in New England had collapsed, and scientists linked the fishes' demise in that area to rapidly warming water in the Gulf of Maine, which has heated far more quickly than the world ocean as a whole (Pershing et al. 2015). Fisheries managers had cut allowable harvests in the area for several years but these reductions did not stem the cod's decline. The increase in temperatures in the Gulf of Maine "was a magnitude . . . that few ocean ecosystems have ever experienced," said Andrew J. Pershing, lead author of a study on the cod's decline (Goode 2015). Part of the increase was attributed to general warming of ocean water that was compounded by a shift in the Gulf Stream that flows into that area from the southwest. The bounty of the cod had sustained people in the area for centuries.

Further Reading

Goode, Erica. "Cod's Failure to Recover Is Linked to Warming Gulf of Maine." *The New York Times*, October 30, 2015, A18.

Lavelle, Marianne. "Collapse of New England's Iconic Cod Tied to Climate Change." Science Online, October 29, 2015. http://news.sciencemag.org/climate/2015/10/collapse-new-england-s-iconic-cod-tied-climate-change.

Pershing, Andrew J., et al. "Slow Adaptation in the Face of Rapid Warming Leads to Collapse of the Gulf of Maine Cod Fishery." *Science* 350 (November 13, 2015): 745.

Wines, Michael, and Jess Bidgood. "Waters Warm, and Cod Catch Ebbs in Maine." *The New York Times*, December 14, 2014. http://www.nytimes.com/2014/12/15/us/waters-warm-in-gulf-of-maine-and-cod-catch-ebbs.html.

"As the world warms, the only way for wildlife species to live in the temperature they prefer is to move their ranges slowly poleward," according to a study published in the *Journal of the Marine Biological Association* (Connor 2002, 12). "Fish are good indicators of temperature change because they are unable to regulate their temperature independently of the surrounding water. They therefore swim to keep themselves in waters of their preferred temperature range. Not only are changes in fish distribution likely to reflect temperature increases, but the arrival of new fish species are well monitored by fishermen, as well as scientists," Stebbing said (Connor 2002, 12).

A series of U.S. Environmental Protection Agency (EPA) reports issued beginning in 1995 anticipated that several cool-water fish species, especially trout, will diminish as waters warm. This study projects that between eight and 10 U.S. states could lose all of their cool-water sports fisheries within 50 to 60 years. Between 11 and 16 other states could lose half of their cool-water sports fisheries, according to the study (U.S. Environmental Protection Agency 1995, ix).

Several Species Move as Water Warms

Cod, a cold-water fish that is similar in some ways to salmon, has been declining as its habitat has warmed. In mid-July 2000, the World Wildlife Fund (WWF) placed North Sea cod, a staple of British fish and chips (contributing to a $60 million annual fishery), on its endangered species list. Cod stocks in the North Sea have been in nearly continuous decline since the early 1970s, falling below safe biological limits after 1984 (Smith 2001).

Populations of North Sea cod have declined 90 percent in 30 years, according to the WWF (Brown 2000). Cod have been overfished, and their population is falling because they do not breed well in warmer water. In 2000, the North Sea was as much as 3°C warmer than it had been in 1970 (Brown 2000). During the 1930s, British fishermen harvested roughly 300,000 tons of cod annually; by 1999, the catch was down to 80,000 tons ("Fished" 2000). Spawning cod in the North Sea fell from about 277,000 English tons during the early 1970s to 54,700 English tons in 2001. Global warming and predation (overfishing) by humans and seals are major factors in the decline of the cod fishery. Another problem advanced by fishermen is dredging and pipe laying (Smith 2001).

Research by Gregory Beaugrand and colleagues supports the idea that cod are declining in the North Sea not only because of overfishing but also because larval cod feed on plankton, which also have declined in part because of rising temperatures. They have written that "variability in temperature affects larval cod survival; [we conclude] that rising temperature since the mid-1980s has modified the plankton ecosystem in way that reduces the survival of young cod" (Beaugrand et al. 2003, 661). Such observations have been supported by North Sea fishermen, who have been dragging rolls of silk behind their boats for 70 years to monitor the density of plankton populations. Given the plankton's decline, even a complete ban on cod fishing is unlikely to restore the fishery. Cod have declined not only in population but also in size. The peak of plankton abundance also now occurs later in the year after cod larvae have already experienced their greatest need for plankton. This mismatch means that fewer larval cod develop into adults.

Cod is only one of several fish species that are moving north as waters warm. William W. L. Cheung, Reg Watson, and Daniel Pauly at the University of British Columbia (UBC) wrote in *Nature* (2013) that they had developed an index, the mean temperature of the catch (MTC), calculated "from the average inferred temperature preference of exploited species weighted by their annual catch," which "increased at a rate of 0.19 [°C] per decade between 1970 and 2006, and nontropical MTC increased at a rate of 0.23 [°C] per decade." The index, which covers 52 large marine ecosystems, including many of Earth's coastal and shelf areas, "shows that ocean warming has already affected global fisheries in the past four decades, highlighting the immediate need to develop adaptation plans to minimize the effect of such warming on the economy and food security of coastal communities, particularly in tropical regions" (Cheung et al. 2013, 365).

Cheung and UBC colleagues studied 968 species of fish and invertebrates and found that most were moving toward the poles. The research is more confirmation that "global change is real and has been real for a long time," said Boris Worm, a professor of marine biology at Dalhousie University in Halifax, Nova Scotia, who was not part of the study. "It's not something in the distant future. It is well underway" (Bernstein 2013). Migration of sea life poses a danger to people in the tropics. "As the subtropical fish go away because it's too warm for them, you don't have hypertropical fish replacing them," said Pauly, a professor of fisheries (Bernstein 2013).

Tropical Fish and Sharks off Maine

Following warmer waters, a few sharks were sighted off the coast of Maine during the summer of 2002. According to an account in the *Boston Herald*,

> Whether it was blue or mako sharks that invaded the crowded Maine beach, banishing swimmers to the sand, the unusual occurrence is the result of either fish chasing food close to shore or warmer waters luring a more tropical variety to New England waters, marine experts say. The Wells, Maine, shark sightings are "a very unusual circumstance," said Greg Skomol, a shark specialist for the state Division of Marine Fisheries. "We wouldn't expect to see that in Maine or in New England, in general." (Richardson 2002)

> The sharks swam near the beach three days in a row in a popular resort called Vacationland as the tide was receding, probably hoping to catch whatever baitfish were being washed out of nearby crevices in a rocky ledge. The sharks' usual food supply (mackerel and some herring) may have been lured to shallow waters by unusually warm air temperatures in the 90s (Richardson

2002). "We had hot weather and sometimes that will change the distribution of fish and bait fish that sharks feed on," Skomol said. "The high temperatures may have increased the mackerel and herring, therefore bringing in the sharks. But with one event, it's hard to predict anything, though it's worth watching" (Richardson 2002).

Bruce Joule, a marine biologist in Boothbay Harbor who advised officials to close the beach when sharks appeared, said he believes water temperature had nothing to do with the sharks. "I don't think the temperature has really anything to do with it," Joule said. "The sharks came in following bait, whether it was natural migration [of bait fish] or the smaller fish were chasing their own food. It's nature" (Richardson 2002). Following Joule's advice, Wells Fire Chief Marc Bellefeuille closed the beach as a precautionary measure, allowing sunbathers into the water only up to their ankles. "This is the first time anyone can remember shark sightings like this," Bellefeuille said (Richardson 2002).

At about the same time, unusually warm water temperatures also brought tropical fish to New England shores, including yellow fin, dolphin fish, white marlin, and sometimes skipjack tuna, Skomol said. "How close they come depends on the water. They're not aggressive. We've been seeing them the last three years" (Richardson 2002). President George W. Bush was photographed during the summer of 2002 landing a large striped bass that his daughter Jenna had caught on a fishing holiday off the coast of Maine. Striped bass is a warmer-water species and only a generation ago would never have been seen so far north along America's Atlantic coast (Connor 2002, 12). Bush did not indicate whether he realized his catch was unusual or influenced by climate change.

Further Reading

Connor, Steve. "Strangers in the Seas; Exotic Marine Species Are Turning Up Unexpectedly in the Cold Waters of the North Atlantic." *The Independent* (London), August 5, 2002, 12–13.
Richardson, Franci. "Sharks Take the Bait: Experts—Sightings in Maine an 'Unusual Circumstance.'" *Boston Herald*, August 11, 2002, 3.

Outside of the tropics, the Maine shrimp catch has been suspended because the crustaceans have been fleeing the warming waters. Lobster habitats are also moving northward, with catches declining in some areas and rising in others, with black sea bass moving in from the south. Blue crabs, well known in Chesapeake Bay, have been caught in waters near Portland, Maine. Those waters warmed an average of 1°F each 21 years until about 2000. After that, water temperature rises (which vary year to year like air temperatures) have risen an average of 1°F every two years (Wines and Bidgood 2014). "What we're experiencing is a warming that very few ocean ecosystems have ever experienced," said Andrew J. Pershing, Gulf of Maine

Research Institute chief scientific officer. Fish and shellfish try to maintain constant temperatures not only for comfort but also because water temperature often influences their ability to reproduce and thus survive. Fish also may seek colder water by descending to lower depths.

Lobsters Moving Northward

The lobster catch in New England varies annually with temperature. During 2002, record high temperatures were a factor in lighter than usual lobster catches offshore. Marine scientists attributed unusually warm waters to seasonal aberrations, including a mild winter, a short spring, and a hot summer (Healy 2002). Bob Glenn, a biologist at the state Division of Marine Fisheries, said he took a temperature reading of 77°F in 20 feet of water off the coast of Bourne, Massachusetts, during the third week of August. The average reading is near 70. "I'm not ready to jump on the global warming bandwagon, but this has been an exceptionally warm summer in many places off the Cape," Glenn said. "And this kind of water has a real impact on just about everything" (Healy 2002).

In 2012, warming waters to the south provoked an explosion in lobster populations off Maine's coast; this collapsed prices to as low as $1.25 a pound. Atlantic lobsters have been moving north at about 10 miles a decade over the last 40 years. South of Maine, off Cape Cod and Long Island, lobsters are being killed by a shell disease caused by warming water ("Maine's Lobster" 2013).

According to one *Boston Globe* account,

> The movements of lobsters, blue-fin tuna, and cod have become difficult to predict this summer [2002] because they are moving out of warmer water, several fishermen said. Even on the ocean side of the Cape, which typically has colder waters than its bay, lobsters haven't been appearing in some of the usual streams this month, and some lobstermen's businesses have taken a beating as a result. "I've lost $8,000 in two weeks because my traps aren't seeing nearly as many lobsters," said Billy Souza, who has 700 traps from Provincetown to Wellfleet and sells his catch out of his garage here. (Healy 2002)

Leaning against his truck after delivering about 150 pounds of lobster to his wife and chief saleswoman, Cheryl, Souza declared the day his best in weeks. His haul fell as low as 25 or 30 pounds on some days, when he mostly netted "eggers"—lobsters bearing eggs that must be thrown back—and found himself baffled by the lobsters' movements. "I've been lobstering for 20 years, and from my own system these lobsters like it in 42-degree water," Souza said. "More and more of my traps are coming up totally empty" (Healy 2002).

Massachusetts state biologists associated the warm water temperatures during the summer of 2002 to temperatures in Cape Cod Bay and Buzzards Bay the previous December, which were the warmest in 12 years, according to the state Division of Marine Fisheries. In one reading in 65 feet of water off Plymouth, the mean temperature was 46°F in December, significantly higher than the usual mean of 42 degrees. In 75 feet of water off Buzzards Bay, December's monthly mean was 50°F, compared to the normal reading of 45 degrees (Healy 2002). Temperatures in New England returned to average or below during the winters of 2002–2003 and 2003–2004 as pervasive cold and snow returned.

Warming Affects Sea Creatures

Roughly 300 surviving North Atlantic right whales may be threatened by a decline in their main food source, plankton, as a result of shifts in ocean circulation, At the same time, the common dolphin, *Delphinus delphis*, a warm-water species, has been increasing its range while the cold water range of the white-beaked dolphin, *Lagenorhynchus albirostris*, is shrinking. Predators are following their prey as species of fish change their latitude or depth in response to a warming climate. Exotic southern fish species such as red mullet, anchovy, sardine, and poor cod are now being found in the North Sea. Fish species are unable to regulate their body temperature, and their distribution and abundance are temperature dependent ("Climate Change" 2006).

As the North Sea has grown warmer, the Pacific oyster, *Crassostrea gigas*, which was brought to Europe for commercial reasons, has been able to breed in the wild and is now displacing native oysters in the Wadden Sea. Previously, these oysters were unable to survive outside artificial pens ("Climate Change" 2006).

In 1984, the jumbo squid (a smaller species that is related to the giant squid) migrated as far north in the eastern Pacific as Point Conception on the California coast. By 2005, some of them were caught by fishermen off the southern coast of Alaska. The range of marlins has also expanded several miles northward to the coast of Washington state. A 2003 meta-analysis of 1,700 bird species (mostly birds and some sea animals) found that over several decades their ranges were expanding an average of 600 meters northward per year (Kintisch 2006, 776).

Fish populations in many areas will also probably be affected by interruptions in their breeding cycles caused by saltwater intrusion into estuaries as the oceans rise. Roughly 70 percent of the world's fish use shoreline waters in their breeding cycles. Writes Robert Gough, "Fish production will thus suffer when such nursery habitats are lost" (Gough 1999, 47). Gough warns that many species in the sea and on the land that have adapted to specific environments may not be able to adapt to the speed of coming temperature changes: "Surviving species may succumb to predatory pressure and competition from more exotic species better adapted to the new conditions" (Gough 1999, 48).

Warming Limits Fishes' Oxygen Intake

The viviparous eelpout (*Zoarces viviparous*) needs more oxygen at higher temperatures. Because warming water in the tidal flats of the Wadden Sea (near the Netherlands) carries less oxygen, the fish there are becoming smaller and scarcer (Portner and Knust 2007).

A mismatch between the demand for oxygen and the capacity of oxygen supply to tissues is the first mechanism to restrict whole-animal tolerance to thermal extremes. We show in the eelpout, *Zoarces viviparus*, a bioindicator fish species for environmental monitoring from North and Baltic Seas that thermally limited oxygen delivery closely matches environmental temperatures beyond which growth performance and abundance decrease. Decrements in aerobic performance in warming seas will thus be the first process to cause extinction or relocation to cooler waters. (Portner and Knust 2007, 95)

Warm summers dramatically reduced populations of the fish the following year. Population declines, long before temperature rises, threaten the fish as individuals.

Further Reading

Beaugrand, Gregory, et al. "Plankton Effect on Cod Recruitment in the North Sea." *Nature* 426 (December 11, 2003): 661–664.

Bernstein, Lenny. "World's Fish Have Been Moving to Cooler Waters for Decades, Study Finds." *Washington Post,* May 15, 2013. http://www.washingtonpost.com/national/health-science/worlds-fish-have-been-moving-to-cooler-waters-for-decades-study-finds/2013/05/15/730292e8-bcd7-11e2-9b09-1638acc3942e_print.html.

Brown, Paul. "Overfishing and Global Warming Land Cod on Endangered List." *The Guardian* (U.K.), July 20, 2000, 3.

Cheung, William W. L., Reg Watson, and Daniel Pauly. "Signature of Ocean Warming in Global Fisheries Catch." *Nature* 497 (May 16, 2013): 365–368.

"Climate Change Dislocates Migratory Animals, Birds." Environment News Service, November 17, 2006. http://www.ens-newswire.com (no longer available).

Connor, Steve. "Strangers in the Seas; Exotic Marine Species Are Turning Up Unexpectedly in the Cold Waters of the North Atlantic." *The Independent* (London), August 5, 2002, 12–13.

"Fished to the Point of Ruin, North Sea Cod Stocks So Low as to Spell Disaster." *The Herald* (Glasgow, Scotland), November 7, 2000, 18.

Gough, Robert. "Stress on Stress: Global Warming and Aquatic Resource Depletion." *Native Americas* 16(3–4) (Fall–Winter 1999): 46–48. http://nativeamericas.aip.cornell.edu (no longer available).

Healy, Patrick. "Warming Waters: Lobstermen on Cape Cod Blame Light Hauls on Higher Ocean Temperatures." *Boston Globe*, August 30, 2002, B1.

Kintisch, Eli. "As the Seas Warm." *Science* 313 (August 11, 2006): 776–779.

"Maine's Lobster Bubble Is Threatening to Burst, Thanks to Global Warming and Canada." quartz.com, August 7, 2013. http://qz.com/112273/maines-lobster-bubble-is-threatening-to-burst-thanks-to-global-warming-and-canada/. Accessed August 30, 2013.

McNeil, Ben I., and Tristan P. Sasse. "Future Ocean Hypercapnia Driven by Anthropogenic Amplification of the Natural CO_2 Cycle." *Nature* 529 (January 21, 2016): 383–386.

"Ocean Acidification (Effects on Marine Animals: Fish)—Summary." *CO_2 Science.* Ocean Acidification Database. November 14, 2014. http://www.co2science.org/subject/o/summaries/acidfish.php.

Portner, Hans O., and Rainer Knust. "Climate Change Affects Marine Fishes through the Oxygen Limitation of Thermal Tolerance." *Science* 315 (January 5, 2007): 95–97.

Smith, Craig S. "One Hundred and Fifty Nations Start Groundwork for Global Warming Policies." *The New York Times*, January 18, 2001, 7.

U.S. Environmental Protection Agency. *Ecological Impacts from Climate Change: An Economic Analysis of Freshwater Recreational Fishing.* Washington, DC: EPA, 1995 (April).

Wines, Michael, and Jess Bidgood. "Waters Warm, and Cod Catch Ebbs in Maine." *The New York Times*, December 14, 2014. http://www.nytimes.com/2014/12/15/us/waters-warm -in-gulf-of-maine-and-cod-catch-ebbs.html.

See also: El Niño and La Niña; Jellyfish; Ocean Acidity, North America; Ocean Acidity, Worldwide; Oceans' Absorption of Heat; Phytoplankton; Salmon and Trout

JELLYFISH

Some sea species thrive on conditions that kill others. For example, jellyfish seem to have a biological affinity for some forms of human pollution as well as warmer habitats. Jellyfish populations have been increasing rapidly in many parts of the world. In some areas, the increase appears to be part of a natural cycle (Jellyfish populations are also declining in a few areas) (Pohl 2002). Jellyfish, have been called "the cockroaches of the open waters," maritime survivors who thrive in damaged environments (Rosenthal 2008). They breed more quickly as waters warm.

By the summer of 2004, reports indicated that jellyfish populations were on the rise in Puget Sound, the Bering Strait, and the harbors of Tokyo and Boston. "Smacks" or swarms of jellyfish shut down fisheries in Narragansett Bay, parts of the Gulf of Alaska, and sections of the Black Sea. In the Philippines, 50 tons of jellyfish shut down a power plant, causing blackouts when they were sucked into the plant's cooling system (Carpenter 2004, 68). In late July 2003, thousands of barrel jellyfish and moon jellyfish washed up on the coast of southern Wales.

Jellyfish Thrive Worldwide

On the shores of the Sea of Cortez, in northwestern Mexico, most fish are gone, but "cannonball jellies" have arrived. Former fishermen are loading their boats, salting them, and shipping their new harvest off to China, "where they are a bland sort of staple" (Vance 2013, 62). Welcome to the future.

Near Barcelona, Spain, fishermen have found their nets slimed with jellyfish as more than 300 beachgoers per day were treated for stings during the summer of 2008. All over the world, jellyfish have been proliferating, a sign of overfishing of their predators (tuna, sharks, and jellyfish), pollution that depletes oxygen levels, and gradually rising water temperatures. These problems are especially severe in the landlocked Mediterranean Sea. Beaches have also been closed because of jellyfish swarms around the world, including the Cote d'Azur in France, the Great Barrier Reef of Australia, Waikiki Beach in Hawaii, and Virginia Beach in the eastern United States. In Australia, where the deadly Irukandji jellyfish is expanding its range in warming waters, the number of jellyfish stings rose to 30,000 treated cases in 2007, double the total in 2005 (Rosenthal 2008).

"Human-caused stresses, including global warming and overfishing, are encouraging jellyfish surpluses in many tourist destinations and productive fisheries," said the U.S. National Science Foundation, which issued a report on the explosion of jellyfish populations in the fall of 2008. The report lists Australia, the Gulf of Mexico, Hawaii, the Black Sea, Namibia, coastal Britain, the Mediterranean, the Sea of Japan, and the Yangtze estuary as problem areas (Rosenthal 2008).

In November 2007, a 16-square-kilometer smack of jellyfish at a depth of 35 feet invaded the Northern Salmon Company, Northern Ireland's only salmon farm, killing more than 100,000 fish at a cost of more than $2 million. Billions of *Pelagia nocticula* (mauve stingers) swarmed into the cages holding the fish about a mile into the Irish Sea off Glenarm Bay and Cushendun. Managing director John Russell said he had never seen anything like it during three decades in the business. "The sea was red with these jellyfish and there was nothing we could do about, it, absolutely nothing," he said ("Jellyfish Attack" 2007). The jellyfish had not been sighted so far north until recent years (their usual range is in the Mediterranean Sea), so warming waters probably played a role. Observers likened the attack to a swarm of locusts on land. The Torness nuclear power station in Scotland shut down its two reactors on June 28, 2011, after a huge smack of jellyfish clogged the water intakes for its cooling towers ("Jellyfish Close" 2011). The French company EDF Energy operates the power station.

Jellyfish as Indicators of Stressed Ecosystems

"Jellies are a pretty good group of animals to track coastal ecosystems," said Monty Graham, a scientist at the University of South Alabama. "When you start to see jellyfish numbers grow and grow, that usually indicates a stressed system" (Pohl 2002). Those stresses include increased water temperature, a rise in nutrients (from fertilizers and sewage), and depleted stocks of other fish, which is often caused by overfishing and the removal of jellyfish competitors. All of these changes are usually human caused, according to Graham.

Otto Pohl of *The New York Times* described the increasing numbers and toxic potency of jellyfish in some areas of Australia that are popular with swimmers:

When Robert King climbed back on the boat after snorkeling off the Great Barrier Reef . . . on March 31 [2002], he knew something was wrong. "I don't feel so good," he said, rubbing his chest. He had been stung by a jellyfish, and his condition deteriorated rapidly. By the time [an] emergency helicopter arrived, he was screaming in agony; a few hours later he was in a coma, eyes frozen wide, bleeding into his brain. He never regained consciousness. King, 44, from Columbus, Ohio, was the second person in Australia to die this year from the sting of a species of jellyfish, *Carukia barnesi*, found only in Australia and never before known to be fatal. More than 200 other victims went to hospitals, several times the number in a normal summer season here. . . . Jellyfish release millions of microscopic harpoons when touched, shooting tiny hypodermic needles into a victim's skin. They are lined with barbs and filled with venom, and they often linger painfully in the skin for months after the toxin has worn off. (Pohl 2002)

"This year [2002] [was] incredibly abnormal" in the numbers of jellyfish stings, said Jamie Seymour, a jellyfish expert at James Cook University (Pohl 2002). Seymour believes that strong and unusual wind patterns helped blow the jellyfish toward the shore, where they flourished in unseasonably warm waters. Seymour, who has spent years analyzing the venom from each sting that receives hospital treatment in the Barrier Reef region, had never seen the type of venom that killed the two tourists in 2002. Across the world, according to a report in *The New York Times*, shrimp fishermen in the Gulf of Mexico are struggling with rising numbers of jellyfish that fill their nets with slimy gelatin that ruins their catches (Pohl 2002).

Smackdown: Hawaii to New England

On Waikiki Beach in Hawaii, lifeguard Landy Blair counted jellyfish stings on more than 900 people during a single day in 2002; the stings were severe enough on some 1 percent of victims to send them to area hospitals. Blair has been keeping track of jellyfish populations near the beach since 1991. The problem has grown steadily worse, he said. "We have see the highest numbers ever over the past year" (Pohl 2002). On the beaches near Auckland, New Zealand, half a dozen sting victims have required hospitalization, said Robert Ferguson of Surf Life Saving New Zealand, an organization of lifeguards. "It's the first time I've ever heard of victims needing hospital care," Ferguson said. "This is a new type of jellyfish with stings that are much more severe, much harsher" (Pohl 2002).

At about the same time, a report appeared in the *Boston Globe* that described a massive infestation of jellyfish in Narragansett Bay and Long Island Sound. A group of fishermen who expected "an array of marine life in their nets . . . got jellyfish, nothing but jellyfish; jellyfish so plentiful that the gelatinous organisms came up dangling through the net like slimy icicles. And with each haul came more" (Arnold 2002). "Eventually it seemed that our deck was coated with vaseline," said Captain Eric Pfirrmann, who works for Save The Bay, a group whose members engage in environmental issues related to Rhode Island's Narragansett Bay. He piloted a research vessel that had taken several high-school teachers on a marine field trip. "I've seen blooms like this before," Pfirrmann said, "but never so early in the summer." The culprit is a nonstinging invertebrate about the size and shape of a tulip blossom and commonly known as the comb jelly. Along with sea squirts (an entirely different organism), these jellyfish were taking over Long Island Sound, thriving in large part because water temperatures have risen some 3°F over the past two decades, according to scientists (Arnold 2002).

The rapid increase in jellyfish has nearly wiped out the winter flounder population in Narragansett Bay at the same time that the non-native sea squirt has overrun local oysters and blue mussels. Both are favored by warmer water temperatures and human pollution. "There is evidence of jellyfish explosions around the world that appear related to the adverse impact of human activities, and those include global warming," said Sarah Chasis, the senior attorney for the New York City–based Natural Resources Defense Council (Arnold 2002).

Comb Jellies Run Riot

According to the *Boston Globe*,

> Historically, Narragansett Bay was the northern limit of the comb jelly, whose domain extends as far south as Argentina. During November of 2000 comb jellies were documented for the first time in Boston Harbor, although in numbers too sparse at that time to affect the harbor's ecology. In Narragansett Bay, according to Barbara Sullivan, an oceanographer from the University of Rhode Island who has been studying the jellyfish infestation, warmer water is changing the rules of who eats whom. (Arnold 2002)

The comb jelly's reproductive cycle adjusts to the warmth of the water in which it lives. Populations usually explode during the warmth of late summer and early autumn. Narragansett Bay has warmed an average of 3.4°F over the past 20 years, while between the late 1970s to 2001, the average temperature of Long Island Sound during for the first three months of each year has increased about 8 percent from 37.4 to 40.2°F (Arnold 2002). As waters warmed, the comb jelly has reproduced and "bloomed" four months earlier than previously, which has enabled it to gobble up the eggs and larvae deposited in the spring by spawning fish to the point of nearly replacing them (Arnold 2002). "We have seen areas of the bay where these things have cleaned out everything edible floating in the water column," Chasis said.

The winter flounder population has nearly vanished in Narragansett Bay. In 1982, approximately 4,200 metric tons of flounder were landed in Rhode Island. Within 20 years, the harvest fell to approximately 600 tons (Arnold 2002). "Comb jellies do indeed eat many winter flounder larvae," said Tim Lynch, a marine biologist with the state Division of Fish and Wildlife, "but I do not believe they are the sole cause of the population decline in Narragansett Bay." Factors such as overfishing and water quality probably also play a role, he said (Arnold 2002).

By 2008, the National Science Foundation (NSF) reported that swarms of stinging jellyfish and jellyfish-like animals were turning many fisheries and vacation spots into "jellytoriums." Along with warming waters, jellyfish populations were exploding because of non-native species, overfishing, and the presence of artificial structures such as oil and gas rigs. In addition to damaging fisheries, jellyfish swarms were inflicting an increasing toll on fish farms, ocean mining, desalination plants, and nuclear power plants by clogging intake pipes. In the Gulf of Mexico, some swarms had reached a density of 100 jellyfish per cubic meter of water ("Jellyfish Swarms" 2008).

Monty Graham of Alabama's Dauphin Island Sea Lab, which is located on a barrier island in the Gulf of Mexico, said that the abnormally large, dense, and frequent jellyfish swarms are "a symptom of an ecosystem that has been tipped off balance by environmental stresses" ("Jellyfish Swarms" 2008). Jellyfish often swarm in dead zones, where they face few predators or competitors. Rising water temperatures accelerate their growth and reproduction. According to the NSF report *Jellyfish Gone Wild!! Environmental Change and Jellyfish Swarms*, marine turtles, which eat jellyfish, have declined in many areas, and jellyfish populations have exploded. One jellyfish-like creature. Australia's *Chironex fleckeri*, is the world's most venomous animal; its toxins can kill a person in less than three minutes.

Jellyfish as Dominant Species?

The United Nations has warned that overfishing, pollution, and warming waters may make jellyfish the dominant species in some waters within a few decades, in "a global regime shift from a fish to a jellyfish ocean" One U.N. report, *Review of Jellyfish Blooms in the Mediterranean and Black Sea,* says that jellyfish blooms flourish in warm polluted waters when overfishing removes predators from the top of the food chain (Boero 2013). Professor Boero, a zoologist and marine biologist, directs the the University of Salento's Laboratory of Zoology and Marine Biology in Lecce, Italy. The jellyfish eat fish eggs and larvae of species that are being overfished, hastening their demise. Jellyfish from tropical climates thrive and expand their ranges as waters warm. Seawalls built to forestall coastal erosion also provide jellyfish ideal habitat. As Boero wrote:

> Gelatinous plankton is formed by representatives of *Cnidaria* (true jellyfish), *Ctenophora* (comb jellies), and *Tunicata* (salps). . . . The life cycles of gelatinous plankters are conducive to bloom events, with huge populations that are occasionally built up whenever conditions are favorable. Such events have been known since ancient times and are part of the normal functioning of the oceans. . . . In the last decade, however, the media are reporting on an increasingly high number of gelatinous plankton blooms. The reasons for these reports is that thousands of tourists are stung, fisheries are harmed or even impaired by jellyfish that eat fish eggs and larvae, coastal [power] plants are stopped by gelatinous masses. ("UN Warns" 2013).

By 2013, power plants had sucked masses of jellyfish into their intake pipes in Japan, China, the Philippines, India, the Gulf of Oman, the Baltic Sea, the Arabian (or Persian) Gulf, Qatar, and the United States. Jellyfish have been invading new habitats such as the Black Sea. where *Mnemiopsis leidyi*, also known as the "sea walnut," from the Atlantic Ocean, invaded and had such an "overwhelming impact on [Black Sea] fish populations that fisheries were put on their knees" ("UN Warns" 2013). This species traveled from coastal U.S. waters to the Black Sea in the ballast waters of oil tankers. Once established in the Black Sea, these jellyfish obliterated a local anchovy nursery by eating the fish larvae and its food sources. However, *Beroe ovata*, another invader (and one that eats *Mnemiopsis*) soon restored some semblance of balance.

Further Reading

Arnold, David. "Global Warming Lends Power to a Jellyfish in Narragansett Bay and Long Island Sound; Non-Native Species Are Taking Over." *Boston Globe*, July 2, 2002, C1.

Boero, Fernando. *Review of Jellyfish Blooms in the Mediterranean and Black Sea.* Rome: Food and Agriculture Organization of the United Nations, 2013. http://www.fao.org/docrep /017/i3169e/i3169e.pdf.

Carpenter, Betsy: "Feeling the Sting: Warming Oceans, Depleted Fish Stocks, Dirty Water—They Set the Stage for a Jellyfish Invasion." *U.S. News and World Report*, August 16, 2004, 68–69.

"Jellyfish Attack Destroys Salmon." British Broadcasting Corporation, November 21, 2007. http://news.bbc.co.uk/2/hi/uk_news/northern_ireland/7106631.stm

"Jellyfish Close Scottish Nuclear Power Plant." Environment News Service, June 30, 2011. http://ens-newswire.com/2011/07/01/jellyfish-close-scottish-nuclear-power-plant/

"Jellyfish Swarms Invade Ecosystems Out of Balance." Environment News Service, December 16, 2008. http://www.ens-newswire.com/ens/dec2008/2008-12-16-01.asp (no longer available).

National Science Foundation. "Jellyfish Gone Wild!!" Arlington, VA: NSF, no date. https:// www.nsf.gov/news/special_reports/jellyfish/textonly/index.jsp.

Pohl, Otto. "New Jellyfish Problem Means Jellyfish Are Not the Only Problem." *The New York Times*, May 21, 2002, F3.

Rosenthal, Elisabeth. "Stinging Tentacles Offer Hint of Oceans' Decline." *The New York Times*, August 3, 2008. http://www.nytimes.com/2008/08/03/science/earth/03jellyfish.html.

"UN Warns of Global Shift 'From a Fish to a Jellyfish Ocean'." Environment News Service, May 30, 2013. http://ens-newswire.com/2013/05/30/un-warns-of-global-shift-from-a -fish-to-a-jellyfish-ocean/.

Vance, Erik. "Letter from the Sea of Cortez: Emptying the World's Aquarium." *Harper's*, August 2013, 53–62.

See also: Coral Reefs; El Niño and La Niña; Jellyfish; Ocean Acidity, North America; Ocean Acidity, Worldwide; Oceans' Absorption of Heat

"MANAGED REALIGNMENT"

Parts of England's coastline are afflicted by the same problems as coasts along the U.S. Eastern Seaboard and the Gulf of Mexico. The land is subsiding as ice melt and thermal expansion slowly raise sea levels. The U.K. Climate Impact Program, a government-funded program at Oxford University, forecasts that the sea could be as much as a meter higher by late in the 21st century. In addition to climate change, "isostatic rebound," the rise of Scotland's coast following the last ice age, is contributing to a subsiding coastline southward along the English coast. As the sea rises 3 millimeters a year abreast of Essex, the land itself is sinking half as rapidly. By 2003, the rising waters were threatening the closed Bradwell Nuclear Power Station ("Rising Tide" 2003). The marks of climate change are everywhere. Cornwall farmers, for example, have adapted to warmer, wetter weather by growing jalapeño peppers.

Anthony Browne commented in the *Times* (London): "A thousand years after King Canute showed that man could not hold back the tide, the Government has come to the same conclusion. The Environment Agency, the government body responsible for flood defenses, is planning a strategic withdrawal from large parts of the English coastline because it believes that it can no longer defend them from rising sea levels, the result of global warming" (Browne October 2002).

The Sea Floods Low-Lying Farmland

The new U.K. strategy is officially called "managed realignment" and will allow the sea to flood low-lying farmland rather than try to fend off invading waters by building ever-higher defenses. The policy, which will allow the encroaching sea to submerge several thousand acres of land, has been welcomed by environmentalists. Farmers, however, said that the strategy was "unviable" and have demanded more concrete flood defenses. The area of coast affected ranges from the Humber Estuary, around East Anglia, to the Thames Estuary and west to the Solent. Strategic withdrawal also has been planned for sections of the Severn Estuary (Browne 2002). The first site surrendered to the sea was in Lincolnshire. About 200 acres of farmland were flooded by seawater at Freiston Shore after diggers broke through the flood defense banks to create a salt-marsh bird reserve (Browne 2002).

Rising Seas Dunk the Saint Andrews Golf Course

Golf has been played at St. Andrews in Scotland since the 15th century, but by 2000 a warming climate, rising tides, and coastal erosion were allowing the sea to swallow some of the game's oldest and most storied links. According to one observer, "One theory even claims that golfers will be lucky to enjoy another 50 years of competition on the links in Fife before the Old Course becomes just another submerged attraction akin to the lost city of Atlantis" (Aitken 2001).

Peter Mason, external relations manager of the St. Andrews Links Trust, explained how a combination of high tides, large waves, and strong winds have battered the land along the Eden Estuary and swept away a coastal path. "Three meters of linksland, just centimeters away from the par-4 eighth hole on the Jubilee Course, were lost to the sea in a single day last year after storms. "It fell away so fast, it frightened the life out of us," Mason recalled (Aitken 2001).

Parts of the St. Andrews links were once reclaimed from the ocean. The original courses were called "links" because they were located on strings of dunes and knolls that linked sea and land (Hale 2001). Today, all golf courses are called "links" after St. Andrews' origins, whether or not they border water. As satellite surveys indicate that the height of waves striking Britain's coasts has increased 25 percent in 20 years, some of these holes may be relocated for the first time in their two-century history.

Sloping gabions (stone-filled wire cages) were installed along a 100-meter stretch where the path ran before the storms. This defense was backed by 4,400 cubic meters of sand; an additional 12,000 cubic meters of sand was added later. If that fails to hold back the sea, more gabions may be needed (Aitken 2001). Roughly 200,000 rounds of golf are played as St. Andrews annually, making the town and course a major tourist draw.

Golfers who drive off the eighth tee of the Jubilee Course now stare down at 400 yards of rock-filled cages that stand between the course and the ocean. The third hole on the Royal Aberdeen course once was 60 yards from dunes that marked the ocean's edge. By 2001, on the eve of the 130th British Open on the course, golfers teed off at this hole two yards from a 40-foot drop into the ocean (Hale 2001). Violent winter storms are responsible for much of the erosion.

Other British golf courses are also eroding. At the Royal West Norfolk Golf Club near Brancaster in England, advancing seas have isolated the clubhouse, forcing use of knee-high rubber boots to play through.

Further Reading

Aitken, Mike. "St. Andrews Stymied by Natural Hazard." *The Scotsman*, April 18, 2001, 20.

Hale, Ellen. "Seas Create Real Water Hazard; Changing Climate at Root of Erosion That's Putting Links Courses in Jeopardy." *USA Today*, July 18, 2001, 3C.

"Managed realignment is in its infancy . . . but it has been an emotive issue; a lot of people are concerned about it. But you have to get the message across that we are defending property," said Brian Empson, flood defense policy manager for the Environment Agency (Browne 2002). The land that was returned to the sea in 2002 had been reclaimed 150 years earlier and protected by human-made banks. A grass-covered wall in Abbots Hall farm country on the east coast of Essex that had held back the sea for almost 400 years also was breached intentionally in 2002 to surrender the area to the sea. The use of breakwaters made of clay, bricks, and then large blocks of concrete had not stopped the rising waters.

Until recently, the Thames River barrier, which was built to protect London and surrounding areas from unusually high tides and storm surges, closed an average of only two or three times a year. Between November 2001 and March 2002, however, the barrier was raised 23 times. A British report released in September 2002 said that 59,000 square miles (home to 750,000 people) in and around London are vulnerable to flooding because they are below high-tide levels, some by as much as 12 feet.

Local Land Losses

Late in 2006, 50 feet of Roger Middleditch's sugar-beet field fell into the North Sea. A tenant farmer, Middleditch once tilled 23 acres of potatoes, but 20 acres have been reclaimed by the sea over several years. "We've lost so much these last few years," he said. "You plant, and by harvest it's fallen into the water" (Rosenthal 2007). Although the area has been sinking slowly for at least 100 years, the pace has accelerated over the last few years. The government no longer maintains local sea-walls, part of its policy of "managed realignment" or "managed retreat." Homes in

Happisburgh, an English village, fall into the sea each year, collapsing over a crumbling cliff. The North Norfolk District Council and Coastal Concern Action Limited have started to shore up Happisburgh's cliff with rocks, financed in part by an Internet campaign, "Buy a Rock for Happisburgh." "The [United Kingdom] won't let London flood," Mr. Viner said, "but the national government's not going to worry about an odd village or farm" (Rosenthal 2007).

Land loss at Benacre "has accelerated dramatically," said Mark Venmore-Roland, the estate's manager. "At first it was like a chap losing his hair—bit by bit, so you'd get used to it." But in the past few years, he said, "it's been really frightening" (Rosenthal 2007). Elisabeth Rosenthal of *The New York Times* described how the shoreline of eastern England is changing:

Walkers and birders who frequent these famous Broads, or salt marshes, will find that the hiking path through Benacre that once gently declined from a low grassy plateau toward the beach, now ends in a precipitous drop of 16 feet to the water; the rest fell into the sea in February. The 6,000-acre Benacre Estate is losing swaths of land 30 feet wide along its entire two miles of coastline each year. Inland trees that were once sold for timber are dying or no longer commercially valuable, because the proximity to the salty sea air has left them stunted. (Rosenthal 2007)

Further Reading

Browne, Anthony. "Canute was Right! Time to Give up the Coast." *Times* (London), October 11, 2002, 8.
"Rising Tide: Who Needs Essex Anyway." London *Guardian*, June 12, 2003, 4.
Rosenthal, Elisabeth. "As the Climate Changes, Bits of England's Coast Crumble." *The New York Times,* May 4, 2007. http://www.nytimes.com/2007/05/04/world/europe/04erode.html.

See also: Sea Level Rise

METHANE BURP

As the study of methane clathrates became more popular in recent years, evidence accumulated that their release, especially from oceans, may occasionally have been a major and sudden driving force in Earth's past climate cycles. James P. Kennett et al. studied climate records for the last 60,000 years off Santa Barbara, California, and parts of Greenland, finding that "surface and bottom temperatures change in concert" (Kennett 2000; Blunier 2000, 68). This findings support assertions by E. G. Nisbet (1990, 148) that massive release of oceanic methane from clathrates has played a significant role in rapid warming in Earth's past, even without added provocations by fossil fuel combustion.

Scientists have yet to reach any sort of consensus on the causes of Earth's paleoclimatic "methane burps." No one yet knows why a trillion tons of methane may have been suddenly released from solid methane hydrates around the world in times past, but evidence suggests that the release was a reaction to and accelerated sudden (in geologic time) global warming of 4 to 10°C. James Cook and Gerald Dickens theorize that the "methane probably oxidized to form carbon dioxide which eventually reached the atmosphere, driving greenhouse warming" (Kerr 1999).

A 10°F Warming in a Short Time

Sediment cores drilled and recovered by Miriam E. Katz et al. contained remnants of small marine organisms called *foraminifera* that preserve a record in their shells of carbon levels in the ocean (Katz et al. 1999). The shells tell a story of an extreme warming (possibly more than 10°F) in the ocean over a short time, an event that killed more than half of all *foraminifera*. The sediment cores also contains evidence of an underwater landslide that scientists believe took place as melting methane clathrates "warmed dramatically, breaking apart into water and methane gas, and [bubbled] ferociously out of the sea floor" (Witze 2000).

This line of reasoning was supported by Richard Norris of the Woods Hole Research Center and Ursula Rohl of Germany's University of Bremen, who wrote in *Nature* that the "methane burp" occurred when an as-yet-unknown natural provocation pumped greenhouse gases into the atmosphere and caused a sudden bout of global warming. Large natural changes in the global carbon cycle have occurred in the past at rates similar to those being induced today by human activity. This surge in global temperatures may have played a crucial role in the evolution of warm-blooded mammals as Earth's dominant species 10 million years after a cataclysmic event (probably the impact of a large asteroid) ended the dominance of dinosaurs. Katz et al. contend that "elevated temperatures quickly opened high latitude migration routes for the widespread dispersal of mammals" (Katz 1999, 1531).

Methane bursts during a past warming transition may provide a guide to future Arctic climate change. Earth's climate can sharply warm or cool rapidly. Massive shifts have taken place in paleoclimates quickly. "About 11,600 years ago, at the end of the Younger Dryas cold period," wrote E. G. Nisbet and J. Chappellaz in *Science*, "the planet warmed very suddenly, with strong increases in atmospheric greenhouse gases, especially methane" (Nisbet and Chappellaz 2009, 477). Petrenko et al. used radiocarbon (^{14}C) data to determine that wetlands were responsible for the rapid increase of atmospheric methane concentrations (Petrenko et al. 2009, 506).

A "Methane Pulse" from the Ocean Floors

Stephen P. Hesselbo and colleagues reported that roughly 140 million to 200 million years ago large quantities of methane were liberated from ocean floors, possibly because of warming global temperatures. This "methane pulse"—a "voluminous and extremely rapid release of methane from gas hydrate contained in marine continental-margin sediments" (Hesselbo et al. 2000, 392)—combined with oxygen in the oceans to form carbon dioxide, accelerating the worldwide warming. Along the way, a large proportion of oceanic animal life (perhaps 80 percent) died for lack of oxygen. "One of the important questions that is debated a lot today is the stability of this methane hydrate reservoir, and how easy it is to release the methane," said Stephen P. Hesselbo, lead author of the paper. "The extinction and the association with lack of oxygen has been fairly well established, but the association with methane release is something that hadn't been realized before," he said ("Prehistoric Extinction" 2000).

During past periods of rapid warming, methane in gaseous form has been released from the sea floor in intense eruptions. An explosive rise in temperatures on the order of approximately 8°C over a few thousand years accompanied a methane release 55 million years ago called the Palaeocene–Eocene Thermal Maximum. According to considerable evidence, "Just a small amount of warming could kick-start a positive feedback loop between hydrate release and further warming, sending global temperatures soaring" (Schiermeier 2003, 681). Following the temperature spike 55 million years ago, Earth eventually recovered its temperature equilibrium— but it took roughly 100,000 to 200,000 years to do so.

This period has assumed a crucial position in climate studies because atmospheric concentrations of carbon dioxide are believed to have reached 800 to 1,000 parts per million at its peak—the level that will prevail by the end of the 21st century at current rates of increase. The warming triggered 55 million years ago was quite spectacular. Scientists drilling in the Arctic have discovered that temperatures in the Arctic Ocean rose suddenly to as high as 68°F (20°C). "What no one expected was how much warmer it really was. That is a huge surprise," said Andy Kingdon of the British Geological Survey ("North Pole" 2004). Scientists discovered fossils of marine plants and animals that died quickly because they could not cope with the surge in temperatures.

Signatures in the Geological Record

Scientists have known for some time about a massive release of methane in the atmosphere about 55 million years ago because they can detect a distinct chemical signature in the geological record. Until recently, however, the cause of this spike in greenhouse gases eluded them. A European team led by Henrik Svensen at the University of Oslo discovered ancient "escape conduits" on the floor of the Norwegian Sea. Svensen and several colleagues suggested that the eruption of carbon-rich sedimentary strata from these vents under the present-day Norwegian Sea may have provided the trigger.

The researchers suggested in *Nature* that such conduits were common throughout the Northeast Atlantic 55 million years ago as the sea floor literally ripped apart between Greenland and northern Europe (Munro 2004). The European scientists suggest that upwellings of molten rock from deep inside Earth heated and cooked organic material in sea sediments, producing excessive amounts of methane gas that bubbled out into the atmosphere. "Similar volcanic and metamorphic processes may explain climatic events associated with other large igneous provinces such as the Siberian Traps (about 250 million years ago)," they wrote in *Nature* (Svensen et al. 2004, 542). Believing that carbon dioxide release alone was insufficient to explain the magnitude of warming, a case has been made by Gabriel J. Bowen and colleagues that rapid increases in humidity (water vapor being a greenhouse gas) occurred at the same time (Bowen et al. 2004, 495).

Crocodiles and Palm Trees in the Arctic

Researchers who have drilled into sediment layers near the east coast of Florida found evidence that melting methane clathrates thawed "suddenly"—that is, over the course of a few thousand years—approximately 55 million years ago. The thawing initiated a sudden episode of global warming that ended with crocodiles and palm trees in the Arctic. At the peak of this episode, greenhouse gas levels in the atmosphere were between two and six times as high as they are now. Lisa Sloan, a paleoclimatologist at the University of California–Santa Cruz, and Gerald Dickens, a paleoceanographer at James Cook University in Queensland, Australia (two of several scientists who conducted the study), reported the findings at a meeting of the American Geophysical Union late in 1999.

Could the warming episode about 55 million years ago have been ignited by different methane releases in different parts of the world? T. L. Hudson and L. B. Magoon have advanced a theory that 52 million to 58 million years ago in the Gulf of Alaska, "large amounts of sediment, eroded from freshly uplifted mountains, were deposited in deep water off the 2,200-kilometer coast of Alaska" (Clift and Bice 2002, 130).

An oceanic tectonic plate subducting below a continental plate generates heat and produces hydrocarbons, including liquid methane, from organic matter in the sediment. Writing in *Nature*, Peter Clift and Karen Bice proposed that the methane may have subsequently turned to gas and bubbled into the atmosphere, "increasing greenhouse gas concentrations [that] might [help to] explain the higher global temperatures 58 [million] to 52 million years ago" (Clift and Bice 2002, 130). The authors assert that while this process may have contributed significantly to a prolonged period of sudden atmospheric warming at that time but that it was probably too geographically isolated to have been the sole cause of worldwide warming (Clift and Bice 2002, 130).

An Analog to Human-Provoked Warming?

Some scientists who have studied past releases of methane clathrates believe the releases could provide a dramatic analog for what might happen if human transport and industry continue to pump massive amounts of carbon dioxide and other greenhouse gases into the air. The rise in temperature that has been associated with this event is indeed extremely close to what the Intergovernmental Panel on Climate Change (IPCC) has projected—given "business as usual" conduct—during the lives of our great-grandchildren.

"The case is getting stronger that global warming is associated with large hydrocarbon releases, and here's an example where it seems to have happened in the past," said Roy Hyndman, a senior scientist at the Pacific Geoscience Center of the

Geological Survey of Canada (Munro 2004). "If you put a lot of methane or other hydrocarbons into the atmosphere you can get abrupt and very large global warming and that's what we're doing now," said Hyndman. "The amount we are putting in now is huge in the geological context, very large" (Munro 2004).

Kennett et al. (2003) make a case that the conversion of clathrates was much more important in driving climate change than increases in wetland emissions, a common explanation. However, Gerald R. Dickens, an earth scientist at Rice University in Houston, wrote, "There are . . . two gaping problems: the authors provide no compelling evidence that methane released from the sea floor passed through the water column or that atmospheric methane initiated climate change. Until these major holes are plugged, their full hypothesis should rightfully be considered highly speculative" (Dickens 2003). Dickens added: "It's the first really nice evidence that hydrocarbons were coming out of the sea floor at this time" (Munro 2004). Dickens and Hyndman said that the warming spike 55 million years ago is an intriguing, if not yet completely solid, analog for what humans are doing to the atmosphere today. "The question is, will it just warm steadily as we add more and more carbon, or is there some point where it will warm a lot? There is a danger of a sudden or catastrophic big change," said Hyndman (Munro 2004).

Hot magma was delivered into organic-rich sediments throughout the North Atlantic at this time, initiating a massive hydrocarbon discharge from the sea floor. The composition of the expelled fluids and the timing of their release has not been explicitly defined, however, leading to questions about how many gigatons (GTs) of carbon were released. Some calculations put that amount at about 3,000 GT, a "stupendous amount of hydrocarbons," considering that the entire world's deposits of oil and gas today total some 5,000 GT (Dickens 2004, 514). Although his explanation for the warming spike 55 million years ago needs refining, Dickens asserted that if cause and effect can be proved here, these events may become "an intriguing but imperfect analog of current fossil fuel emissions" (Dickens 2004, 515).

"It is the first time geology has isolated an individual methane release event in the distant past," said Santo Bains of Oxford University's Department of Earth Sciences. "Now we can see just how it played its part in the global warming process of that era" (Keys 2001). Bains led geologists during three years of research through the badlands of Wyoming, parts of Antarctica, and off Florida's east coast. "By studying that event, we may well be able to understand the effect of future global warming on the Arctic," said paleoclimatologist Professor Euan Nisbet of Royal Holloway College, University of London.

Further Reading

Blunier, Thomas. "'Frozen' Methane Escapes from the Sea Floor." *Science* 288 (April 7, 2000): 68–69.

Bowen, Gabriel J., et al. "A Humid Climate State during the Palaeocene/Eocene Thermal Maximum." *Nature* 432 (November 25, 2004): 495–499.

Clift, Peter, and Karen Bice. "Earth Science: Baked Alaska." *Nature* 419 (September 12, 2002): 129–130.

Dickens, Gerald R. "A Methane Trigger for Rapid Warming?" *Science* 299 (February 14, 2003): 1017. Washington, DC: American Geophysical Union, 2002.

Dickens, Gerald R. "Global Change: Hydrocarbon-driven Warming." *Nature* 429 (June 3, 2004): 513–515.

Hesselbo, Stephen P., et al. "Massive Dissociation of Gas Hydrate during a Jurassic Oceanic Anoxic Event." *Nature* 406 (July 27, 2000): 392–395.

Katz, Miriam E., et al. "The Source and Fate of Massive Carbon Input during the Latest Paleocene Thermal Maximum." *Science* 286 (November 19, 1999): 1531–1533.

Kennett, James P., et al. "Carbon Isotopic Evidence for Methane Hydrate Instability during Quaternary Interstadials." *Science* 288 (April 7, 2000): 128–133.

Kennett, James P., et al. *Methane Hydrates in Quaternary Climate Change: The Clathrate Gun Hypothesis.* Washington, DC: American Geophysical Union, 2003.

Kerr, Richard A. "A Smoking Gun for an Ancient Methane Discharge." *Science* 286 (November 19, 1999): 1465.

Keys, David. "Global Warming: Methane Threatens to Repeat Ice-Age Meltdown." *The Independent* (London), June 16, 2001, n.p. (LEXIS).

Munro, Margaret. "Earth's 'Big Burp' Triggered Warming: Prehistoric Release of Methane a Cautionary Tale for Today." *Edmonton Journal*, June 3, 2004, A10.

Nisbet, E. G. "The End of the Ice Age." *Canadian Journal of Earth Sciences* 27(1) (1990): 148–157.

Nisbet, E. G., and J. Chappellaz. "Shifting Gear, Quickly." *Science* 324 (April 24, 2009): 477–478.

"North Pole Had Sub-Tropical Seas Because of Global Warming." Agence France Presse, September 7, 2004 (LEXIS).

Petrenko, Vasilii V., et al. "$^{14}CH_4$ Measurements in Greenland Ice: Investigating Last Glacial Termination CH_4 Sources." *Science* 324 (April 24, 2009): 506–508.

"Prehistoric Extinction Linked to Methane." *Omaha World-Herald.* July 27, 2000, 9.

Schiermeier, Quirin. "Gas Leak: Global Warming Isn't a New Phenomenon." *Nature* 423 (June 12, 2003): 681–682.

Svensen, Henrik, et al. "Release of Methane from a Volcanic Basin as a Mechanism for Initial Eocene Global Warming." *Nature* 429 (June 3, 2004): 542–545.

Witze, Alexandra. "Evidence Supports Warming Theory." *Omaha World-Herald*, January 12, 2000, 4.

See also: Methane Hydrates; Oceans' Absorption of Heat; Permafrost; Sea Level Rise; Thermal Inertia

METHANE HYDRATES TODAY

Warming oceans eventually could cause intense eruptions of methane from the sea floor, accelerating global warming caused by humankind's industries and transportation. Approximately 10,000 billion tons of solid methane clathrates, described as "a lattice of ice crystals rather like a honeycomb" (Pearce 2007, 91) are stored beneath the ocean and on the continents. By comparison, the contribution of humans to the atmosphere's inventory of greenhouse gases via burning of fossil fuels has amounted to about 200 billion tons of carbon. If even a small portion of the oceans' stored methane were to escape into the atmosphere, greenhouse warming could greatly accelerate. Among scientists, this mechanism has come to be called the *methane burp* or *clathrate gun hypothesis.*

Submarine hydrates total several thousand petagrams (Pg) worldwide. A petagram is a billion metric tons, and several thousand would equal at least 1,000 times the methane now in the atmosphere. In addition, the top three meters of Arctic permafrost contains roughly 1,000 Pg of carbon as organic matter that could be converted to methane that would equal 300 times the methane in the atmosphere (Kerr 2010, 620).

Whether and when methane hydrates pump up the atmosphere's greenhouse gas load has become a matter of intense debate in scientific circles. Most of the hydrates are several hundred meters deep in the oceans and under varying depths of mud in areas that probably will not enter the atmosphere as gas for centuries after much of the world's surface ice has melted. However, once warmed, they would react at a temperature similar to that of water ice, so any reaction could be rapid and massive. In the Arctic, writes David Archer in *The Long Thaw: How Humans Are Changing the Next 100,000 Years of Earth's Climate,* methane hydrates have formed within 200 meters of the surface. Because Arctic warming is intense, some of them could melt in decades to centuries (Archer 2009, 132). The potential is compounded by the complicated nature of hydrate chemistry (methane turns to carbon dioxide as it liquefies) as well as by the extremely rapid speed of anthropogenic warming compared to natural changes observed in paleoclimate. However, ultimately, writes Archer, "Methane hydrates could release as much carbon as the CO_2 released from fossil fuels, doubling the long-term impact of global warming" (Archer 2009, 136).

Methane "Hot Spots" Today

Even today, melting permafrost is producing "hot spots" across the Arctic where ejections of methane bubble out so strongly that ice never forms, even in the coldest winter weather. "It could be 10 or 30 liters of methane per day, from one little hole," said Katey Walter Anthony, an ecologist at the University of Alaska Fairbanks, who has observed this phenomenon in Alaska, Greenland, and Siberia. "And it does that all year. And then you realize there are hundreds of spots like that, and millions of lakes" (Lavelle 2012, 94). Some of this methane comes not only from near-surface mud but also from deeper geologic formations that were capped by permafrost, "and that contain hundreds of times more methane than is in the atmosphere now" (Lavelle 2012, 95). These may be harbingers of the theorized methane burp.

By 2010, scattered emissions of methane hydrates had been reported in parts of the oceans, but they did not seem to be adding up—yet—to the kind of steady increase that would raise worldwide atmospheric levels markedly. Scientists were detecting destabilization of some clathrates in the vicinity of the Gulf Stream off the eastern coast of the United States. "A changing Gulf Stream has the potential to thaw and convert hundreds of gigatons of frozen methane hydrate trapped below the sea floor into methane gas, increasing the risk of slope failure and methane release," wrote Benjamin Phrampus and Matthew Hornbach in *Nature* (2012, 527). Using thermal models and seismic data, Phrampus and colleagues associated this destabilization with warming water and attempted to gauge what effect these changes may have on stability of methane hydrates. According to their reasoning,

> recent changes in intermediate-depth ocean temperature associated with the Gulf Stream are rapidly destabilizing methane hydrate along a broad swathe of the North American margin. The area of active hydrate destabilization covers at least 10,000 square kilometers of the United States eastern margin, and occurs in a region prone to kilometer-scale slope failures. Previous hypothetical studies postulated that an increase of five degrees Celsius in intermediate-depth ocean temperatures could release enough methane to explain extreme global warming events like the Palaeocene–Eocene Thermal Maximum (PETM) and trigger widespread ocean acidification. Our analysis suggests that changes in Gulf Stream flow or temperature within the past 5,000 years or so are warming the western North Atlantic margin by up to [8°C] and are now triggering the destabilization of 2.5 gigatons of methane hydrate (about 0.2 percent of that required to cause the PETM). This destabilization . . . may continue for centuries. It is unlikely that the western North Atlantic margin is the only area experiencing changing ocean currents. Our estimate of 2.5 gigatons of destabilizing methane hydrate may therefore represent only a fraction of the methane hydrate currently destabilizing globally. The transport from ocean to atmosphere of any methane released—and thus its impact on climate— remains uncertain. (Phrampus and Hornbach 2012)

Methane Bubbles Worldwide

A group of British and German research scientists using sonar has found that at least 250 sources of methane bubbles have been rising from hydrates in the sea bed off Norway, perhaps an early indication of a "methane gun." The group reported in *Geophysical Research Letters* (Westbrook et al. 2009) that the methane was rising from between 150 and 4,500 meters near West Spitsbergen. Temperature records indicate that this area has warmed by about 1°C during the last 30 years, destabilizing some of the hydrates. Professor Tim Minshull of the National Oceanography Center at Southampton told BBC News, "We already knew there was some methane hydrate in the ocean off Spitsbergen and that's an area where climate change is happening rather faster than just about anywhere else in the world. . . . Our survey was designed to work out how much methane might be released by future ocean warming; we did not expect to discover such strong evidence that this process has already started," Minshull said (Burns 2009).

This research indicates that the methane release may be part of a long-term pattern that reaches, in some cases, to the last years of the most recent ice age. Most of the methane ejected from the hydrates reacts with oxygen in the ocean, forming carbon dioxide as carbonic acid and contributing to ocean acidification. "If this process becomes widespread along Arctic continental margins, tens of teragrams of methane per year could be released into the ocean," the scientists said (Westbrook et al. 2009).

Approximately 8 million tons of methane a year have been bubbling from a frozen seabed north of Siberia, according to a March 5, 2010, article in *Science*. Investigators are not sure whether this is a sign of global warming or merely the first detection of methane from stores trapped under permafrost that has been going on for decades. "Subsea permafrost is losing its ability to be an impermeable cap," said study coauthor Natalia Shakhova, a scientist at the University of Fairbanks, Alaska. Shakhova said that the sea floor was believed to be an impermeable barrier. The scientists measured methane levels at 5,000 sites on the East Siberian Arctic Shelf between 2003 and 2008 (Doyle 2010; Heimann 2010, 1211). Was the venting caused by global warming or by natural factors? Shakhova says, "No one can answer this question." A future rise in temperatures could accelerate the thaw, however. "It's good that these emissions are documented. But you cannot say they're increasing," said Professor Martin Heimann, an expert at the Max Planck Institute for Biogeochemistry in Germany. "These leaks could have been occurring all the time" since the last ice age 10,000 years ago, he said (Doyle 2010). Monitoring could resolve whether the venting was "a steadily ongoing phenomenon or signals the start of a more massive release period," according to the scientists, who are based at research institutions in the United States, Russia, and Sweden. The release of just a "small fraction of the methane held in [the] East Siberian Arctic Shelf sediments could trigger abrupt climate warming," they wrote (Shakhova et al. 2010).

In the December 8, 2009, issue of the *Proceedings of the National Academy of Sciences*, David Archer of the University of Chicago and colleagues reported a model that suggests that a 3°C warming of the oceans could melt half of existing hydrates. Thus far, the deep ocean has warmed only a tiny fraction of that amount in most places, although one location near Norway's Svalbard archipelago, where methane plumes have been reported, has warmed 1°C in 30 years. Calculations indicate that rises in atmospheric methane levels stem thus far mainly from other sources such as wetlands in northern latitudes, the Amazon River valley, and Indonesia as well as from human emissions. However, according to Archer, tipping points for submarine hydrates and permafrost will come slowly, driven by thermal inertia that will take time to transfer ocean or Earth-bound warming to the air (Kerr 2010, 621).

Linking Methane Releases and the Carbon Cycle

Writing in *Science*, Kai-Uwe Hinrichs and colleagues Laura Hmelo and Sean Sylva of the Woods Hole Oceanographic Institution provided a direct link between methane reservoirs in coastal marine sediments and the global carbon cycle, an indicator of global warming and cooling (Hinrichs et al. 2003). Molecular fossils of methane-consuming bacteria found in the Santa Barbara Basin off California deposited during the last glacial period (70,000 to 12,000 years ago) indicate that large quantities of methane were emitted from the sea floor during warmer phases of Earth's climate in the recent past. Preserved molecular remnants found by the Woods Hole team were analyzed; the researchers determined that the bacteria fed exclusively on methane and that large quantities of this powerful greenhouse gas were probably present in coastal waters off California. The team studied samples that were deposited 37,000 to 44,000 years ago.

"For the first time, we are able to clearly establish a connection between distinct isotopic depletions in forams and high concentrations of methane in the fossil record," said Hinrichs, an assistant scientist in the Woods Hole Institution's Geology and Geophysics Department. "The large amounts of methane presumably released during one event about 44,000 years ago suggest a mechanism different from those underlying the emissions at warmer periods, [that is,] slow decomposition of methane hydrate triggered by warming of bottom waters," Hinrichs continued. "The sudden release of these enormous quantities of methane was probably caused by landslides and melting of the methane hydrate" (Hinrichs et al. 2003).

Further Reading

Archer, David. *The Long Thaw: How Humans Are Changing the Next 100,000 Years of Earth's Climate.* Princeton, NJ: Princeton University Press, 2009.

Burns, Judith. "Methane Seeps from Arctic Sea-Bed." BBC News. August 18, 2009. http://news.bbc.co.uk/go/pr/fr/-/2/hi/science/nature/8205864.stm.

Doyle, Alister. "Methane Bubbles." Reuters in Australian Broadcasting Corp. March 5, 2010. http://www.abc.net.au/science/articles/2010/03/05/2837443.htm.

Heimann, Martin. "How Stable Is the Methane Cycle?" *Science* 327 (March 5, 2010): 1211–1212.

Hinrichs, Kai-Uwe, Laura R. Hmelo, and Sean P. Sylva. "Molecular Fossil Record of Elevated Methane Levels in Late Pleistocene Coastal Waters." *Science* 299 (February 21, 2003): 1214–1217.

Kerr, Richard A. "'Arctic Armageddon' Needs More Science, Less Hype." *Science* 329 (August 6, 2010): 620–621.

Lavelle, Marianne. "Good Gas, Bad Gas." *National Geographic*, December 2012, 90–109.

Pearce, Fred. *With Speech and Violence: Why Scientists Fear Tipping Points in Climate Change.* Boston: Beacon Press, 2007.

Phrampus, Benjamin J., and Matthew J. Hornbach. "Recent Changes to the Gulf Stream Causing Widespread Gas Hydrate Destabilization." *Nature* 490 (October 25, 2012): 527–530.

Shakhova, Natalia, et al. "Extensive Methane Venting to the Atmosphere from Sediments of the East Siberian Arctic Shelf." *Science* 327 (March 5, 2010): 1946–1950.

Westbrook, Graham K., et al. "Escape of Methane Gas from the Seabed along the West Spitsbergen Continental Margin." *Geophysical Research Letters 36* (August 6, 2009): L15608. doi: 10.1029/2009GL039191.

See also: Methane Burp; Oceans' Absorption of Heat; Permafrost; Sea Level Rise; Thermal Inertia

OCEAN ACIDITY, NORTH AMERICA

Ocean pH is now lower than it has been in 20 million years, and it is heading lower, according to marine chemist Richard Feely of the Marine Environmental Laboratory run by the National Oceanic and Atmospheric Administration (NOAA) in Seattle. Models by Feely and colleagues anticipate that ocean pH will decline from 8.2

before the industrial revolution to 7.7 to 7.8 by 2100, increasing acidity by some 150 percent (Kerr 2010, 1501).

Corals are damaged by acidification independently of temperature. Taryn Foster and colleagues, describing their use of high-resolution, three-dimensional x-ray microscopy, wrote in *Science Advances* (2016):

> Successful coral recruitment and high rates of juvenile calcification are critical to the replenishment and ultimate viability of coral reef ecosystems. . . . We show that ocean acidification not only causes reduced overall mineral deposition but also a deformed and porous skeletal structure in newly settled coral recruits. In contrast, elevated temperature (+3°C) had little effect on skeletal formation except to partially mitigate the effects of elevated PCO_2 [PCO_2 (partial pressure of CO_2)]. The striking structural deformities we observed show that new recruits are at significant risk, being unable to effectively build their skeletons in the PCO_2 conditions predicted to occur for open ocean surface waters under a "business-as-usual" emissions scenario . . . by the year 2100. (Foster et al. 2016)

Ocean Acidity and Absorption of Carbon Dioxide

Carbon dioxide is being injected into the oceans much more quickly than nature can neutralize it. Seawater is usually alkaline and around 8.2 pH. The pH scale is logarithmic, so a 0.1 decrease in pH—the average worldwide change by 2015 since the beginning of the Industrial Revolution—indicates a 30-percent increase in the concentration of hydrogen ions. Under a business-as-usual scenario, the pH will fall enough by the year 2100 to increase the level of hydrogen ions to three times the preindustrial "baseline" concentration (Henderson 2006, 30).

Since the Industrial Revolution began, human beings have infused roughly 120 billion tons of carbon dioxide into the oceans. By 2006, the seas were absorbing an additional 2 billion tons of CO_2 per year. Every day, each citizen of the United States adds an average of 40 pounds of carbon dioxide to the world's oceans (Kolbert 2006, 68–69). In 1800, the carbon dioxide level of the atmosphere was 280 parts per million (ppm), and ocean pH averaged 8.16. Today, atmospheric CO_2 is around 400 ppm, and ocean pH averages 8.00. Some estimates suggest a fall of pH to about 7.9 by the year 2100 (Ruttimann 2006, 978).

Acidity Limits Absorption of CO_2

As the oceans absorb more carbon dioxide, their chemistry has been changing. "The more carbon dioxide the ocean absorbs, the more acidic it becomes and the less carbon dioxide it can absorb," said the lead author of the study published in *Nature*, Samar Khatiwala of the Lamont-Doherty Earth Observatory at Columbia University and professor at the Georgia Institute of Technology. Although the ocean's ability to absorb carbon dioxide kept pace with rising emissions through the 1950s, the rate of absorption declined in the 1980s and has gone down even more since 2000. "It's a small change in absolute terms" (Khatiwala et al. 2009, 346). "What I think is fairly clear and important in the long term is the trend toward lower [pH] values, which implies that more of the emissions will remain in the atmosphere" (Bhanoo 2009).

Khatiwala and colleagues used a mathematical model that utilized 20 years of seawater temperature measurements. They also included measures of salinity and chlorofluorocarbons. They then estimated these measures for 1765, the beginning of the Industrial Age. The oceans' uptake rate dropped most sharply, approximately 10 percent, from 2000 to 2007, according to this study (Khatiwala et al. 2009, 346).

Analogs: Millions of Years

The worst mass extinction known to scientists, the End Permian extinction, took place 250 million years ago, largely provoked by acidifying ocean waters. Ninety percent of Earth's species died, as Eric Hand wrote in *Science* (2015, 165), after "volcanoes dumped trillions of tons of carbon dioxide into the atmosphere and some of it dissolved in the oceans, leading to an acidity that can weaken sea creatures' ability to make calciferous shells." A sharp drop in pH, wrote Hand, is "a cautionary lesson: due to the burning of fossil fuels, today's oceans are acidifying at an even faster rate than they were at the time of the extinctions, although it hasn't yet persisted nearly as long."

During the Paleocene–Eocene Thermal Maximum, between 55 million and 56 million years ago, massive amounts of carbon dioxide oxidized from methane clathrates and surged into the atmosphere from the oceans, raising sea surface temperatures by 5°C in the tropics to approximately 9°C at high latitudes. An initial rapid rise in temperatures over 1,000 years was followed by a slower increase during the next 30,000 years (Zachos et al. 2005, 1612; Clarkson et al. 2015, 229).

Scientists have been studying the acidification of the oceans during this long-ago epoch as an analog to similar conditions anticipated in response to human-provoked increases in carbon dioxide. One study of the problem, published in *Science*, made this pointed conclusion:

What, if any, implications might this have for the future? If combustion of the entire fossil fuel reservoir (about 4,500 gigatons of carbon) is assumed, the impacts on deep-sea pH [acidity] and biota will likely be similar to those in the Paleocene–Eocene Thermal Maximum. However, because the anthropogenic carbon input will occur within just 300 years, which is less than the mixing time of the ocean, the impacts on ocean surface pH and biota will probably be more severe. (Zachos et al. 2005, 1614)

Although the PETM presents something of an analog to the changes taking place in the atmosphere and oceans today with respect to pH change and rise in CO_2 level, the *rate* of change with human-propelled emissions is occurring at least

10 times as quickly, according to paleoceanographer James Zachos of the University of California–Santa Cruz. Nature took several thousand years to vent 2,000 to 7,000 gigatons of carbon dioxide and methane (which oxidized to CO_2) from volcanoes, methane hydrates, and peat—or perhaps all of these. Human emissions of greenhouse gases are doing the same thing within a few centuries (Kerr 2010, 1500).

"What does well in disturbed environments are invasive generalists," said Ken Caldeira, a climate expert at Stanford's Carnegie Institution for Science, who helped popularize the term *ocean acidification*. "The ones that do poorly are the more highly evolved specialists. Yes, there will be winners and losers, but the winners will mostly be the weeds" (Welch 2013b).

Zero Emissions the Only Solution?

Calderia said that the only way to save the oceans is to aim for zero human-generated carbon-dioxide emissions. "People laugh at this," he said, but the oceans naturally absorb only 0.1 gigatons more carbon dioxide per year than they release. Now they are soaking up an extra 2 gigatons a year, more than 20 times the natural rate. "Even if we halve emissions," he said, "that will merely double the time until we kill off your favorite plant or animal" (Henderson 2006, 32).

The oceans have reached only one-third of their capacity for absorption of humankind's excess carbon dioxide, but even at this level the rising level of carbon in ocean water is impeding sea animals' ability to grow protective shells because of increasing acidity. Scientists who project this trend into the future find ample reason to worry about the sustainability of ocean life (Feely et al. 2004, 362; Sabine et al. 2004, 367; "Report Says" 2004).

Scientific Meetings and Declarations

In June 2009, 70 science academies around the world sounded a call to include ocean acidity on the international climate-change agenda. Their collective assertion was that the oceans already had become more acidic than at any time in at least the last 800,000 years. "Unless global CO_2 emissions can be cut by at least 50 percent by 2050 and more thereafter, we could confront an underwater catastrophe, with irreversible changes in the makeup of our marine biodiversity," said Martin Rees, president of the United Kingdom's Royal Society. "The effects will be seen worldwide, threatening food security, reducing coastal protection and damaging the local economies that may be least able to tolerate it" ("Climate Deal" 2009).

Concern about rising ocean acidity was reinforced in January 2009 as 155 scientists from 26 countries organized by several international groups under the aegis of the United Nations released a report of its deliberations. "Severe damages are imminent," the group said in the summary to the Monaco Declaration arrived at during a symposium in Monaco in October 2008 (Dean 2009).

Acidification Disorients Fish

A study led by Danielle Dixson of Australia's James Cook University in Queensland (published in *Ecology Letters*) reports that increasing ocean acidification leaves some fish (orange clown fish in this study) unable to sense chemical signals given off by predators (Dixson et al. 2009). Acidification also disrupts fish behavior by causing confusing sensory cues (hearing, smell, sight) as illustrated by American biologist Danielle Dixson's study of clownfish. Dixson found that clownfish were drawn back to particular spots on reefs adjacent to New Guinea's island rain forests by the scents of the trees' leaves that had fallen into the water. Elevated carbon dioxide destroyed the fishes' ability to use smell as an environmental cue. They failed to be able tell one smell from another—including the scent, in the case of clownfish, of predators such as dottybacks and rock cod. The clownfish soon swam directly toward predators and were eaten.

Paradoxically, wrote Craig Welch in the *Seattle Times* (2013), "Another American living in Australia, marine scientist Jodie Rummer, was learning that high carbon dioxide boosted aerobic capacity for some tropical fish, transforming them into super athletes." The same fish also sometimes exhibited behavioral problems. They also got eaten more often despite their agility. "You might expect that a more athletic fish can be better at chasing food, or be better at getting away from a predator, or finding a mate," she said. "But if their cognitive function—or their brain—is compromised under these high-CO_2 conditions, they might make bad choices. They might turn the wrong direction and end up right in a predator's mouth" (Welch 2013).

Further Reading

Dixson, Danielle L., Philip L. Munday, and Geoffrey P. Jones. "Ocean Acidification Disrupts the Innate Ability of Fish to Detect Predator Olfactory Cues." *Ecology Letters*, November 16, 2009. http://www3.interscience.wiley.com/journal/1226849 40/abstract?CRETRY=1&SRETRY=0] DOI: 10.1111/j.1461-0248.2009.01400.x.

Welch, Craig. "Sea Change: The Pacific's Perilous Turn." *Seattle Times*, September 11, 2013. http://apps.seattletimes.com/reports/sea-change/2013/sep/11/pacific-ocean-perilous-turn-overview/.

"The chemistry is so fundamental and changes so [rapidly and severely] that impacts on organisms appear unavoidable," said James Orr, a chemical oceanographer at the Marine Environmental Laboratory in Monaco who headed the symposium's scientific committee. According to the declaration, "ocean acidification may render most regions chemically inhospitable to coral reefs by 2050" (Dean 2009). The group said that ocean acidification will continue to rise unless atmospheric levels of carbon dioxide stop rising. Some proposed solutions, including "fertilizing" the oceans with iron to absorb CO_2, may make acidification worse.

The impacts of rising acidity in the oceans can also affect the birthrates of birds such as gulls and murres, which eat fish that ate other fish that ate plankton. Scientists have been using Tatoosh Island off the northwest coast of Washington state (at the intersection of the Pacific Ocean and the Strait of Juan de Fuca) to study ocean acidification's effects on flora and fauna of both land and water (Solie 2012).

Effects of increasing acidity of the world's oceans will continue for "tens of thousands of years," according to an analysis in *Science* on June 18, 2010. Nothing in geologic records describes effects as sudden and severe as the current plunge in pH, so effects on marine wildlife are unknown. Researchers are seeing signs that "coral growth does slow, oyster larvae suffer, and plankton with calcareous skeletons lose mass" (Kerr 2010, 1500).

Ocean Acidity: The North American West Coast

In upwelling zones such as those seen along the U.S. West Coast, prevailing onshore winds drive water. They are notably vulnerable to amplified levels of ocean acidification. "Near-shore waters of the California Current System already have a low carbonate saturation rate," wrote Nicolas Gruber and colleagues, who used eddy-resolving model simulations in a study that projects acidification along the U.S. West Coast through the year 2050. They believe that "[h]abitats along the sea floor will become exposed to year-round under-saturation within the next 20 to 30 years. These projected events have potentially major implications for the rich and diverse ecosystem" (Gruber et al. 2012, 220).

In general, rising atmospheric levels of carbon dioxide are combining with industrial discharges, septic runoff, and exhaust from motor-vehicle traffic to accelerate the acidity of Washington state's coastal waters and Puget Sound. The waters have become so acidic that wild oysters have not reproduced since 2005. Acidity problems are worst in Hood Canal, site of most of the area's shellfish industry. Scientists from the University of Washington and the National Oceanic and Atmospheric Administration (NOAA) warned on July 12, 2010, that western Washington's waters had become a hot spot for marine acidity.

In addition to eroding calcium shells, by 2016 scientists at Seattle's NOAA Northwest Fishery Science Center had learned that rising acidity harms animals with chiton shells (in this instance, Dungeness crabs, a $221 annual industry in the Pacific Northwest), because it alters their metabolism. Rising acidity also reduces the survival rate of crab larvae, an important food source for many other sea creatures, including salmon.

Acidic water also kills oyster larvae. On the oceanic coast, the highest acidity is found near the coast, the area richest in marine life, Coastal acidity is being carried by ocean currents into the enclosed waters of Hood Canal and Puget Sound, raising levels from industrial sources there even higher. Hood Canal is becoming a dead zone as pollution and runoff from septic tanks kills fish, octopus, crab, and other sea life. Measured pH in southern Puget Sound was as low as 7.4 by 2010 (Welch 2010).

Alaska's King Crabs Endangered

Ocean acidification is damaging the crab fishery in Alaska, most notably the huge Bristol Bay red king crab. With arms the size of baseball bats, it has become the star of restaurant surf and turf offerings around the world, a $100 million a year industry and "the showboat of the Northwest's billion-dollar fishing industry . . . a television sensation and a marketer's dream, its image emblazoned on bumper stickers, mugs, caps, and T-shirts throughout the Pacific Northwest and Alaska" (Welch 2013a).

"Crabbing attracts tough adrenaline junkies who disappear for weeks into the storm-buffeted frontier of the Bering Sea," wrote Craig Welch in the *Seattle Times*. "They lounge in cramped quarters watching bad movies and wait for crab to fill their cages. Then workers scramble day or night on icy decks through stomach-churning swells, amped on coffee and nicotine" as seen in the Discovery Channel's *Deadliest Catch* (Welch 2013a). When cruise ships stop in Ketchikan, on the Alaskan panhandle, passengers pay $159 each for a half-day side excursion to watch fishermen haul the monster crabs out of the water. At the Pike Street Market on Seattle's waterfront, the same crabs retail for $40 or more per pound.

Snow crab, another major harvest, also may be damaged. "With red king crab, it's all doom and gloom," said Robert Foy, who oversaw the crab research for NOAA in Kodiak, Alaska. "With snow crab, there's so little is known we just can't say. But we don't see anything from our experience that's good for any of these crab. Some is just not as bad as others" (Welch 2013a). Rising ocean acidification is damaging the crabs' shells.

The problems in the crab industry also illustrate the paradoxes of ocean acidification on sea life. Similar species in the same habitat react differently. "The real issue here is unpredictability," said Richard Aronson, a Florida-based marine scientist who has tracked king crab in Antarctica. "There are all these unanticipated collateral impacts. The problem is, most of them are nasty surprises" (Welch 2013a).

"Many crab species appear hardy in the face of souring seas, or at least not so frail," reported Welch (2013a). Exceedingly corrosive waters actually pump up Maryland blue crabs to three times their usual size and turn them into voracious predators. Sour waters kill Dungeness crab, but far less often than Alaska red king crab. Laboratory tests with anticipated mid-century acidity levels show intensified damage, especially to Alaska red king crab. The models were developed with the future in mind, but in some areas acidity had reached danger levels by 2015.

Ocean Acidity Perils for Periwinkles, Urchins, Oysters, Algae, Fishes, and Squid

In addition to corals, many other calcifying organisms that grow calcium carbonate skeletons or shells suffer when pH falls significantly below 8.2 in surface ocean seawater, reported geochemist Justin Ries of the University of North Carolina–Chapel Hill and colleagues at the 2008 American

Geophysical Union (AGU) meeting. In the laboratory, they tested 18 different species (including periwinkles, urchins, oysters, and some forms of algae) in a range of carbon dioxide concentrations, including some that are expected in the oceans by the end of the 21st century. One species of mussels was not harmed, and some crustaceans (such as an edible shrimp and an American lobster) reacted by growing thicker shells even at high levels of acidification (Kerr 2009).

High concentrations of carbon dioxide dissolved in seawater also inhibits respiration of some fishes and squid, according to marine geochemists Peter Brewer and Edward Peltzer of the Monterey Bay Research Institute in Moss Landing, California, who also spoke at the 2009 AGU meeting. They described watching a squid that had been immersed in a high CO_2 test tank sink motionless to the bottom. "We can expect multiple impacts as we go forward into this strange CO_2 world," Brewer said (Kerr 2009).

Further Reading

Kerr, Richard A. "The Many Dangers of Greenhouse Acid." *Science* 323 (January 23, 2009): 459.

Crabs react differently, depending on the water chemistry of their habitat. Little has yet been proven, but some species may quickly evolve tolerance to some level of acidity. Golden king crabs, which live in extremely deep water with high levels of carbon dioxide, may be evolving a tolerance, but acidity levels may be rising more quickly than they can change. "It's almost like an arms race," said Gretchen Hofmann, a marine biologist at the University of California–Santa Barbara. "We can see that the potential for rapid evolution is there. The question is, will the changes be so rapid and extreme that it will outstrip what they're capable of?" (Welch 2013b).

Golden king crab, for example, live below 1,000 feet, where waters already are naturally rich in CO_2, making them highly tolerant of changes in sea chemistry. Crabs that live near the water's surface and close to shore may evolve tolerance from swings in acidity experienced from rising and falling tides and even photosynthesis.

Eventually, crabs that cannot survive acidic water caused by human emissions of carbon dioxide may be raised in aquatic farms under managed conditions. One such farm, the Alutiiq Pride Shellfish Hatchery, already exists in Seward, Alaska, and provides seed stock for new species, which supports both shellfish enhancement and other research ("Alutiiq Pride" 2015).

By 2008, scientists surveying waters along North America's western coast found rising levels of acidified ocean water within 20 miles of the shoreline, raising concern for marine ecosystems from Canada to Mexico. Researchers on the *Wecoma*, then Oregon State University's research vessel, discovered that the acidified upwelling from the deeper ocean is probably 50 years old. Future ocean acidification levels probably will rise as atmospheric levels of carbon dioxide increase (Feely et al. 2008, 1490; "Pacific Coast" 2008).

"When the upwelled water was last at the surface, it was exposed to an atmosphere with much lower CO_2 (carbon dioxide) levels than today's," said Burke Hales, an associate professor in the College of Oceanic and Atmospheric Sciences at Oregon State University, a coauthor of the study. "The water that will upwell off the coast in future years already is making its undersea trek toward us, with ever-increasing levels of carbon dioxide and acidity." The researchers found that the 50-year-old upwelled water had CO_2 levels of 900 to 1,000 parts per million, placing it "right on the edge of solubility" for calcium carbonate-shelled aragonites, Hales said ("Pacific Coast" 2008). Continued carbon dioxide overload in the oceans could make ocean water more acidic than it has been "for tens of millions of years and, critically, at a rate of change 100 times greater than at any time over this period" (Riebesell et al. 2007, 545).

Isaac Kaplan, a computer modeler at NOAA in Seattle, anticipates declines in flounder, sole, sharks, skates, Pacific whiting (hake), and rays. "Some species will go up, some species will go down," said Phil Levin, ecosystems leader for NOAA's Northwest Fisheries Science Center in Seattle. "On balance, it looks to us like most of the commercially caught fish species will go down" (Welch 2013b). The North Pacific pollock catch (which supplies a range of products from imitation crab to fish sticks) is hauled in by ocean-going fishermen and factory trawlers. By itself, it is a $1 billion annual industry. The pollock have problems recognizing their food sources in water at enhanced carbon dioxide levels. However, over several generations they might evolve a tolerance.

Washington state has been among the first governmental entities to monitor ocean acidity—to tally levels in water and emissions sources. This is acknowledged as just a first step. "The only long-term solution to acidification is for the world to reduce industrial emissions of carbon dioxide, allowing the ocean to reach a less acidic equilibrium" ("Marine Life" 2012). The state will next

seek to reduce carbon pollution from land based sources, including agricultural and urban runoff. There will also be practical, site based steps to offset carbon, like planting sea grasses (which themselves are endangered globally) in shellfish hatcheries. And there will be an extensive campaign to educate the public, business leaders and policy makers about the risks of increasing acidification. ("Marine Life" 2012)

Further Reading

"The Alutiiq Pride Shellfish Hatchery." Web page. http://alutiiqpridehatchery.com. Accessed January 31, 2015.

Bhanoo, Sindya N. "Seas Grow Less Effective at Absorbing Emissions." *The New York Times*, November 19, 2009. http://www.nytimes.com/2009/11/19/science/earth/19oceans.html.

Clarkson, M. O., et al. "Ocean Acidification and the Permo-Triassic Mass Extinction." *Science* 348 (April 10, 2015): 229–232.

"Climate Deal Must Cover Acid Oceans, World Scientists Urge." Environment News Service, June 1, 2009. http://www.ens-newswire.com/ens/jun2009/2009-06-01-01.asp (no longer available).

Dean, Cornelia. "Rising Acidity Is Threatening Food Web of Oceans, Science Panel Says." *The New York Times*, January 31, 2009. http://www.nytimes.com/2009/01/31/science/earth/31ocean.html.

Feely, Richard A., et al. "Impact of Anthropogenic CO_2 on the $CaCO_3$ System in the Oceans." *Science* 305 (July 16, 2004): 362–366.

Feely, Richard A., et al. "Evidence for Upwelling of Corrosive 'Acidified' Water onto the Continental Shelf." *Science* 320 (June 13, 2008): 1490–1492.

Foster, Taryn, et al. "Ocean Acidification Causes Structural Deformities in Juvenile Coral Skeletons." *Science Advances* 2(2) (February 19, 2016): e1501130. doi: 10.1126/sciadv.1501130.

Gruber, Nicolas, et al. "Rapid Progression of Ocean Acidification in the California Current System." *Science* 337 (July 13, 2012): 220–223.

Hand, Eric. "Acid Oceans Cited in Earth's Worst Die-off." *Science* 348 (April 10, 2015): 165–166.

Henderson, Casper. "The Other CO_2 Problem." *New Scientist*, August 5, 2006, 28–33.

Kerr, Richard A. "Ocean Acidification Unprecedented, Unsettling." *Science* 328 (June 18, 2010): 1500–1501.

Khatiwala S., F. Primeau, and T. Hall. "Reconstruction of the History of Anthropogenic CO_2 Concentrations in the Ocean." *Nature* 462 (November 19, 2009): 346–349.

Kolbert, Elizabeth. "The Darkening Sea: What Carbon Emissions Are Doing to the Oceans." *The New Yorker*, November 20, 2006, 66–75.

"Marine Life on a Warming Planet" (Editorial). *The New York Times*, December 2, 2012. http://www.nytimes.com/2012/12/03/opinion/marine-life-on-a-warming-planet.html.

"Pacific Coast Turning More Acidic." Earth Observatory. May 22, 2008. http://earthobservatory.nasa.gov/Newsroom/MediaAlerts/2008/2008052226903.html (no longer available).

"Report Says Oceans Hit by Carbon Dioxide Use." *Omaha World-Herald*, July 17, 2004, 5A.

Riebesell, U., et al. "Enhanced Biological Carbon Consumption in a High CO_2 Ocean." *Nature* 450 (November 22, 2007): 545–548.

Ruttimann, Jacqueline. "Oceanography: Sick Seas." *Nature* 442 (August 31, 2006): 978–980.

Sabine, Christopher L., et al. "The Oceanic Sink for Anthropogenic CO_2." *Science* 305 (July 16, 2004): 367–371.

Solie, Stacey. "Scientists Adopt Tiny Island as a Warming Bellwether." *The New York Times*, October 6, 2012. http://www.nytimes.com/2012/10/07/us/scientists-in-washington-state-adopt-tiny-island-as-climate-change-bellwether.html.

Welch, Craig. "Puget Sound Waters Becoming More Corrosive." *Seattle Times*, July 13, 2010, B1, B2.

Welch, Craig. "Sea Change: Scientists Fear Ocean Acidification Will Drive the Collapse of Alaska's Iconic Crab Fishery." *Seattle Times*, September 11, 2013a. http://apps.seattletimes.com/reports/sea-change/2013/sep/11/alaska-crab-industry/.

Welch, Craig. "Sea Change: The Pacific's Perilous Turn." *Seattle Times*, September 11, 2013b. http://apps.seattletimes.com/reports/sea-change/2013/sep/11/pacific-ocean-perilous-turn-overview/.

Zachos, James C., et al. "Rapid Acidification of the Ocean During the Paleocene–Eocene Thermal Maximum." *Science* 308 (June 10, 2005): 1611–1615.

See also: Oceans' Absorption of Heat; Ocean Acidity, Worldwide; Phytoplankton

OCEAN ACIDITY, WORLDWIDE

Scientists have been investigating the effects of rising ocean acidification on marine animals with calcium shells, including phytoplankton. These animals form the base of the oceanic food chain and provide half of the world's oxygen. Gretchen Hofmann of the University of California–Santa Barbara reported that rising ocean temperatures and acidification could be fatal to the purple sea urchin (*Stronylocentrotus purpuratus*). At an ambient pH of 7.8, the larvae of the purple sea urchin had great difficulty building their skeletons. Warming the water in which the sea urchins lived compounded the effect (Kintsch and Stokstad 2008).

The limits of acidity tolerance for many corals is now expected at about 550 parts per million carbon dioxide, a level that will be reached at current rates of accretion around the middle of the 21st century. As CO_2 rises above that level, the oceans will encounter a "reef gap" at which point "these magnificent, diverse structures disappear from the world" (Zalasiewicz and Williams 2016, 189). The last reef gap occurred 55 million years ago during the Paleocene–Eocene Thermal Maximum. The corals eventually recovered after several million years. Life will continue during such a gap, and coral reefs "will be replaced by 'slime-rock' systems dominated by algal and microbial mats and jellyfish" (Zalasiewicz and Williams 2016, 191).

A report on ocean acidification by Britain's Royal Society concluded that, "without significant action to reduce CO_2 emissions," there may be "no place in the future oceans for many of the species and ecosystems we know today" (Kolbert 2006, 74). In other words, increasing ocean acidification because of rising carbon dioxide levels may threaten large numbers of ocean species with extinction, including most with calcium shells, which break down as acidity rises.

Acidic Water Kills Krill

The shrimplike krill, one of Earth's most fundamental species and a key link in the Antarctic food chain, may suffer significant depletion as carbon dioxide levels rise in the oceans. Krill are so abundant in the ocean around Antarctica that they can form schools several miles long with thousands of them in a cubic foot. Commercial fishing boats can harvest tons of krill in a few months.

Dr. So Kawaguchi, a wildlife biologist with the Australian federal government's Antarctic Division, studied krill for a quarter century. Two decades into his career, he began to suspect that rising CO_2 levels were harming krill by reducing their hatch rates. This reduction will affect the many marine animals that prey on krill, including whales, squid, seals, several species of fish, penguins, and many types of seabirds. "Higher levels of carbon dioxide in the water mean greater levels of ocean acidification," said Kawaguchi. "This interrupts the physiology of krill. It stops the eggs hatching, or the larvae developing" (Innis 2015). "If we continue with business as usual, and we don't act on reducing carbon emissions, in that case, there could be a 20 to 70 percent reduction in Antarctic krill by 2100," Kawaguchi said. "By 2300, the Southern Ocean might not be suitable for krill reproduction."

Further Reading

Innis, Michelle. "Warming Oceans May Threaten Krill, a Cornerstone of the Antarctic Ecosystem." *The New York Times*, October 19, 2015. http://www.nytimes.com/2015/10/20/science/australia-antarctica-krill-climate-change-ocean.html?_r=0.

The effects of acidity in the oceans also will continue long after the burning of fossil fuels peaks on land. Ken Caldeira modeled ocean acidification for fossil fuel burning with an assumed CO_2 emissions peak in 2100 and found that the oceans would continue to become more acidic for several centuries after that. Under this model, surface acidity would peak at about 2,750 parts per million. A kilometer down in the ocean, acidification would rise for a thousand years. "People should know that the consequences of what we're doing in the next decade will last for thousands of years," said Caldeira (Ruttimann 2006).

Growth of Corals Reduced since the Dawn of Industrialization

A team of scientists recently replicated oceanic acidic levels at the beginning of the Industrial Revolution and traced them to the present day and learned that rising acidity has been stunting coral growth since humankind began reducing oceanic pH. "Acidification-induced reductions in calcification are projected to shift coral reefs from a state of net accretion to one of net dissolution this century," they wrote. "When ocean chemistry is restored closer to pre-industrial conditions, net community calcification increases. In providing results from the first seawater chemistry manipulation experiment of a natural coral reef community, we provide evidence that net community calcification is depressed compared with values expected for pre-industrial conditions, indicating that ocean acidification may already be impairing coral reef growth" (Albright et al. 2016).

Corals raised in water near Australia with a pH matching that of the preindustrial world grew approximately 7 percent more than those in water with contemporary acidity. Previous studies had been completed in laboratories. "Our work provides the first strong evidence from experiments on a natural ecosystem that ocean acidification is already slowing coral reef growth," said study lead author Rebecca Albright of the Carnegie Institution for Science. "Ocean acidification is already taking its toll on coral reef communities. This is no longer a fear for the future; it is the reality of today," she added (Rice 2016). This research was released as an intense El Niño event raised ocean temperatures to record highs around the world, contributing to the longest episode of coral bleaching (death) on record—from 2014 to 2016.

Corals at Risk

By the end of the 21st century, according to Ken Caldeira, surface acidity around Antarctica will be roughly double preindustrial levels (a 0.2 decrease in pH), which will threaten the ability of aquatic life forms to maintain their shells (Kolbert 2006, 70). Such a level would put about two-thirds of cold-water corals in corrosive waters (Kintsch and Stokstad 2008). Acidification will affect corals acutely, dissolving their shells at a time when warming temperatures are already threatening their survival. "While bleaching . . . is an acute stress that's killing them off . . . acidification is a chronic stress that's preventing them from recovering," said Joanie Kleypas, a coral-reef scientist at the National Center for Atmospheric Research in Boulder, Colorado (Kolbert 2006, 72).

Ilsa B. Kuffner of the U.S. Geological Survey in St. Petersburg, Florida, reported in *Nature Geoscience* that an increase in ocean acidity harms crustose coralline algae, the builders of coral reefs, as well as the reefs themselves. Kuffner and colleagues manipulated the pH of ocean water near reefs in Hawaii for conditions expected in the year 2100 and found after seven weeks "far less algae encrusted on clear plastic cylinders inside the more acidic tanks" (Fountain 2008). Instead, the space was taken by "soft" algae that do not secrete calcium carbonate. The secretion of calcium by crustose coralline algae acts as a mortar to help maintain the structure of reefs.

Coral reefs are among the richest biological areas of the oceans, so acidification is a major long-term threat to aquatic life. Thomas Lovejoy, who coined the term *biological diversity* in 1980, compared the effects of ocean acidification to "running the course of evolution in reverse." The two most important biological factors for organisms in the ocean are temperature and acidity, Lovejoy said. The effects of changes in both provoked by human-induced carbon dioxide emissions reach to the base of the oceanic food chain with profound long-term implications that favor "lower" forms of life such as jellyfish and other invertebrates. In the extreme long run, some scientists fear human intervention in the oceanic system may favor the return to slime as a predominant life form there, according to German marine biologist Ulf Riebesell (Kolbert 2006, 75).

Deepwater corals are at risk from increasing ocean acidification for several reasons. First, they are made of aragonite, a carbonate material that is more soluble than the calcite used by corals closer to the surface. The vulnerabilty of carbonates to dissolution also increases in colder water at greater pressure. By the end of this century, two-thirds of deepwater corals could be exposed to seawater that is corrosive to aragonite; none are exposed today.

Ocean Acidity and Oceanic Dead Zones

In addition to increased nitrogen-based fertilizer use and sewage runoff, accelerating emissions of carbon dioxide into the atmosphere (and its absorption by the oceans) is causing "dead zones" to expand rapidly; in these desolate regions, pollution-fed algae deprive other marine life of oxygen. The number of such areas worldwide increased to approximately 200 by 2008, a 34 percent rise in two years. Given carbon dioxide's long life in the atmosphere, this trend could persist for centuries or even longer after greenhouse gas emissions end. A 100,000-year computer simulation indicates that severe ocean oxygen depletion could last for thousands of years. Dead zones that cover about 2 percent of the ocean surface today could expand to 20 percent by that time (Shaffer et al. 2009, 105). In the simulation, business-as-usual carbon emissions increase to current rates until 2100; atmospheric levels increase 400 percent and the climate warms by approximately 5°C.

The scientists' simulation, led by Gary Shaffer of the Niels Bohr Institute, University of Copenhagen, led to the conclusion that there would be:

> severe, long-term ocean oxygen depletion, as well as a great expansion of ocean oxygen-minimum zones for scenarios with high emissions or high climate sensitivity. . . . Climate feedbacks within the Earth system amplify the strength and duration of global warming, ocean heating, and oxygen depletion. Decreased oxygen solubility from surface-layer warming accounts for most of the enhanced oxygen depletion in the upper 500 meters of the ocean. Possible weakening of ocean overturning and convection lead to further oxygen depletion, also in the deep ocean. We conclude that substantial reductions in fossil fuel use over the next few generations are needed if extensive ocean oxygen depletion for thousands of years is to be avoided. (Shaffer et al. 2009)

Further Reading

Shaffer, Gary, Steffen Malskær Olsen, and Jens Olaf Pepke Pedersen. "Long-Term Ocean Oxygen Depletion in Response to Carbon Dioxide Emissions from Fossil Fuels." *Nature Geoscience* 2 (2009): 105–109.

The death of corals (coccolithophores) near the surface of the oceans could amplify global warming because of albedo. When they bloom, these organisms lighten the surface, reflecting sunlight. They also produce dimethylsulphide, which accounts for much of the aerosolized sulfate in the air above the oceans that "seed" cloud droplets. Without them, oceanic cloud cover could decline, allowing more sunlight and heat to reach the surface (Ruttimann 2006, 980).

"If you look at the business-as-usual scenario for emissions and its impact with respect to aragonite in surface waters, by the end of the [21st] century there [will be] no place left with the kind of chemistry where corals grow today," said Ken Caldeira (Henderson 2006, 31). Corals could become rare because temperatures and acidity levels are rising at the same time. In addition, pteropods, which form an important part of the food chain for cod, salmon, and whales in colder water, could find their shells dissolving at the pH levels anticipated by 2050 (Henderson 2006, 31).

An experiment conducted between 1996 and 2003 at Columbia University's Biosphere 2 lab in Tucson, Arizona, concluded that the growth of corals was reduced by half in aquariums where they were exposed to a level of carbon dioxide projected for 2050. Coupled with the warmer sea temperatures that climate change produces, corals may not survive by the end of the century, said Christopher Langdon, a marine biology professor at the University of Miami. Ocean acidification is "going to be on a global scale and it's also chronic," said Langdon. "Twenty-four/seven, it's going to be stressing these organisms. . . . These organisms probably don't have the adaptive ability to respond to this new onslaught" (Eilperin 2006). Stanford University marine biologist Robert B. Dunbar has studied the effect of increased carbon dioxide on coral reefs in Israel and Australia's Great Barrier Reef. "What we found in Israel was [that] the community is dissolving," Dunbar said (Eilperin 2006).

Ken Caldeira has mapped areas where corals grow today and the pH levels of the water in which they live. He maintains that by the end of the 21st century, no seawater will be as alkaline as corals' current habitats. If carbon dioxide emissions continue at their current levels, he said, "It's say 'goodbye' to coral reefs" (Eilperin 2006). "We're taking a huge risk," said Ulf Riebesell, a marine biologist at the Leibniz Institute of Marine Sciences in Kiel, Germany. "Chemical ocean conditions 100 years from now will probably have no equivalent in the geological past, and key organisms may have no mechanisms to adapt to the change" (Schiermeier 2004).

No Precedent in Recorded History

Scientists investigated 328 colonies of massive *Porites* corals on the Great Barrier Reef off Australia that grew to more than six meters tall in a period spanning from decades to centuries. Results from 69 sections of the reef found that calcification had declined 14.2 percent between 1990 and 2005, impeding the reefs' growth by 13.3 percent (De'ath et al. 2009, 116). Such a sudden and massive decline in the reef's calcification has no precedent in recorded history, some 400 years of observation. Increasing temperature stress and a rising carbon dioxide level in the water around the reef are the probable causes. "This study has provided the first really vigorous snapshot of how calcification might be changing [worldwide]," said marine biologist Ove Hoegh-Guldberg of Australia's University of Queensland. "The results are extremely worrying" (Pennisi 2009, 27). During the 2015–2016 El Niño, more than half of the corals in the Great Barrier Reef's northernmost 700 miles died, but 95 percent of those farther south survived, according to Australia's ARC Center of Excellence for Coral Reef Studies.

Analog of a Future Acidic Ocean

As an analog for a future acidic ocean, consider volcanic vents near Normanby Island (off Papua New Guinea) that spout water with high levels of carbon dioxide. The water chemistry near these vents resembles what scientists believe may be typical ocean water near the end of the 21st century. As Craig Welch wrote in the *Seattle Times*, "That makes this isolated splash of coral reef a chilling vision of our future oceans" (Welch 2013b) as the oceans worldwide absorb about 20 trillion pounds of carbon dioxide a year. The changing chemistry of the oceans was expected to inflict damage on living organisms in the far future a few decades ago, but in some areas ocean habitats are already turning toxic as acidity causes billions of oysters to go sterile along the coast of Washington state (as well as within a hatchery on Hood Canal across Puget Sound from Seattle). Some mussels have been damaged in the same areas, along with clams and baby scallops, as acidity also "is dissolving a tiny plankton species eaten by many ocean creatures, from auklets and puffins to fish and whales—and that had not been expected for another 25 years" (Welch 2013b). The U.S. West Coast has become an acidification hot spot as strong prevailing winds bring water rich in carbon dioxide to the surface and into shellfish habitats.

"I used to think it was kind of hard to make things in the ocean go extinct," said James Barry of the Monterey Bay Aquarium Research Institute in California. "But this change we're seeing is happening so fast it's almost instantaneous. I think it might be so important that we see large levels, high rates, of extinction" (Welch 2013b).

According to Katharina Fabricius, an ecologist from the Australian Institute of Marine Science who has been exploring the seascape near the vents rich in carbon dioxide, "Instead of tiered jungles of branching, leafy reefs or a watery Eden of delicate corals arrayed in fans . . . saw mud, stubby spires, and squat boulder corals. Snails and clams were mostly gone, as were most of a reef's usual residents: worms, colorful sea squirts, and ornate feather stars" (Welch 2013b).

Another analog for an acidified ocean lies near Castello Aragonese, a volcanic island near Naples, Italy. The coral reefs there provide a riot of life and color with several species of fish, sea urchins, and smaller animals. A few hundred yards away, volcanic vents shoot streams of silver bubbles into the air—carbon dioxide—that combine with seawater to form carbonic acid, which raises the CO_2 level to what scientists, projecting current trends, predict will be the worldwide average in 2100. The seascape there is a barren and dull green, hosting little except algae, some sea grass, and jellyfish. Anything that lives in a shell, from the corals to the sea urchins, has disappeared. This may be the ocean of the future because seawater, once carbonated, is nearly impossible to reclaim. Restoring the usual 8.1 pH of ocean water would require two tons of lime for every ton of carbon dioxide absorbed by the oceans from the atmosphere, and the oceans are now absorbing a million tons of CO_2 each hour (Kolbert 2011, 102, 108).

Further Reading

Kolbert, Elizabeth. "The Acid Sea." *National Geographic*, April 2011, 100–121.
Welch, Craig. "Sea Change: The Pacific's Perilous Turn." *Seattle Times*, September 11, 2013b. http://apps.seattletimes.com/reports/sea-change/2013/sep/11/pacific-ocean -perilous-turn-overview/.

Computer models forecast that, at the rate that acidity has been increasing, polar waters will no longer sustain viable populations of pteropods. Once acidity reaches levels that dissolve calcium shells in the tropics, "It's a doomsday scenario for coral reefs," said Caldeira—that is, for corals not already killed by rising water temperatures. He anticipates that coral reefs will survive only in walled off enclosures where acidity can be controlled by humankind. "Our emissions are huge compared with natural fluxes," said Caldeira. "If you could stop emissions and wait 10,000 years, natural processes would probably take care of most of it" (Holland 2007, 111). Emissions, however, are not being curtailed.

Effects on Plankton

Increasing ocean acidity is already reducing the weight of some phytoplankton, which form the base of the marine food chain ("Acidic Oceans" 2014). As early as 2011, scientists found severe damage to shells of a sea snail (*Limacina helicina*) that plays an important role in the food web. A report in *Science* documented the effects on an estimated one-fifth of these plankton, also called "sea butterflies," some of which had damaged shells even in preindustrial times because of natural water acidity. Now "anthropogenic emissions of carbon dioxide made the water more acidic, more than doubling the number of individuals with damaged shells" (Kintisch 2014). Birds, pollack, baby salmon, and whales (among many other species) eat these sea butterflies in enormous numbers. Ocean acidification may ultimately imperil 1 billion people worldwide who acquire most of their protein from the oceans because stocks of fish consumed by people will be reduced by a lack of smaller organisms that make up their diets.

Shells of microscopic oceanic animals already have thinned by as much as one-third because of rising carbon dioxide levels. With funding from the Australian Government Department of Climate Change, Will Howard of the Antarctic Climate & Ecosystems Cooperative Research Center in Australia. and his colleagues collected microscopic marine animals—called *planktonic foraminifera*, or *forams*—from the South Tasman Rise region of the Southern Ocean. They compared the weights of the shells of these modern forams to those trapped in ocean sediments before the Industrial Revolution and the buildup of carbon dioxide and found that today's shell are 30 percent to 35 percent lighter than older fossils. This is nearly equal to the amount by which the proportion of CO_2 has risen (Thompson 2009). The findings were described in the March 8, 2009, issue of the journal *Nature Geoscience*.

The depletion of forams has significant implications for oceanic food chains. These findings "are a worrying signal of what we can expect to see elsewhere in the future," Howard said. "The Southern Ocean is giving us a strong indication of an acidification process that will spread throughout the global ocean" (Thompson 2009). Andrew D. Moy and colleagues wrote:

> Planktonic foraminifera are single-celled calcite-secreting organisms that represent between 25 and 50 percent of the total open-ocean marine carbonate flux [that influences] the transport of organic carbon to the ocean interior. Here we compare the shell weights of the modern foraminifer *Globigerina*

bulloides collected from sediment traps in the Southern Ocean with the weights of shells preserved in the underlying Holocene-aged sediments. We find that modern shell weights are 30–35 percent lower than those from the sediments, consistent with reduced calcification today induced by ocean acidification. We also find a link between higher atmospheric carbon dioxide and low shell weights in a 50,000-year-long record obtained from a Southern Ocean marine sediment core. (Moy et al. 2009)

Jason M. Hall-Spencer and colleagues studied the effects of acidification in eco-systems at shallow coastal sites where CO_2 from volcanic vents lowers the pH of the water and found that calcareous organisms such as corals and sea urchins were adversely affected. This probably was the first time that the effects of elevated acidity had been tested in ocean water; Most other studies have been done under laboratory conditions. The species populating the vent sites make up a suite of organisms that are resilient to naturally high concentrations of CO_2 and indicate that ocean acidification may benefit highly invasive non-native algal species (Hall-Spencer et al. 2008, 96).

As scientists learn more about acidification of the oceans, the reality of the threat becomes more evident. The date when increasing carbon dioxide levels in the oceans are expected to change acidity enough to dissolve the calcium carbonate shells of corals, plankton, and other marine animals has now advanced to the next few decades, sooner than previously projected. A team of scientists recently writing in *Nature* said, "In our projections, Southern Ocean surface waters will begin to become undersaturated with respect to aragonite, a metastable form of calcium carbonate, by the year 2050. By 2100, this undersaturation could extend throughout the entire Southern Ocean and into the subarctic Pacific Ocean" (Orr et al. 2005, 681).

Although the fate of plankton and marine snails may not seem as compelling as vibrantly colored coral reefs, they are critical to sustaining marine species such as salmon, redfish, mackerel, and baleen whales. "These are groups everyone depends on, and if their numbers go down there are going to be reverberations throughout the food chain," said John Guinotte, a marine biologist at the Marine Conservation Biology Institute. "When I see marine snails' shells dissolving while they're alive, that's spooky to me" (Eilperin 2006).

Further Reading

"Acidic Oceans Shrink Plankton." *Nature* 510 (June 12, 2014): 190.

Albright, Rebecca, et al. "Reversal of Ocean Acidification Enhances Net Coral Reef Calcification." *Nature* 530 (released online February 24, 2016). http://www.nature.com/nature/journal/vaop/ncurrent/full/nature17155.html.

De'ath, Glenn, Janice M. Louygh, and Katharina E. Fabricius. "Declining Coral Calcification on the Great Barrier Reef." *Science* 323 (January 2, 2009): 116–119.

Eilperin, Juliet. "Growing Acidity of Oceans May Kill Corals." *Washington Post,* July 5, 2006, A1. http://www.washingtonpost.com/wp-dyn/content/article/2006/07/04/AR2006070400772_pf.htmla.

Fountain, Henry. "More Acidic Ocean Hurts Reef Algae as Well as Corals." *The New York Times,* January 8, 2008, D3.

Hall-Spencer, Jason M., et al. "Volcanic Carbon Dioxide Vents Show Ecosystem Effects of Ocean Acidification." *Nature* 454 (July 3, 2008): 96–99.

Henderson, Casper. "The Other CO_2 Problem." *New Scientist*, August 5, 2006, 28–33.

Holland, Jennifer S. "The Acid Threat: As CO_2 Rises, Shelled Animals May Perish." *National Geographic*, November 2007, 110–111.

Kintisch, Eli. "'Sea Butterflies' Are a Canary for Ocean Acidification." *Science* 344 (May 9, 2014): 569.

Kintisch, Eli, and Erik Stokstad. "Ocean CO_2 Studies Look beyond Coral." *Science* 319 (February 22, 2008): 1029.

Kolbert, Elizabeth. "The Darkening Sea: What Carbon Emissions Are Doing to the Oceans." *The New Yorker*, November 20, 2006, 66–75.

Moy, Andrew D., et al. "Reduced Calcification in Modern Southern Ocean Planktonic Foraminifera." *Nature Geoscience*, March 8, 2009 (online). doi: 10.1038/ngeo460.

Orr, James C., et al. "Anthropogeenic Ocean Acidification over the Twenty-First Century and Its Impact on Calcifying Organisms." *Nature* 437 (September 29, 2005): 681–686.

Pennisi, Elizabeth. "Calcification Rates Drop in Australian Reefs." *Science* 323 (January 2, 2009): 27.

Rice, Doyle. "Study: Ocean Acidification Stunting Growth of Coral Reefs." *USA Today*, February 24, 2016. http://www.usatoday.com/story/weather/2016/02/24/ocean-acidification-coral-reefs/80857670/.

Ruttimann, Jacqueline. "Oceanography: Sick Seas." *Nature* 442 (August 31, 2006): 978–980.

Schiermeier, Quirin. "Researchers Seek to Turn the Tide on Problem of Acid Seas." *Nature* 430 (August 19, 2004): 820.

Thompson, Andrea. "Growing Acid Problem Thins Shells of Ocean Creatures." LiveScience, March 8, 2009. http://www.livescience.com/environment/090308-acidification-shells.html.

Zalasiewicz, Jan, and Mark Williams. *Ocean Worlds: The Story of Seas on Earth and Other Planets*. New York: Oxford University Press, 2016.

See also: Oceans' Absorption of Heat; Ocean Acidity, North America; Phytoplankton

OCEAN CIRCULATION

Warming seas could affect patterns of oceanic circulation around the world (the thermohaline circulation) that replenish the world's oceans with oxygen, leading to possible extinction of several sea creatures. A debate also has developed regarding whether changes in ocean circulation could cause some areas of the world (most notably, Western Europe) to experience a marked cooling trend by suppressing the warm waters of the Gulf Stream, which help to supply that area with a winter climate that is unusually warm for such a northerly latitude.

By 2015, however, scientific consensus was leaning away from this possibility because changes earlier considered long-term were being regarded instead as part of cyclical variation and probably unrelated to a warming climate (Schiermeier 2007). Stuart Cunningham and colleagues at the National Oceanography Center at Southampton, United Kingdom, found that the thermohaline circulation varied by 25 percent in one year (Cunningham et al. 2007).

Warming and Inhibition of Thermohaline Circulation

The term *thermohaline* is used because "water density in the ocean is determined by both temperature and salinity" (Rahmstorf 2003, 699). According to Peter U. Clark and colleagues writing in *Nature*, most but not all coupled general circulation model projections of 21st century climate anticipate a reduction in the strength of the Atlantic overturning circulation. The result would be warming seas provoked by increasing concentrations of greenhouse gases (Clark et al. 2002). This flow is part of what marine scientists called the *global conveyor*, a vast submarine flow of water south from the Arctic. The Gulf Stream that keeps Britain 5°C warmer than expected for such a northerly latitude. The Gulf Stream delivers 27,000 times more heat to British shores than all of that nation's power stations combined (Radford 2001).

Warming Water off the U.S. West Coast

Western North America has recently experienced record drought and warmth, which seems to be connected with a mass of warm ocean water off the West Coast, as the eastern portion of the continent has experienced a wet, cold cycle. Are these being provoked by distortion in the jet stream or is that just one of several possible influences? During the winters of 2013–2014 and 2014–2015, an exceptionally large area of seawater that was at least 2.5°C (4.5°F) above average persisted along the U.S. West Coast. This are was 1,600 kilometers (1,000 miles) wide and as much as 100 meters (300 feet) deep and sat under a long-lasting dome of high pressure that kept surface waters from mixing with cooler layers below. Most years, such a pressure pattern is prevalent during summer and early fall but recedes southward in lieu of stormier weather in winter (Bond et al. 2015).

According to NASA's Earth Observatory ("Warm Water" 2015),

> effects have rippled through the marine environment. This warm "blob," as researchers began to call it, saw significantly reduced chlorophyll levels—fewer phytoplankton. With less food in the region, Pacific Northwest and California current fish species started appearing in the waters off Alaska. Seabirds and sea lions were dying off or becoming emaciated because of a lack of food.

Dennis Hartmann, also at the University of Washington, compiled historical data and computer models to link recent warming of the ocean off the U.S. West Coast to recent El Niños, the Pacific Decadal Oscillation, the North Pacific Mode, all of which alter atmospheric circulation (Hartmann 2015). "It's an interesting question if that's just natural variability happening or if there's something changing about how the Pacific Ocean decadal variability behaves," said Hartmann. "I don't think we know the answer. Maybe it will go away quickly and we won't talk about it anymore, but if it persists for a third year, then we'll know something really unusual is going on" ("Warm Water" 2015).

Further Reading

Bond, Nicholas A., et al. "Causes and Impacts of the 2014 Warm Anomaly in the NE Pacific." *Geophysical Research Letters*, 42 (April 6, 2015). http://onlinelibrary.wiley .com/doi/10.1002/2015GL063306/abstract?campaign=wlytk-41855.52 82060185.
Hartmann, D. L. "Pacific Sea Surface Temperature and the Winter of 2014." *Geophysical Research Letters*, March 19, 2015, 42. http://onlinelibrary.wiley.com/doi/10.10 02/2015GL063083/abstract?campaign=wlytk-41855.5282060185.
"Warm Water and Strange Weather May Be Connected." NASA Earth Observatory, April 18, 2015. http://earthobservatory.nasa.gov/IOTD/view.php?id=85714&src =eoa-iotd.

A report presented at the annual meeting of the American Association for the Advancement of Science during February 2005 by Ruth Curry, a scientist at the Woods Hole Oceanographic Institute, indicated that massive amounts of freshwater from melting Arctic ice are seeping into the Atlantic Ocean. According to Curry's research, between 1965 and 1995, enough freshwater to fill Lake Superior, Lake Erie, Lake Ontario, and Lake Huron combined, melted from the Arctic region and poured into the northern Atlantic. Curry projected that if this pattern continued at the same rate, the thermohaline circulation may have begun to shut down in about two decades. The changes proved not to be linear, however, throwing any such projections into doubt. However, the problem of circulation breakdown is still salient in the long run.

Curry and colleagues also projected that Greenland's ice, which has been been melting quickly, affects ocean circulation when its relatively cold freshwater flows into the ocean. "We are taking the first steps," Curry said in a news conference. "The system is moving in that direction" (Borenstein 2005). In the longer range, according to calculations by Curry and colleagues, "at the observed rate, it would take about a century to accumulate enough freshwater (e.g., 9,000 cubic meters) to substantially affect the ocean exchanges across the Greenland–Scotland Ridge, and nearly two centuries of continuous dilution to stop them" (Curry and Mauritzen 2005, 1774).

Ocean Salinity and Circulation

The salinity of the North Atlantic Ocean is closely related to world ocean circulation. If the North Atlantic becomes too fresh (most likely from melting Arctic ice), its waters could stop sinking, and the global conveyor could slow and even perhaps stop. In *The Discovery of Global Warming*, Spencer R. Weart provides a capsule description of possible changes in the ocean's thermohaline circulation under the influence of sustained and substantial global warming:

If the North Atlantic around Iceland should become less salty—for example, if melting ice sheets diluted the upper ocean layer with freshwater—the surface layer would no longer be dense enough to sink. The entire circulation that drove cold water south along the bottom could lurch to a halt. Without the vast compensating drift of tropical waters northward, a new glacial period could begin. (Weart 2003, 64)

Atmospheric winds move heat quickly across large distances, but fundamental changes in ocean temperatures and circulation may require centuries. Thermohaline circulation involves warm surface currents that distribute tropical heat, as well as deeper currents that carry cold water back toward the equator. Together, these currents form a system that circulates ocean water, heat, oxygen, and nutrients through the North and South Atlantic Oceans and then into the Indian Ocean and the Pacific Ocean. The North Atlantic is one of only two areas in the world's oceans where the presence of ice aids formation of salty, dense water that then sinks and helps to drive ocean circulation, carrying oxygen with it. The other is the Weddell Sea in the Antarctic.

According to one expert observer, "If the warming is strong enough and sustained long enough, a complete collapse [of thermohaline circulation] cannot be excluded" (Clark et al. 2002, 863). "What is fairly clear is that if the ocean circulation patterns that now warm much of the North Atlantic were to slow or stop, the consequences could be quite severe," Clark said. "This might also happen much quicker than many people appreciate. At some point the question becomes how much risk do we want to take?" (Cowen 2002). This circulation is vital to the circulation of oxygen and nutrients in much of the Atlantic Ocean—and thus necessary to marine life as we know it. Similar circulation systems perform the same function in other oceans around the world. Without the thermohaline circulation, the oceans could become largely stagnant with large areas bereft of life.

Conditions May Change Quickly

Thomas Stocker and colleagues also argue that these changes may be nonlinear; "they may have large amplitudes and may occur as surprises." Inherent in such changes is their "reduced predictability" (Stocker et al. 2001, 277). "Among such nonlinear changes," they have written, "are the collapse of large Antarctic ice masses and rapid sea level rise, the desertification of entire land regions, the thawing of permafrost and associated release of large amounts of radiatively active gases, and the collapse of the large-scale Atlantic thermohaline (i.e., the temperature and salinity driven) circulation" (Stocker et al. 2001, 277).

Carsten Ruhlmann et al. wrote in *Nature* that Earth's emergence from its last glaciation was punctuated by several short, sharp periods during which glacial conditions returned. These variations seem to have been triggered by the type of changes in the North Atlantic thermohaline circulation that are considered possible in a world warmed by increasing emissions of greenhouse gases: "The thermohaline circulation was the important trigger for these rapid climate changes" (Ruhlmann et al.

1999, 512). Climate changes that diminish thermohaline circulation tend to divert the Gulf Stream southward, cooling the North Atlantic Ocean (as well as Greenland and Western Europe), but "warming . . . the western tropical North Atlantic . . . and most of the Southern Hemisphere" (Ruhlmann et al. 1999, 512).

Thermohaline Circulation Changes Measured

A study of North Atlantic ocean circulation published in late 2005 (Quadfasel 2005) reported a 30-percent reduction in the meridional overturning (thermohaline) circulation at 26.5° north latitude based on readings taken in 1957, 1981, 1992, and 1998. Harry Bryden at the Southampton Oceanography Center in the United Kingdom, whose group carried out the analysis, said he was not sure whether the change was temporary or part of a long-term trend. This analysis is, however, "the first observational evidence that such a decrease of the oceanic overturning circulation is well underway" (Quadfasel 2005, 565).

"We don't want to say the circulation will shut down," Bryden told the *New Scientist*. "But we are nervous about our findings. They have come as quite a surprise" (Pearce 2005). Bryden's team measured north–south heat flow during 2004 using a set of instruments in various locations in the Atlantic between the Canary Islands and the Bahamas, then compared results to surveys in surveys in 1957, 1981, and 1992. They then calculated that the amount of water flowing north had fallen by around 30 percent (Bryden et al. 2005, 655). The area that was surveyed is limited and could be influenced by regional variations.

Although the role of global warming as direct provocation of changes in thermohaline circulation is open to debate, at least one global climate model relates aerosol emissions in the atmosphere to temperature variability in the North Atlantic Ocean, "suggesting that human activity influences extreme weather events" (Evan 2012). These influences range from effects on hurricane activity to the intensity of drought in the African Sahel and the Amazon River valley. "A number of studies have provided evidence that aerosols can influence long-term changes in sea surface temperatures," wrote Ben Booth and colleagues in *Nature* (2012), "but climate models have so far failed to reproduce these interactions, and the role of aerosols in decadal variability remains unclear."

Desiccating Winds and Storm Tracks

Research conducted at the University of Arizona (UA) in Tucson indicates that hotter, drier springs in the American Southwest are caused at least in part by desiccating winds aggravated by human-abetted climate change. The winds are part of upper-level wind patterns that have contributed to a northward shift in the upper-level winds that steer storms since the 1970s, most often in late winter. Fewer storms have been bringing rain and snow to Southern

California, Nevada, Arizona, Utah, western New Mexico, and western Colorado during that season.

"When you pull the storm track north, it takes the storms with it," said Stephanie A. McAfee, a UA doctoral candidate in geosciences. "During the period it's raining less, it also tends to be warmer than it used to be," McAfee said. "We're starting to see the impacts of climate change in the late winter and early spring, particularly in the Southwest. It's a season-specific kind of drought." Drier, warmer conditions earlier in the year affect snowpack, hydrological processes, and water resources, McAfee said. Other researchers, including Tom Swetnam, director of the UA's Laboratory of Tree-Ring Research, have linked warmer, drier springs to more and larger forest fires ("Drier, Warmer" 2008).

The study associates a general pole-ward movement of storm-steering westerly winds to changes in the West's winter storm pattern for the first time. This change has been provoked by the atmospheric effects of global warming and depleted stratospheric ozone. Climate models suggest that similar movements will continue as the atmosphere warms.

Further Reading

"Drier, Warmer Springs in U.S. Southwest Stem from Human-Caused Changes in Winds." NASA Earth Observatory, August 19, 2008. http://earthobservatory.nasa .gov/Newsroom/MediaAlerts/2008/2008081927359.html (no longer available).

Booth and colleagues used an Earth-system climate model "to show that aerosol emissions and periods of volcanic activity explain 76 percent of the simulated multidecadal variance in 1860–2005 North Atlantic sea surface temperatures. . . . Our findings suggest that anthropogenic aerosol emissions influenced a range of societally important historical climate events such as peaks in hurricane activity and Sahel drought" (Booth et al. 2012, 228).

Worldwide Effects of Circulation Changes

Oceanographers also have found similar trends in parts of the oceans bordering Antarctica. Marine geochemist Wallace Broecker and colleagues at Columbia University's Lamont-Doherty Earth Observatory have found that the "renewal of deep waters by sinking surface waters near Antarctica has slowed to only one-third of its flow a century or two ago" (Kerr 1999, 1062). The work of Broecker, Stewart Sutherland, and Tsung-Hung Peng suggests that the slowing of deepwater formation may be related to warming attending the end of Europe's Little Ice Age (roughly 1400 to 1800 CE). Furthermore, they posit that a "see-sawing of deep water production between the northern Atlantic and the Southern oceans may lie at the heart of the 1,500-year ice-rafting cycle" (Broecker et al. 1999, 1132).

Variations in the thermohaline circulation of the North Atlantic may have shaped climate, including the course of civilizations, in Mesoamerica as well as in Europe. In *The Great Maya Droughts* (2000), for example, Richardson B. Gill postulates that a southward shift in the Gulf Stream forced by cold deep-water flow from the north cooled climate in northwestern Europe between 800 and 1000 CE. The same shift is said by Gill to have displaced the North Atlantic high-pressure system southwestward, causing many years of severe drought in regions populated by the classic Maya civilization in Mesoamerica and perhaps contributing to its collapse.

Further Reading

Booth, Ben B., et al. "Aerosols Implicated as a Prime Driver of 20th-Century North Atlantic Climate Variability." *Nature* 484 (April 12, 2012): 228–232.

Borenstein, Seth. "Scientists Worry about Evidence of Melting Arctic Ice." *Seattle Times*, February 18, 2005, A6.

Broecker, Wallace S., Stewart Sutherland, and Tsung-Hung Peng. "A Possible 20th-Century Slowdown of Southern Ocean Deep Water Formation." *Science* 286 (November 5, 1999): 1132–1135.

Bryden, Harry L., Hannah R. Longworth, and Stuart A. Cunningham. "Slowing of the Atlantic Meridional Overturning Circulation at 25° North." *Nature* 438 (December 1, 2005): 655–657.

Clark, P. U., et al. "The Role of the Thermohaline Circulation in Abrupt Climate Change." *Nature* 415 (February 21, 2002): 863–868.

Cowen, Robert C. "Into the Cold? Slowing Ocean Circulation Could Presage Dramatic—and Chilly—Climate Change." *Christian Science Monitor*, September 26, 2002, 14.

Cunningham, Stuart A., et al. "Temporal Variability of the Atlantic Meridional Overturning Circulation at 26.5°N." *Science* 317 (2007): 935–937. doi: 10.1126/science.1141304.

Curry, Ruth, and Cecilie Mauritzen. "Dilution of the Northern North Atlantic Ocean in Recent Decades." *Science* 308 (June 17, 2005): 1772–1774.

Evan, Amato. "Climate Science: Aerosols and Atlantic Aberrations." *Nature* 484 (April 12, 2012): 170–171.

Gill, Richardson Benedict. *The Great Maya Droughts: Water, Life, and Death.* Albuquerque: University of New Mexico Press, 2000.

Kerr, Richard A. "Oceanography: Has a Great River in the Sea Slowed Down?" *Science* 286 (November 5, 1999): 1061–1062.

Pearce, Fred. "Failing Ocean Current Raises Fears of Mini Ice Age." *New Scientist*, November 30, 2005. https://www.newscientist.com/article/dn8398-failing-ocean-current-raises -fears-of-mini-ice-age/.

Quadfasel, D. "Oceanography: The Atlantic Heat Conveyor Slows." *Nature* 438 (December 1, 2005): 565–566.

Radford, Tim. "As the World Gets Hotter, Will Britain Get Colder? Plunging Temperatures Feared after Scientists Find Gulf Stream Changes." *The Guardian* (U.K.), June 21, 2001, 3.

Rahmstorf, Stefan. "Thermohaline Circulation: The Current Climate." *Nature* 421 (February 13, 2003): 699.

Ruhlmann, Carsten, et al. "Warming of the Tropical Atlantic Ocean and Slowdown of Thermohaline Circulation during the Last Deglaciation." *Nature* 402 (December 2, 1999): 511–514.

Schiermeier, Quirin. "Ocean Circulation Noisy, Not Stalling." *Nature* 448 (August 23, 2007): 844–845.

Stocker, Thomas F., Reto Knutti, and Gian-Kasper Plattner. "The Future of the Thermohaline Circulation—A Perspective." Pp. 277–293 in Dan Seidov, Bernd J. Haupt, and Mark Maslin, eds., *The Oceans and Rapid Climate Change: Past, Present, and Future*. Washington, DC: American Geophysical Union, 2001.

Weart, Spencer R. *The Discovery of Global Warming*. Cambridge, MA: Harvard University Press, 2003.

See also: Oceans' Absorption of Heat; Sea Level Rise; Thermohaline Circulation

OCEANS' ABSORPTION OF HEAT

The oceans have been soaking up much of the excess radiation that Earth retains from its elevated levels of atmospheric greenhouse gases, a major reason why warming on land lags emissions by some 50 years. Kevin Trenberth and colleagues at the National Center for Atmospheric Research analyzed ocean temperature records for 1958 to 2009 and "found that about 30 percent of the extra heat has been absorbed by the oceans and mixed by winds and currents to a depth below about 2,300 feet"; this may be part of the as much as 90 percent excess heat that oceans as a whole have been absorbing. "It turns out there is a spectacular change in the surface winds which then get reflected in changing ocean currents that help to carry some of the warmer water down to this greater depth," Trenberth said. "This is especially true in the tropical Pacific Ocean and subtropics" where patterns that scientists call the Pacific decadal oscillation govern heat retention related to the El Niño (positive) and La Niña (negative) cycle (Roach 2013).

This oscillation switched from positive to negative after an intense El Niño phase in 1997 and 1998, which played a major role in record global temperatures. For a decade after that, global temperatures backed and filled before setting new records in 2010 and 2014 in the absence of an El Niño phase. World temperatures resumed their rapid ascent in 2015 and 2016 with a new El Niño phase that became the most powerful in recorded history. By the end of 2015, temperatures in the central Pacific Ocean had soared to 3°C (5.4°F) above average—a record.

In the alternate phase, La Niña, the tropical Pacific Ocean is cooler and absorbs heat more readily, Kevin Trenberth explained. "So, some of this heat may come back in the next El Niño event . . . but some of it is probably contributing to the warming of the overall planet, the warming of the oceans. . . . It means that the planet is really warming up faster than we might have otherwise expected" (Roach 2013).

Magdalena A. Balmaseda and colleagues wrote in 2013:

The elusive nature of the post-2004 upper ocean warming has exposed uncertainties in the ocean's role in the Earth's energy budget and transient climate sensitivity. Here we present the time evolution of the global ocean heat content for 1958 through 2009 from a new observational based reanalysis of the ocean. Volcanic eruptions . . . are identified as sharp cooling events punctuating a long-term ocean warming trend, while heating continues during the recent upper-ocean-warming hiatus, but the heat is absorbed in the deeper ocean. In the last decade, about 30 percent of the warming has occurred below 700 meters, contributing significantly to an acceleration of the warming trend. The warming below 700 meters remains even when the Argo observing system is withdrawn although the trends are reduced. Sensitivity experiments illustrate that surface wind variability is largely responsible for the changing ocean heat vertical distribution. (Balmaseda et al. 2013, 1754)

Giant Squid Love Warming Water

In addition to jellyfish, one species seems to benefit from a warming environment: the giant squid. Squids have reproduced with such success, increasing their physical size and numbers to a point where, by the end of the 20th century, on a worldwide scale, they had overtaken humans in terms of total planet-wide biomass, according to some estimates. Along with a warmer habitat, human overfishing of other species has favored the squid (reducing competition for food). Australian scientist George Jackson said he believed that global warming is causing squid to grow larger ("Giant Squid" 2002). Although squid may thrive in warmer waters, they face another problem: rising carbon dioxide levels in the oceans affect their respiration rates, which inhibits their ability to swim.

A report in the Australian science journal *Australasian Science* said that most marine researchers now agree that warming habitats have given cephalopods an advantage not available to any other large sea creature and that they have flourished as a result. Their numbers have been increasing, along with their body size (Benson 2002). In July 2002, according to a report by Agence France Presse, the body of a giant squid weighing about 200 kilograms (440 pounds) washed up on a beach in the Australian state of Tasmania. Within days of that event, hundreds of dead squid washed ashore on the coast of California.

George Jackson, with the Institute of Antarctic and Southern Ocean studies in Tasmania, said squid thrived during environmental disasters such as global warming. The animal ate anything in that came their way, bred whenever possible and kept growing (Benson 2002). "This trend has been suggested to be due both to the removal of cephalopod predators such as toothed whales and tuna and an increase of cephalopods due to removal of finfish competitors," said Jackson. "The fascinating thing about squid is that they're short-lived. . . . I haven't found any tropical squid in Australia older than 200 days" (Benson 2002). Jackson continued: "Many of the species have exponential growth, particularly during the juvenile stage so if you increase the water temperature by even a degree it has a tremendous snowballing effect of rapidly increasing their growth rate and their ultimate body size. They get much bigger and they can mature earlier and it just accelerates everything" (Benson 2002).

During July 2009, several thousand 100-pound Humboldt jumbo flying squid, which have been described as "aggressive 1.5-metre-long . . . with razor-sharp beaks and toothy tentacles," invaded shallow waters near San Diego, California, driving scuba divers from the water before many of the squid died on local beaches ("Jumbo Squid" 2009). Some of the squid covered divers' masks with tentacles and swiped cameras. Some divers call the squid, who are native to deep waters off western Mexico, "red devils" for their color and nasty disposition. The squid taste with their tentacles and may be sizing up divers as food. The squid have been sighted outside their usual range as far north as the Alaska panhandle in recent years. The same species was seen in 2005 off the coast of San Diego.

Further Reading

Benson, Simon. "Giant Squid 'Taking over the World.'" *Daily Telegraph* (Sydney), July 31, 2002, 4.
"Giant Squid Film Team Makes Spectacular Catch." September 14, 2002. Agence France Presse, September 14, 2002.
"Jumbo Squid Invade San Diego Shores, Spook Divers." *New Zealand Herald*, July 17, 2009 (LEXIS).

Balmaseda and colleagues extended earlier work along the same lines, indicating that the oceans have been soaking up a large proportion of human-provoked warming. A team of scientists writing in *Nature* in May 2010 calculated that the upper 300 meters of the world's oceans warmed 0.64 watts per square meter between 1993 and 2008, an amount the characterized as "robust" (Lyman et al. 2010, 334).

They also wrote, however, that the "underlying uncertainties in ocean warming are unclear, limiting our ability to assess closure of sea-level budgets, the global radiation imbalance, and climate models. For example, several teams have recently produced different multi-year estimates of the annually averaged global integral of upper-ocean heat content anomalies. . . . Patterns of inter-annual variability, in particular, differ among methods" (Lyman et al. 2010, 334). Although data describing heat absorption in the upper 300 meters of world oceans from 1993 to 2008 provide clear evidence of warming, not all of the methods of measurement are consistent, Kevin E. Trenberth asserted in *Nature* in 2010 (Trenberth 2010, 304).

Xianyao Chen (Key Laboratory of Physical Oceanography, Ocean University of China, Qingdao, China) and Ka-Kit Tung (Department of Applied Mathematics, University of Washington) supported the work of Balmaseda and colleagues:

A vacillating global heat sink at intermediate ocean depths is associated with different climate regimes of surface warming under anthropogenic forcing: The latter part of the 20th century saw rapid global warming as more heat stayed near the surface. In the 21st century, surface warming slowed as more heat moved into deeper oceans. In situ and reanalyzed data are used to trace the pathways of ocean heat uptake. In addition to the shallow La Niña–like patterns in the Pacific that were the previous focus, we found that the slowdown is mainly caused by heat transported to deeper layers in the Atlantic and the Southern oceans, initiated by a recurrent salinity anomaly in the subpolar North Atlantic. Cooling periods associated with the latter deeper heat-sequestration mechanism historically lasted 20 to 35 years. (Chen and Tung 2014, 897)

Oceans Absorb Excess Heat

After setting record highs by large margins during 1998, global surface temperatures on land appeared to back and fill for the next decade and a half. Global warming contrarians asserted that warming's back had been broken, even as indicators of climate change—such as rises in sea level, reduced Arctic ice cover, and rising atmospheric carbon dioxide levels—indicated continued warming. By 2015, scientists had reached widespread consensus on what had caused the "hiatus" in air temperatures. The excess heat had been absorbed, for the most part, by the oceans, according to research by Veronica Nieves and colleagues, published in *Science* (Nieves et al. 2015), who compiled 20 years of ocean temperature records from underwater floats and other instruments.

The Pacific Ocean covers nearly one-third of Earth's surface, so it has an outsized impact on the global thermostat. "As the top 100 meters of the Pacific goes, so goes the surface temperatures of the planet," said William C. Patzert, a coauthor of the study. After 2013, surface air and ocean temperatures resumed their rapid rise, with an assist in 2015 and 2016 from a major El Niño event, which raises equatorial ocean temperatures across much of the Pacific Ocean.

"Natural, decadal variability has been with us for centuries, and it continues to have big regional impacts on society," said Nieves. "We can expect to have more hiatuses in the future, but unless future hiatuses are stronger than usual, they will be less visible due to fast rising greenhouse gases. Right now, the combined effect of the human-caused warming and the Pacific changing to a warm phase can play together and produce warming acceleration" ("New Study" 2015).

Further Reading

"New Study: Heat Is Being Stored Beneath the Ocean Surface." NASA Earth Observatory, July 10, 2015. http://earthobservatory.nasa.gov/IOTD/view.php?id=86184&src =eoa-iotd.

Nieves, Veronica., Josh K. Willis, and William C. Patzert. "Recent Hiatus Caused by Decadal Shift in Indo-Pacific Heating." *Science*, July 9, 2015. doi: 10.1126/science. aaa4521.

Also in 2014, Wniu Cai and colleagues elaborated on the important role greenhouse warming will play in raising the number of severe weather events over Pacific Island nations in coming years:

> The South Pacific convergence zone (SPCZ) is the Southern Hemisphere's most expansive and persistent rain band, extending from the equatorial western Pacific Ocean southeastward towards French Polynesia. Owing to its strong rainfall gradient, a small displacement in the position of the SPCZ causes drastic changes to hydroclimatic conditions and the frequency of extreme weather events—such as droughts, floods, and tropical cyclones—experienced by vulnerable island countries in the region. The SPCZ position varies from its climatological mean location with the El Niño/Southern Oscillation (ENSO), moving a few degrees northward during moderate El Niño events and southward during La Niña events. During strong El Niño events, however, the SPCZ undergoes an extreme swing—by up to ten degrees of latitude toward the Equator—and collapses to a more zonally oriented structure with commensurately severe weather impacts·· Understanding changes in the characteristics of the SPCZ in a changing climate is therefore of broad scientific and socioeconomic interest. *Here we present climate modeling evidence for a near doubling in the occurrences of zonal SPCZ events between the periods 1891–1990 and 1991–2090 in response to greenhouse warming, even in the absence of a consensus on how ENSO will change. . . . The change is caused by a projected enhanced equatorial warming in the Pacific and may lead to more frequent occurrences of extreme events across the Pacific island nations most affected by zonal SPCZ events.* (Cai et al. 2012, 365, emphasis added)

In addition, as the climate warms and sea levels rise, the frequency of local extremes in storm surges will increase along much of the Eastern coastline of the United States. To assess changes in local flood risk, Claudia Tebaldi and her colleagues at Climate Central in Princeton, New Jersey, combined projections from a model of global sea level rise with long-term records from 55 tidal gauges around the United States. The team estimated that by 2050, one-third of gauge locations will see an increase in the frequency of extreme high-water levels that are currently expected to occur only about once a century. Some locations can expect to see these extremes every 10 years on average and others even annually.

Further Reading

Balmaseda, Magdalena A., Kevin E. Trenberth, and Erland Källén. "Distinctive Climate Signals in Reanalysis of Global Ocean Heat Content." *Geophysical Research Letters* 40(9) (May 16, 2013): 1754–1759.

Cai, Wniu, et al. "More Extreme Swings of the South Pacific Convergence Zone Due to Greenhouse Warming." *Nature* 488 (August 16, 2012): 365–369.

Chen, Xianyao, and Ka-Kit Tung. "Varying Planetary Heat Sink Led to Global-warming Slowdown and Acceleration." *Science* 345 (August 22, 2014): 897–903.

Lyman, John M., et al. "Robust Warming of the Global Upper Ocean." *Nature* 465 (May 20, 2010): 334–337.

Roach, John. "Where Did Global Warming Go? The Deep Ocean, Experts Say." NBC News. April 11, 2013. http://science.nbcnews.com/_news/2013/04/11/17708881-where-did -global-warming-go-the-deep-ocean-experts-say?lite.
Trenberth, Kevin E. "The Ocean Is Warming, Isn't It?" *Nature* 465 (May 20, 2010): 304.

See also: Ocean Circulation; Sea Level Rise; Thermohaline Circulation

PHYTOPLANKTON

Phytoplankton are the basis of the maritime food chain, as well as a key consumer of carbon dioxide and producer of oxygen, but rising acidity in the oceans has been devastating these microscopic plants. Surveys by satellites and ships have confirmed that phytoplankton productivity is declining, most notably from the North Pacific to the high Arctic. Mike Toner wrote in the *Atlanta Journal-Constitution*: "Plankton are also as important to the long-term health of the atmosphere as the world's forests. The photosynthesis of the ocean's tiny green plants account for about half of the carbon dioxide that plants remove from the atmosphere each year" (Toner 2002). "The less phytoplankton you have, the less carbon is taken up by the oceans," said Margarita Conkright of the National Oceanic and Atmospheric Administration (Toner 2002).

Warmer ocean surface temperatures are related to lower oceanic phytoplankton biomass and productivity, the source of half the photosynthesis ("net primary production") on Earth, according to a survey of nearly a decade's worth of satellite data compiled by Michael J. Behrenfeld and colleagues in *Nature* (2006). They argue that reduced phytoplankton biomass is a result of changes induced in ocean circulation by warming that reduces supplies of nutrients required for photosynthesis. Many of these nutrients are conducted through the ocean by upwelling of cold, nutrient-rich water.

According to another researcher, "Extrapolating the satellite observations into the future suggests that marine biological productivity in the tropics and mid-latitudes will decline substantially, in agreement with climate-model simulations" (Doney 2006, 696). Productivity probably will increase at higher latitudes. As carbon dioxide levels in the atmosphere have risen for the last century, phytoplankton density in the oceans has declined, according to a study by Daniel G. Boyce and others who "find a strong correspondence between [the] chlorophyll record and changes in both leading climate indices and ocean thermal conditions . . . in which increases in indices of phytoplankton productivity are mirrored by increases in ocean warming" (Siegel and Franz 2010, 569).

With historical data and satellite observations, Boyce and colleagues make the following case:

In the oceans, ubiquitous microscopic phototrophs (phytoplankton) account for approximately half the production of organic matter on Earth. Analyses of satellite-derived phytoplankton concentration (available since 1979) have suggested decadal-scale fluctuations linked to climate forcing, but the length of this record is insufficient to resolve longer-term trends. Here we combine available ocean transparency measurements and in situ chlorophyll observations to estimate the time dependence of phytoplankton biomass at local, regional and global scales since 1899. We observe declines in eight out of ten ocean regions, and estimate a global rate of decline of ~1% of the global median per year. Our analyses further reveal inter-annual to decadal phytoplankton fluctuations superimposed on long-term trends. These fluctuations are strongly correlated with basin-scale climate indices, whereas long-term declining trends are related to increasing sea surface temperatures. We conclude that global phytoplankton concentration has declined over the past century; this decline will need to be considered in future studies of marine ecosystems, geochemical cycling, ocean circulation and fisheries. (Boyce et al. 2010)

Zooplankton Move Northward

Gregory Beaugrand and colleagues reported in *Science* that several sea species, including several types of plankton, have moved northward in the Northern Hemisphere between 1960 and 1999 in response to warming water temperatures: "We provide evidence of large-scale changes in the biogeography of calanoid copepod crustaceans in the Eastern North Atlantic Ocean and European shelf seas. . . . Strong biogeographical shifts in all copepod assemblages have occurred with a northward extension of more than 10 degrees latitude of warm-water species associated with a decrease in the number of colder-water species," they wrote (Beaugrand et al. 2002, 1692).

This study was based on analysis of 176,778 samples collected by the Continuous Plankton Recorder survey taken monthly in the North Atlantic since 1946. The scientists wrote, "The observed bio-geographical shifts may have serious consequences for exploited resources in the North Sea, especially fisheries. If these changes continue, they could lead to substantial modifications in the abundance of fish, with a decline or even a collapse in the stock of boreal species such as cod, which is already weakened by overfishing" (Beaugrand et al. 2002, 1693–1694).

Studying nitrogen balances, Canadian and American oceanographers found evidence that warmer oceans will lose nitrate, which will deplete phytoplankton and boost atmospheric carbon dioxide levels (Leggett 2001, 226). Satellite surveys have detected a sharp decline in plankton in several of the world's oceans, a potential threat to the marine food chain and one that could undercut one of the world's natural buffers to global warming (Toner 2002).

Plankton Decline Varies Ocean to Ocean

The mass of phytoplankton can vary by a factor of more than 100 in various parts of the ocean, depending on local conditions, including the degree of mixing and deposit of wind-borne iron from the continents. Ocean mixing is inhibited by warming water. Satellites have made possible surveys of plankton biomass over large areas of the world ocean. The amount of plankton biomass is also sensitive to the El Niño–La Niña cycle, decreasing as waters warm and increasing as they cool.

The greatest decline has been in the northern Pacific Ocean, where summer levels have dropped by more than 30 percent since the 1980s. Comparing sets of satellite data from early 1980 to data from the late 1990s, researchers reported in *Geophysical Research Letters* that sharp declines in plankton had taken place in *both* the North Pacific and North Atlantic Oceans, where their abundance decreased by 14 percent. In equatorial regions, plankton levels increased. Worldwide plankton stocks decreased more than 8 percent (Gregg and Conkright 2002).

The researchers were not certain whether the decline of phytoplankton is part of a natural cycle in the oceans, a reflection of regional changes, a result of Earth's gradual warming, or all three. They did, however, find a close correlation between the decline of plankton and increasing ocean surface temperatures that indicates "[c]limate change could be a cause as well as an effect of plankton declines" (Toner 2002). Plankton need two things to grow: sunlight and nutrients. The researchers said warmer sea surface temperatures interfere with upwelling of colder water that is rich in nutrients from the oceans' depths.

Scientists in Australia have found similar declines in phytoplankton populations, and have asserted that global warming is starving the depths of the Southern Ocean of oxygen. Research scientists working with the Australian Commonwealth Scientific and Industrial Research Organization (CSIRO), the nation's climate-research agency, said they have found a significant drop in the Southern Ocean's oxygen levels over the past 30 years and that they anticipate this situation will worsen in the future.

Scientist Richard Matear, who is based in Hobart, Australia, said that oxygen-starved oceans may lead to long-term devastation of marine life. Matear said that the decline in oxygen levels was most likely caused by global warming because natural variability could not explain the reduction (Barbeliuk 2002). "The interpretation is that less oxygen-rich water is penetrating into the ocean and this in turn gives additional credibility to climate change models," Matear said. According to Matear, the Southern Ocean was considered by oceanographers to be the "lungs" of the world's oceans, creating 55 percent of the water that regenerates the deep ocean. Any decrease in the Southern Ocean's capacity could have consequences for oceans around the world, he said (Barbeliuk 2002).

Matear said that the Southern Ocean itself is unlikely to be greatly impacted by the change, since it is so oxygen-rich. Other oceans that already have limited oxygen could suffer dire consequences, with the variety of marine life greatly reduced, he said. "Most organisms that live in the ocean require oxygen, only a very few don't," Matear said (Barbeliuk 2002). Matear said that warmer waters are unable to carry as much oxygen or sink as deeply into the ocean's depths. The colder the water, the faster it sinks, acting as an oxygen pump. The Southern Ocean's current oxygen level is about 200 micro-moles per kilogram, representing a decline of about 15 micro-moles during 30 years to 2000.

Matear and fellow scientists Tony Hirst (also of the CSIRO) and Ben McNeil (from the Antarctic Cooperative Research Center), used chemical data gathered during oceanographic research voyages south of Australia to look for changes in the ocean conditions. "Having demonstrated that oxygen is a valuable indicator of climate change in our models, we now have a quantity to monitor to detect future changes," he said (Barbeliuk 2002).

Plankton Decline and Changes in Thermohaline Circulation

The decline of plankton also may be related to the changing nature of the thermohaline circulation (the Atlantic meridional overturning circulation). Writing in *Nature*, Andreas Schmittner reported on results of models simulating disruption of the Atlantic meridional overturning circulation that led to a collapse of North Atlantic plankton stocks to less than half of their initial biomass, "owing to rapid shoaling of winter mixed layers and their associated separation from the deep ocean nutrient reservoir." Schmittner wrote, "These model results are consistent with the available high-resolution paleo record, and suggest that global ocean productivity is sensitive to changes in the Atlantic meridional overturning circulation" (Schmittner 2005, 628).

Adding more support to the idea that phytoplankton populations are declining around the world at least partly because of changing ocean circulation, British scientists examining the Atlantic Ocean south of Iceland found that populations of zooplankton, which feed many larger ocean species, have declined by as much as 90 percent in four decades. This may portend population reductions for larger species from cod and haddock to whales and dolphins. "This is deeply worrying," said marine biologist Dr. Phil Williamson of East Anglia University. "We don't know why zooplankton numbers have plummeted, though global warming looks [like] the best candidate. What is certain is that removing the bottom link from the ocean food chain could have profound and unpleasant results" (McKie 2001; Knutti et al. 2004, 851–854).

Updating a major survey of zooplankton levels in the North Atlantic from 1963, a team of British scientists in November 2001 set out in the marine research vessel *Discovery* to measure changes in these levels. Using automated equipment, the scientists sampled concentrations of *Calanus finmarhicus*, the principal type of Atlantic zooplankton. Having carried out 800 samplings in an area 1,000 miles south of Iceland, the scientists found 5,000 to 10,000 zooplankton per square meter instead of the 50,000 average found in the earlier survey (McKie 2001). The scientists believe that gradually increasing sea temperatures are playing a major role in the zooplankton's decline.

Whether plankton decline is directly related to increased ocean temperatures is not yet certain, according to Watson W. Gregg, a NASA biologist at the Goddard Space Flight Center in Greenbelt, Maryland. Other factors such as the availability of iron also affect the productivity of phytoplankton. According to Gregg, the greatest loss of phytoplankton has occurred where ocean temperatures have risen most significantly between the early 1980s and late 1990s.

In the North Atlantic, summer sea surface temperatures rose about 1.3°F during that period, Gregg said, while in the North Pacific the ocean's surface temperatures rose about 0.7°F (Perlman 2003). "This research shows that ocean primary productivity is declining, and it may be the result of climate changes such as increased temperatures and decreased iron deposition into parts of the oceans," Gregg said. "This has major implications for the global carbon cycle" (Perlman 2003).

Warming may have aided phytoplankton growth in some areas (unlike northern latitudes of the Atlantic Ocean, where it is declining with rising temperatures).

Declining winter and spring snow cover over Eurasia has contributed to an increasing land and ocean temperature gradient that enhances summertime monsoon winds over the western Arabian Sea near the coasts of Somalia, Yemen, and Oman. According to one study, increased upwelling in the area contributes to an increase of more than 350 percent in average summertime phytoplankton biomass along the coast and 300 percent offshore, making the area more productive of aquatic life (Goes et al. 2005, 545).

Further Reading

Barbeliuk, Anne. "Warmer Globe Choking Ocean." *The Mercury* (Hobart, Australia), March 16, 2002, n.p. (LEXIS).

Beaugrand, Gregory, et al. "Reorganization of North Atlantic Marine Copepod Biodiversity and Climate." *Science* 296 (May 31, 2002): 1692–1694.

Behrenfeld, Michael J., et al. "Climate-Driven Trends in Contemporary Ocean Productivity." *Nature* 444 (December 7, 2006): 752–755.

Boyce, Daniel G., Marlon R. Lewis, and Boris Worm. "Global Phytoplankton Decline over the Past Century." *Nature* 466 (July 29, 2010): 591–596.

Doney, Scott C. "Plankton in a Warmer World." *Nature* 444 (December 7, 2006): 695–696.

Goes, Joaquim I., et al. "Warming of the Eurasian Landmass Is Making the Arabian Sea More Productive." *Science* 308 (April 22, 2005): 545–547.

Gregg, Watson W., and Margarita E. Conkright. "Decadal Changes in Global Ocean Chlorophyll. *Geophysical Research Letters* 29(15) (2002): 10. doi: 1029/2002GL014689.

Knutti, R., et al. "Strong Hemispheric Coupling of Glacial Climate through Freshwater Discharge and Ocean Circulation." *Nature* 430 (August 19, 2004): 851–856.

Leggett, Jeremy. *The Carbon War: Global Warming and the End of the Oil Era.* New York: Routledge, 2001.

McKie, Robin. "Dying Seas Threaten Several Species; Global Warming Could Be Tearing Apart the Delicate Marine Food Chain, Spelling Doom for Everything from Zooplankton to Dolphins." *Observer* (London), December 2, 2001, 14.

Perlman, David. "Decline in Oceans' Phytoplankton Alarms Scientists; Experts Pondering Whether Reduction of Marine Plant Life Is Linked to Warming of the Seas." *San Francisco Chronicle*, October 6, 2003, A6.

Schmittner, Andreas. "Decline of the Marine Ecosystem Caused by a Reduction in the Atlantic Overturning Circulation." *Nature* 434 (March 31, 2005): 628–633.

Siegel, David A., and Bryan A. Franz. "Oceanography: Century of Phytoplankton Change." *Nature* 466 (29 July 2010): 569–571.

Toner, Mike. "Microscopic Ocean Life in Global Decline; Temperature Shifts a Cause or an Effect?" *Atlanta Journal-Constitution*, August 9, 2002, 3A.

See also: Coral Reefs; Fisheries; Ocean Acidity, North America; Ocean Acidity, Worldwide; Oceans' Absorption of Heat

SALMON AND TROUT

Rising water temperatures may drive trout and salmon from many U.S. waterways, according to a study compiled by Defenders of Wildlife and the Natural Resources Defense Council. This study included eight species of salmon and trout and suggested that their cold-water habitats could shrink by more than 40 percent during the 21st century, given a "business as usual" emissions of greenhouse gases. Salmon and trout are highly sensitive to the temperature of their aquatic habitats. In many areas, these fish are already living at the margin of their temperature tolerances, meaning that even modest warming could render streams uninhabitable ("Warming Streams" 2002). Habitats for some of these species could shrink as much as 17 percent by 2030, 34 percent by 2060, and 42 percent by 2090 ("Warming Streams" 2002). This analysis includes four species of trout (brook, cutthroat, rainbow, and brown) and four species of salmon (pink, coho, Chinook, and chum). Researchers looked at air and water temperature data from more than 2,000 sites across the United States ("Warming Streams" 2002).

Anticipated increases in water temperatures vary by location, averaging 0.7 to 1.4°F by 2030, 1.3 to 3.2°F by 2060, and 2.2 to 4.9°F by 2090, depending on future levels of greenhouse gases in the atmosphere ("Warming Streams" 2002). In addition to warming waters, wild trout and salmon populations also are under pressure from habitat loss because of human infrastructure development, competition with hatchery fish, and invasive exotic species, among other reasons. "Now we must add climate change to the list of challenges they face," said Mark Shaffer, senior vice president for programs at Defenders of Wildlife. "If we don't address the cumulative impact of all these factors, we will see more of these populations switching from a recreational resource to being listed as threatened or endangered" ("Warming Streams 2002").

"Blob" of Warm Pacific Water Devastates Fisheries

In 2015, an unusually warm mass of relatively warm water nicknamed the "blob" appeared in the Pacific Ocean off the West Coast of the United States. Observers began to notice that the mass was throwing the aquatic food chain into chaos as it

> quelled upwelling that typically delivers nutrients to coastal waters where migratory salmon, tuna, and whales fatten themselves on "forage species" such as anchovies, sardines and krill. . . . The nutrient shortage comes at a time when many of those forage species are already at historic lows. With a strengthening El Niño—warmth in the eastern equatorial Pacific that affects weather patterns worldwide—fisheries managers face a good deal more uncertainty than usual as they prepare to set catch limits for next year. (Gewin 2015)

The "blob" also may be contributing to unusually dry and warm weather along the U.S. West Coast. The water is some 4°F warmer than average and may play a role in dislocating many fish species (for example, tropical sunfish appeared off the Alaska coast), as well as toxification of shellfish associated with algal blooms that, according to a report in *The New York Times*, have rendered shellfish toxic and shut down shellfish fisheries in Washington, Oregon, and California. "A single clam can have enough toxins to kill a person," said Vera L. Trainer, manager of the marine biotoxin program at the Northwest Fisheries Science Center run by the National

Oceanic and Atmospheric Administration (NOAA) in Seattle. Washington state officials also ordered the largest ever shutdown of the state's Dungeness crab fishery (Schwartz 2015).

One observer added:

Warm winter temperatures in the mountains of the western United States this past winter [2014–2015] sharply reduced the region's snowpack. That snow typically serves as a vital water storage reservoir that is slowly released as the snow melts over the dry summer months. Without this snowmelt, stream flows are expected to drop sharply this summer, which in turn is expected to cause water temperatures to rise to a level unhealthy for migrating salmon. As a result, fisheries biologists expect a looming calamity for endangered salmon stocks [in 2015]. (Service 2015, 268)

Beginning in November 2015 and continuing into early 2016, a strong El Niño pattern in the Pacific played a major role in increasing atmospheric winds and surface circulation in this area. The "blob" dissolved. Thus, one warming influence played itself out against another ("The Demise" 2016).

A World Wildlife Fund (WWF) report issued in 2001 asserted that stocks of wild Atlantic salmon had been cut in half in two decades because of global warming and infections spread by hatchery-bred fish. According to the WWF report, wild salmon already had disappeared from 309 of their 2,000 usual breeding areas around the world. Elizabeth Leighton, WWF senior policy officer for Great Britain, said: "When a river loses its salmon stock, that population is gone forever. The miracle of [salmon] returning to its spawning ground cannot be repeated" (Milmo and Nash 2001).

The report further noted that, even in 2001, 90 percent of healthy wild Atlantic salmon populations returning to Europe could be found in only three countries: Norway, Iceland, and the Irish Republic. Salmon have nearly disappeared from Germany, Switzerland, the Netherlands, and Belgium. Wild salmon are nearly extinct in Estonia, Portugal, Poland, parts of the United States, and the warmer parts of Canada. Wild Atlantic salmon had nearly vanished by 2002 from many rivers in Scotland. Wild stocks were extinct in the rivers Shieldaig, Garvan, Attadale, and Sguord on the west coast of Scotland, with only a handful left in another 10 rivers. The genetic integrity of the few remaining wild fish in Scotland's rivers also had been damaged through interbreeding between farmed and wild fish. A wide range of factors was thought to be affecting salmon stocks, including overfishing at sea and global warming. Director of the Spey Fishery Board James Butler said, "In 14 rivers we found that stocks ranged from no fish to just a handful" (Cramb 2002).

Marine Parasites Spread

A marine parasite that began spreading among Chinook salmon in the Yukon River in 2002 may be traced in part to warmer stream waters that allow it to thrive. The parasite has ruined many of the fish, making them inedible, giving them a fruity odor and ruined meat. The illness is caused by a common microorganism that targets ocean fish and was detected in some 35 percent of king salmon sampled in 2002 and 2003, said Richard Kocan, a fish pathologist overseeing a study for the U.S. Department of Interior's Subsistence Management Program. That is a significant increase over the number of infected fish found in 1999, 2000, and 2001 (O'Harra 2004). Kocan said that some Yukon Chinook now spend June and July migrating upstream through water warmed to 59 degrees or higher, temperatures that allow the parasite to spread faster and kill its hosts more quickly. "Some of these fish are swimming within a few degrees of lethal temperatures for healthy salmon, and we know many of them are infected," he said (O'Harra 2004).

The parasite was first noticed in a few salmon during the middle 1980s, but the infection became much worse during the late 1990s. The culprit is a common genus of protozoans called *Ichthyophonus*; the organisms probably enter the fish through its food supply and then spreads into organs and flesh. The pathogen also has been detected in salmon in the Kuskokwim and Taku Rivers, but it is not clear whether the precise same species is found in the Yukon Chinook, according to Kocan's report. *Ichthyophonus* has also been found in a few Yukon burbot, raising the possibility that it is now inside the freshwater system. The fish are not toxic and can be eaten safely. "They taste bad, they smell bad, and they look bad, but if you're starving, go ahead and eat them," Kocan said (O'Harra 2004). The parasite's worst effect, however, is that it inhibits fish breeding and kills salmon before they spawn. An estimated 60 percent of the fish sick with the disease do not make it to spawning grounds.

Salmon Can Adapt—to a Point

According to research published in *Nature Climate Change*, some salmon can adapt to water as much as 2°C warmer than their usual habitat. At a level of 4.4°C, however, they experience a high rate of heart failure.

According to the *Nature Climate Change* report, Pacific salmon provide critical sustenance for millions of people worldwide and changes to their population will have far-reaching impacts on the productivity of ecosystems. Rising temperatures are threatening the salmon's survival. Like many species, however, they can adapt, within limits, displaying "both physiological and genetic capacities to increase . . . thermal tolerance in response to rising temperatures." The researchers continued:

> In juvenile chinook salmon (*Oncorhynchus tshawytscha*), a 4°C increase in developmental temperature was associated with a 2°C increase in key measures of the thermal performance of cardiac function. Moreover, additive genetic effects significantly influenced several measures of cardiac capacity, indicative of heritable variation on which selection can act. However, a lack of both plasticity and genetic variation was found for the arrhythmic temperature of the heart, constraining this upper thermal limit to a maximum of 24.5±2.2°C. Linking this constraint on thermal tolerance with present-day river temperatures and projected warming scenarios, we predict a 17 percent

chance of catastrophic loss in the population by 2100 based on the average warming projection, with this chance increasing to 98 percent in the maximum warming scenario. Climate change mitigation is thus necessary to ensure the future viability of Pacific salmon populations. (Muñoz et al. 2015, 1673)

Salmon Populations Crash

According to a report by the World Wildlife Fund, "While salmon can withstand higher temperatures in summer when food is abundant, in the winter their tolerance drops considerably. As cold-blooded creatures, their metabolism increases in warmer water and keeping up with this high metabolism requires large amounts of food. If sufficient food is not available, salmon can starve" (Mathews-Amos and Berntson 1999).

For Canada's western regions, climate models forecast an increase in precipitation, water runoff, and flooding in winter and a decrease in precipitation and runoff during summer. Higher winter river flows are expected to damage salmon spawning grounds, reduce survival and growth of fish because of increased stream temperatures, and damage Fraser River salmon because of increased predation by warm-water species (Rolfe 1996). In 1995, the Canadian Department of Fisheries and Oceans blamed a collapse of Fraser River salmon runs on predation by mackerel, which invaded the salmon spawning grounds along with warmer-than-average ocean waters provoked by El Niño conditions. At the same time and for the same reasons (according to the Canadian government), Queen Charlotte Chinook salmon runs declined some 80 percent.

The salmon catch in Scotland was 35 percent less in 1999 than during 1998, which itself was the second worst year since comprehensive records began in 1952. The reduced catch, partly the result of warming water temperatures, is threatening the livelihood of river proprietors, country hotels and bed and breakfasts, as well as ghillies (guides). Warmer seawater decreases the population of krill that the salmon eat. Anglers also say that sea lice from salmon reared on fish farms are infesting and killing wild salmon. The fish farms also produce polluting slurry. In eight of 32 rivers on the west coast of Scotland, salmon were virtually extinct by 2000. Seals are also killing salmon in some rivers (Buxton 2000).

Salmon Fishing Shut Down in British Columbia

Rising temperatures in the Fraser River, British Columbia's largest sockeye salmon spawning river, provoked a government shutdown of commercial salmon fishing at the height of the season late in September 1998. According to an Environmental News Service dispatch from Victoria, British Columbia (Thomas 1998), above-average temperatures in the river impaired the salmon's swimming and jumping abilities and afflicted them with "proliferating pathogens and an invasion of warm-water predators such as squaw fish" (Thomas 1998), The article also described fears that "seawater temperatures 1.5°C above average in the adjacent Georgia Strait could trigger toxic algae blooms" (Thomas 1998).

Between one-quarter and two-thirds of salmon returning to various locations in the Fraser River system were dying before they could spawn, many of them from conditions related to rapid warming of their aquatic environment. The water warmed because of below-average snowpack and above-normal temperatures in the interior of British Columbia, the Fraser's drainage basin. A large aluminum smelter also was delivering large amounts of heated water to an upstream tributary of the Fraser, adding yet another human provocation to the salmon's warming environment. Clear-cutting of upstream forests is also blamed for some of the warming of river waters.

Above-average water temperatures in this area had also caused salmon to die in unusually large numbers in 1992 and 1994. If the temperature rises two degrees more (from a late-summer peak of 21°C to 23°C) none of the salmon will survive the swim upstream to spawn. At the higher temperature, most salmon will die of heat prostration. The warming of the river and resulting fishing shutdown in September 1998 brought out angry fishermen, who threatened to block cruise ships in Vancouver's busy harbor.

By late July 2006, the snow pack in the Fraser River watershed, which in most years would be feeding into the river and cooling it, had almost entirely melted. Salmon thus were threatened by a combination of warm water and low river levels.

Further Reading

Buxton, James. "Suspects in the Mystery of Scotland's Vanishing Salmon: Fish Farms, Seals, and Global Warming Are All Blamed for What Some See as a Crisis." *Financial Times* (London), June 13, 2000, 11.

Cramb, Auslan. "Highland River Salmon 'On Verge of Extinction.'" *Daily Telegraph* (London), July 15, 2002, 7.

"The Demise of the Warm Blob." NASA Earth Observatory, February 16, 2016. http://earthobservatory.nasa.gov/IOTD/view.php?id=87513&src=eoa-iotd.

Gewin, Virginia. "North Pacific 'Blob' Stirs Up Fisheries Management." *Nature* 524 (August 27, 2015): 396.

Mathews-Amos, Amy, and Ewann A. Berntson. *Turning Up the Heat: How Global Warming Threatens Life in the Sea.* World Wildlife Fund and Marine Conservation Biology Institute, 1999. http://www.worldwildlife.org/news/pubs/wwf_ocean.htm.

Milmo, Cahal, and Elizabeth Nash. "Fish Farms Push Atlantic Salmon Towards Extinction." *Independent* (London), June 1, 2001, 11.

Muñoz, Nicolas J., et al. "Adaptive Potential of a Pacific Salmon Challenged by Climate Change." *Nature Climate Change* 5 (February 2015): 163–166. doi: 10.1038/nclimate2473.

O'Harra, Doug. "Marine Parasite Infects Yukon River King Salmon; Fish Are Left Inedible; Scientists Study Overall Impacts." *Anchorage Dispatch News*, January 28, 2004, A1.

Rolfe, Christopher. "Comments on the British Columbia Greenhouse Gas Action Plan." West Coast Environmental Law Association. April 17, 1996. http://www.wcel.org/wcelpub/11026.html (no longer available).

Schwartz, John. "The Pacific Ocean Becomes a Caldron." *The New York Times*, November 2, 2015. http://www.nytimes.com/2015/11/03/science/global-warming-pacific-ocean-el-nino-blob.html.

Service, Robert F. "Meager Snows Spell Trouble Ahead for Salmon." *Science* 348 (April 17, 2015): 268–269.

Thomas, William. "Salmon Dying in Hot Waters." Environment News Service, September 22, 1998. http://www.econet.apc.org/igc/en/hl/9809244985/hl11.html (no longer available).
"Warming Streams Could Wipe Out Salmon, Trout." Environment News Service, May 22, 2002. http://ens-news.com/ens/may2002/2002L-05-22-06.html (no longer available).

See also: Biodiversity; Extinctions; Fisheries; Ocean Acidity, North America; Ocean Acidity, Worldwide; Oceans' Absorption of Heat; Temperatures, Global

SEA LEVEL RISE

The projected rise of sea levels through worldwide climate change may be the major controversy of global warming science. Many models project sea level rise to the end of the 21st century not because it will stop at that time but because various projections become vague after that. Even modest estimates spell trouble for hundreds of millions of people who live close to Earth's coastlines. "Sea level rise isn't going to stop in 2100," one observer said. "I think that's something that people don't really take on board. Eventually, they will. Projections of sea level rise far into the future jump from tens of centimeters to tens of meters" (Jones 2013).

Rapid Sea Level Rises in the Past

The oceans have risen and fallen throughout Earth's history. Some 20,000 years ago, New York City's current site sat at the edge of an ice sheet that covered half of North America, and seas were 400 feet below today's level. Research reported in 2009 (Blanchon et al. 2009) indicates that sea levels can rise rapidly—6.5 to 10 feet in less than a century—from natural causes. During the last interglacial period (125,000 years ago), polar temperatures were 3 to 5°C higher than today, and sea levels were also higher. Robert E. Kopp and colleagues use this analysis "as a partial analog for anthropogenic warming scenarios" (Clark and Huybers 2009, 856). As Kopp et al. wrote, "We find a 95 percent probability that global sea level peaked at least 6.6 [meters] higher than today during the last interglacial" (Kopp et al. 2009, 863).

Paul Blanchon and colleagues wrote that their evidence from the past abets "warnings that modern ice sheets will deteriorate owing to global warming and initiate a rise of similar magnitude by AD 2100. . . . Knowing the rate at which sea level reached its highstand during the last interglacial period is fundamental in assessing if such rapid ice-loss processes could lead to future catastrophic sea level rise" (Blanchon et al. 2009).

Sea Level Rise May Accelerate

Writing in the March 2004 edition of *Scientific American*, James Hansen, then director of the NASA Goddard Institute for Space Studies in New York City, warned that catastrophic sea level increases could arrive much sooner than anticipated by the Intergovernmental Panel on Climate Change (IPCC) (Holly 2004). The IPCC has estimated sea level increases of roughly half a meter over the next century if global warming reaches several degrees Celsius above temperatures seen in the late 1800s. Hansen warned that if recent growth rates of carbon dioxide emissions and other greenhouse gases continue during the next 50 years, the resulting temperature increases could provoke large increases in sea levels with potentially catastrophic effects.

Sea Level Rise: A Cake That Is Already Being Baked

With carbon dioxide levels in the atmosphere already as high as the Pliocene Epoch 2 million to 3 million years ago, scientists have been asking a question that will become more important in coming decades: how long will it be before enough ice melts to raise sea levels to reflect these levels?

Sea level rise is not a matter of "if" but "when," according to a study published in *Science* in July 2015 (Dutton et al. 2015). This study documented sea level rises of at least 20 feet (six meters) several times during the last 3 million years and came to the conclusion that current levels do not reflect already surpassed carbon dioxide and other greenhouse gas levels. Generally, because of thermal inertia, temperatures lag any given atmospheric level by 50 years in the air and about 150 to 200 years in the oceans.

Phrased concisely, this cake is already being baked. The scientists found that a rise in temperatures of one to two degrees Celsius virtually guaranteed a rise of 20 feet, which is critical to the several hundred million people around the world who live within 20 feet of high tide. Much of the sea level rise has (and will) come from melting ice sheets in Greenland and Antarctica, said lead author Andrea Dutton, a University of Florida geochemist. "This evidence leads us to conclude that the polar ice sheets are out of equilibrium with the present climate," she said ("Global Sea Levels" 2015).

The team that compiled this study used computer models and the geologic record to gauge the global ice pack's sensitivity to climate change. They determined that when average temperatures rose 1 to 3°C (1.8 to 5.4°F) above levels prevailing in preindustrial times (before about 1850), sea levels peaked at least 20 feet higher than today's levels. "As the planet warms, the poles warm even faster, raising important questions about how ice sheets in Greenland and Antarctica will respond," Dutton said. "While this amount of sea level rise will not happen overnight, it is sobering to realize how sensitive the polar ice sheets are to temperatures that we are on path to reach within decades" ("Global Sea Levels" 2015).

"It takes time for the warming to whittle down the ice sheets," said Anders Carlson, a coauthor of the study and professor at Oregon State University's College of Earth, Ocean, and Atmospheric Sciences. "But it doesn't take forever.

There is evidence that we are likely seeing that transformation begin to take place now."

Another paper by James Hansen et al. (2015) supports the idea that a 2°C rise in temperatures will nearly certainly guarantee a large-scale sea level rise. They wrote in *Atmospheric Chemistry and Physics Discussions*:

There is evidence of ice melt, sea level rise of five to nine meters, and extreme storms in the prior interglacial period that was less than 1 degree Celsius warmer than today. Human-made climate forcing is stronger and more rapid than paleo forcings. . . . We conclude that 2 degrees C global warming above the preindustrial level, which would spur more ice shelf melt, is highly dangerous. (Hansen et al. 2015)

In an e-mail post, Hansen said,

Yet as the evidence accumulates at some point a scientist must say it is time to stop waffling so much and say that the evidence is pretty strong. In my opinion we have reached that point on the sea level issue. My conclusion, based on the total information available, is that continued high emissions would result in multi-meter sea level rise this century and lock in continued ice sheet disintegration such that building cities or rebuilding cities on coast lines would become foolish. (Hansen 2015)

Further Reading

Dutton, A., et al. "Sea Level Rise Due to Polar Ice-Sheet Mass Loss during Past Warm Periods." *Science* 349 (July 10, 2015). doi: 10.1126/science.aaa4019.

"Global Sea Levels Could Soon Rise 20 Feet as Climate Warms." Environment News Service, July 13, 2015. http://ens-newswire.com/2015/07/12/global-sea-levels -could-soon-rise-20-as-climate-warms/.

Hansen, James E. Personal correspondence, July 27, 2015.

Hansen, James E., et al. "Ice Melt, Sea Level Rise and Superstorms: Evidence from Paleoclimate Data, Climate Modeling, and Modern Observations That 2 Degrees C. Global Warming Is Highly Dangerous." *Atmospheric Chemistry and Physics Discussions* 15 (July 2015) :20059–20179. www.atmos-chem-phys-discuss.net/15/20059 /2015/. doi: 10.5194/acpd-15-20059-2015.

Hansen warned that because so many people live on coastlines within a few meters of sea level, a relatively small rise could endanger trillions of dollars of infrastructure. Additional warming already "in the pipeline" could take us halfway to paleoclimatic levels that raised the oceans five to six meters above current levels during the Eemian interglacial period some 120,000 to 130,000 years ago

(Hansen 2004, 73). Past interglacials have been initiated with enough ice melt to raise sea levels roughly a meter every 20 years, "which was maintained for several centuries" (Hansen 2004, 73).

An important issue in global warming, wrote Hansen, is sea level change as related to "the question of how fast ice sheets can disintegrate" (Hansen 2004, 73). "In the real world," wrote Hansen, "Ice-sheet disintegration is driven by highly nonlinear processes and feedbacks. For example, higher sea levels can physically lift marine ice shelves that prevent land ice sheets from sliding into the ocean. This effect accelerates the breakup of the land ice" (Holly 2004). In addition, melting glacier water flows downward through holes in the ice to the bottom of the ice mass, where it serves as a lubricant that further accelerates the disintegration of the land ice and its flow into the sea.

Early in the 21st century, James Hansen explored the relative speed with which ice accumulates, compared to the pace that it melts (2004, 74-785). Hansen's calculations have been confirmed by later studies which conclude that Greenland and West Antarctic ice will suffer major erosion, The main debate now revolves around how quickly this melting will take place. Hansen even in 2004 was projecting that major world-wide sea-level rise will occur with about 1°C of additional worldwide warming, which Is less than even the most conservative IPCC estimates for the next half-century. Given thermal inertia, this melting is now in the pipeline. Hansen recommended restricting methane and soot emissions to balance slow growth in carbon dioxide. That measure plus improved energy efficiency and increased use of renewable energy sources could buy time. He added that new technologies may be developed "that we have not imagined." The question, he concludes, is "Will we act soon enough?" (Hansen 2004, 77).

Hansen described a world in which great ice sheets accumulate slowly and disintegrate quickly: "The great ice sheets on Greenland and Antarctica require millennia to grow, because their rate of growth depends on the snowfall rate in a cold, relatively dry, place. Ice sheet disintegration, on the other hand, is a wet process that can proceed rapidly" (Hansen 2006, 19).

An "Unfolding Planetary Disaster of Monstrous Proportions"

With a temperature increase of 2 to 3°C on average (more in polar regions), Hansen sees human contributions to the Earth's greenhouse gas overload provoking massive sea level increases of approximately 25 meters within a few centuries.

> If additional human-made global warming (above that in 2000) is so large, say 2–3°C, that the expected equilibrium (long-term) sea level rise is of the order of 25 meters, there would be the potential for a continually unfolding planetary disaster of monstrous proportions. If additional warming is kept less than 1 [°C] there may still be the possibility of initiating ice sheet response that begins to run out of our control. However, the long term change that the system would be aiming for would be "only" several meters, at most, and, because the energy imbalance would be much less, the time required to reach a given sea level change would be longer, thus yielding a situation with better opportunities for both adaptation and mitigation. (Hansen 2006, 21)

Hansen further develops his case:

If humanity follows a business-as-usual course with global warming of at least 2–3°C, we should anticipate the likelihood of an eventual sea level rise of 25 meters ± 10 meters. It is not possible to say just how long it would take for sea level to change, as ice sheet disintegration begins slowly until feedbacks are strong enough to evoke a highly nonlinear cataclysmic response. Global warming of 2–3°C would cause larger polar warming, leaving both Greenland and West Antarctica dripping in summer meltwater. It is my opinion that 2–3°C global warming would likely cause a sea level rise of at least ~6 [meters] within a century. Although ice sheet inertia may prohibit large change for a few decades, it is plausible that rapid change would begin this century under the business-as-usual climate forcing scenario. The Earth's history reveals numerous cases in which sea level increased several meters per century. Although the paleoclimate cases may have involved disintegration of ice sheets at slightly lower latitudes, the driving forces were far weaker than the presumed anthropogenic forcings later this century. With global warming of 2–3 [°C], the Greenland and West Antarctic ice sheets would be at least as vulnerable as the paleoclimate ice sheets. (Hansen 2006, 21–22)

In particular, according to Hansen,

The East Coast of the United States, including many major cities, is particularly vulnerable, and most of Florida would be under water with a 25 meter sea level rise. Most of Bangladesh and large areas in China and India also would be under water. . . . The population displaced by a 25 meter sea level rise, for the population distribution in 2000, would be about 40 million people on the East Coast of the United States and 6 million on the West Coast. More than 200 million people in China occupy the area that would be under water with a sea level rise of 25 m. In India it would be about 150 million and in Bangladesh more than 100 million. (Hansen 2006, 22–23)

The effects of sea level rise would be felt most acutely in times of weather emergencies, according to Hansen: "The effects of a rising sea level would not occur gradually, but rather they would be felt mainly at the time of storms. Thus, for practical purposes, sea level rise being spread over one or two centuries would be difficult to deal with. It would imply the likelihood of a need to continually rebuild above a transient coastline" (Hansen 2006, 23).

How Much Sea Level Rise after 2100?

Michael Schaeffer and colleagues (2012) used a semi-empirical model to calibrate sea level data of the past millennium in an attempt to estimate how much the sea will rise even if warming is held to 1.5 to 2.0°C, what is widely regarded today as a level that would avoid catastrophic damage. They found that even at that level, seas will rise an average of five feet.

Limiting warming to these levels with a probability larger than 50 percent produces 75–80 centimeters of sea level rise above the year 2000 by 2100. . . . By 2300 a 1.5°C scenario could peak sea level at a median estimate of 1.5 m above 2000. The 50 percent probability scenario for 2°C warming would see sea level reaching 2.7 m above 2000 and still rising at about double the present-day rate. (Schaeffer et al. 2012)

In 2014, roughly 6 million people in the United States lived on land less than five feet above the high-tide line.

"Worse," wrote Benjamin Strauss and Robert Kopp in *The New York Times* (2012),

Rising seas raise the launching pad for storm surge, the thick wall of water that the wind can drive ahead of a storm. In a world with oceans that are five feet higher, our calculations show that New York City would average one flood as high as Hurricane Sandy's about every 15 years, even without accounting for the stronger storms and bigger surges that are likely to result from warming. (Strauss and Kopp 2012)

Once thermal inertia is factored in, sea levels worldwide will rise about 2.3 meters (7.5 feet) for each degree Celsius (1.8 degrees Fahrenheit) of warming, according to a study sponsored by the U.S. National Science Foundation and the German Federal Ministry of Education and Research that was published in 2013 in the *Proceedings of the National Academy of Sciences*. "The study did not seek to estimate how much the planet will warm, or how rapidly sea levels will rise," said coauthor Peter Clark, an Oregon State University paleoclimatologist. "Instead, we were trying to pin down the sea level commitment of global warming on a multi-millennial time scale. In other words, how much would sea levels rise over long periods of time for each degree the planet warms and holds that warmth?" ("Each Degree" 2013).

Greenland's ice sheet is vulnerable not only because of warming temperatures worldwide but also because "the temperature will increase because of the elevation loss," said Peter U. Clark (Clark and Huybers 2009, 856). "For every 1,000 meters of elevation loss, Greenland warms about six degrees Celsius. That elevation loss would accelerate the melting of the Greenland ice sheet" ("Each Degree" 2013).

Levermann and colleagues project degree of sea level rise well past the year 2100:

We combine paleo-evidence with simulations from physical models to estimate the future sea level commitment on a multimillennial time scale and compute associated regional sea level patterns. Oceanic thermal expansion and the Antarctic ice sheet contribute quasi-linearly, with 0.4 m°C$^{-\circ}$ and 1.2 m°C^{-a} of warming, respectively. The saturation of the contribution from glaciers is overcompensated by the nonlinear response of the Greenland ice sheet. As a consequence we are committed to a sea level rise of approximately 2.3 m°C^{-e} within the next 2,000 y. Considering the lifetime of anthropogenic greenhouse gases, this imposes the need for fundamental adaptation strategies on multicentennial time scales. (Levermann et al. 2013, 13745)

Sea Level Rises in the Past

Sea levels have risen quickly in the past even without human provocation. During the last deglaciation, for example, a large and rapid injection of glacial meltwater from the Laurentide ice sheet to the North Atlantic (called *meltwater pulse 1a*, or MWP-1A) raised sea level by about 20 meters over only two centuries.

"Even a foot rise [of world sea levels] is a pretty horrible scenario," said Stephen P. Leatherman, director of the Laboratory for Coastal Research at Florida International University in Miami. "On low-lying and gently sloping land like coastal river deltas, a sea level rise of only one foot could send water thousands of feet inland. Hundreds of millions of people worldwide make their homes in such deltas; virtually all of coastal Bangladesh lies in the delta of the Ganges River. Over the long term, much larger sea level rises would render the world's coastlines unrecognizable, creating a whole new series of islands" (Rudolf 2007). "Here in Miami," Dr. Leatherman said, "we're going to have an ocean on both sides of us" (Rudolf 2007).

Among the important results of warming seas will be coastal erosion, shoreline inundation because of higher tide levels, higher storm surges, and saltwater intrusion into coastal estuaries and groundwater supplies. As stated by a World Wildlife Fund report, "Scientific evidence strongly suggests that global climate change already is affecting a broad spectrum of marine species and ecosystems, from tropical coral reefs to polar ice-edge communities" (Mathews-Amos and Berntson 1999).

By 2015, many oceanographers were revising their estimates of sea level rise during the 21st century upward. Richard B. Alley and several other climate-change specialists who specialize in the future of ice and oceans wrote that although sea levels may rise about half a meter during the century because of warming climate around the world, "[r]ecent observations of startling changes at the margins of the Greenland and Antarctic ice sheets indicate that dynamical responses to warming may play a much greater role in the future mass balance of the ice sheets than previously considered [and] sea level projections may need to be revised upward" (Alley et al. 2005, 456).

Ice shelves near the edges of Greenland and Antarctica are shrinking most quickly, even as areas of eastern Antarctic increase from heavier snowfalls, which are also caused by warming temperatures, at least in part. Parts of inland Greenland also have been experiencing heavier snowfalls, causing the ice sheet there to grow by 6 centimeters to 7 centimeters a year between 1992 and 2003 (Johannessen et al. 2005, 1013). Alley et al. commented: "Ice shelves are susceptible to attack by warming-induced increases of meltwater ponding in crevasses that cause hydrologically driven fracturing and by warmer sub-shelf waters that increase basal melting" (Alley et al. 2005, 458).

The 2007 assessment of the Intergovernmental Panel on Climate Change (IPCC) was severely criticized for underestimating potential sea level rise during the 21st century. The report reduced its worst-case sea level rise from 88 centimeters to 59 centimeters but excluded real-world evidence that had not been included in models. Bob Correll, chairman of the Arctic Climate Impact Assessment, said that any prediction of less than 1 meter would "not be a fair reflection of what we know" (Pearce 2007, 8). The IPCC also failed to include warnings by the British Antarctic Survey that the Antarctic peninsula is warming more quickly than almost any place on Earth and that the Western Antarctic ice sheet is "unstable and contributing significantly to sea level rise" (Pearce 2007, 8).

The level of the world's seas and oceans rose slowly for much of the 20th century, a millimeter or two a year, enough to produce noticeable erosion on 70 percent of the world's sandy beaches, including 90 percent of sandy beaches in the United States (Edgerton and the Natural Resources Defense Council 1991, 18). In some areas, such as the United States Gulf Coast between New Orleans and Houston, the withdrawal of underground water and oil is causing some areas to sink as the oceans rise.

Sea Level Rise Estimates, 2006

In late March 2006, reports were appearing that melting ice, principally from Greenland and the West Antarctic ice sheet, could contribute to a rise in sea levels of one to three meters or more within a century—a startling upward revision of previous estimates. Jonathan T. Overpeck and colleagues wrote in *Science*,

> Sea level rise from melting of polar ice sheets is one of the largest potential threats of future climate change. Polar warming by the year 2100 may reach levels similar to those of 130,000 to 127,000 years ago that were associated with sea levels several meters above modern levels; both the Greenland ice sheet and portions of the Antarctic ice sheet may be vulnerable. The record of past ice-sheet melting indicates that the rate of future melting and related sea level rise could be faster than widely thought. (Overpeck et al. 2006, 1747)

"The question is: Can we predict sea level? And the answer is no," said David Holland, who directs New York University's Center for Atmosphere Ocean Science. Holland, an oceanographer, added that this could mean researchers will just have to watch the oceans to see what happens. "We may observe the change much more than we ever predict it" (Eilperin 2007). Thus, the wide range in estimates for sea level rise by the end of the century—7.8 inches to two feet from the IPCC's 2007 assessment. How quickly will temperature rise convert into melting ice and sea level rise? No one really knows. A three-foot increase in sea level could turn at least 60 million people into refugees, the World Bank estimates (Eilperin 2007).

Michael Oppenheimer, a Princeton University professor of geosciences and international affairs, says that if either Greenland (which would raise the world ocean about 23 feet) or the West Antarctic ice sheet (17 feet) collapses (it is unlikely that only one would melt), it "would destroy coastal civilization as we know it" (Eilperin 2007). The topography of the land under the thick sheets of ice (which affects its flow) is largely unknown, and few measurements are taken of the remote oceans in polar regions, factors that also add to uncertainty.

Forecasting Uncertainty

Even in 2007, a sizable number of scientists were suggesting the IPCC estimates of sea level rise were too low. Coastlines worldwide could be devastated by a sea level rise of more than five feet by the end of the century, according to a study published December 16, 2007, in *Nature Geoscience*. The team led by Eelco Rohling of Britain's National Oceanography Center, asserted that the sea level will rise during this century could match the rise 124,000 years ago when Earth's climate warmed because its orbit changed. In Great Britain, for example, an increase of that magnitude could flood London and low-lying lands such as the Fens and the marshes of Essex and North Kent. Almost every beach in Britain would be under water. Several million people would lose their homes because of coastal flooding ("Britain's Sea Levels" 2007). The New York City area, home to 20 million people, has 2,400 kilometers of coastline and more than 2,000 bridges and tunnels, most with entrances less than three meters above sea level (Lynas 2007, 157–158).

According to E. J. Rohling et al., "The last interglacial period, Marine Isotope Stage (MIS) 5e, was characterized by global mean surface temperatures that were at least 2°C warmer than present, Mean sea level stood 4 to 6 meters higher than the modern sea level." Much of the sea level rise was a result of the Greenland ice sheet's melting. The most astounding finding of this group was that short-term fluctuations in sea level may have varied as much as 10 meters, although "so far it has not been possible to constrain the duration and rates of change of these shorter-term variations" (Rohling et al. 2007). The average rate of sea level rise was 1.6 meters per century, with temperature rises "comparable to projections for future climate change under the influence of anthropogenic greenhouse gas emissions" (Rohling et al. 2007). The authors added that in the interglacial period, 124,000 to 119,000 years ago, Greenland was 5.4 to 9°F (3 to 5°C) warmer than now, which is similar to the warming period expected in the next 50 to 100 years unless drastic measures are taken to control pollution ("Britain's Sea Levels" 2007).

In 2007, even the IPCC realized that its system was counting only about 60 percent of likely sea level rise using "process" models based on data from 1961 to 2003 that attempt "to represent the physics of every contributing factor" (Jones 2013). "The whole was bigger than the sum of its parts," said John Church, an oceanographer at the Australian Commonwealth Scientific and Industrial Research Organization, and co-lead author of the 2014 IPCC assessment's chapter on sea level rise. "The two biggest effects—the expansion of water as it warms, and the addition of water to the oceans from melting glaciers—each accounted for about one-quarter of the total. A little extra was added in from the melting of the Antarctic and Greenland ice sheets. That left a gaping hole" (Jones 2013).

In 2007, Stefan Rahmstorf, a physical oceanographer at the Potsdam Institute for Climate Impact Research in Germany, realized the problems with existing models and set out to formulate a semi-empirical model that examined both air temperatures and annual rate of sea level rise from the 1880s to the present. "He found a simple relationship," reported Nicola Jones in *Nature* (2013). "The warmer it got, the faster the sea level rose. In 2007, too late to be considered by that year's IPCC assessment, his model predicted up to 1.4 meters of sea level rise by 2100—more than twice the IPCC number" (Rahmstorf 2007). A major question, however, is, how long will a model based on the past maintain any degree of accuracy as melting rates change in the future?

The difference between one meter of sea level rise in 2100 and two meters on the ground means 187 million homes worldwide using current population statistics (Jones 2013). Semi-empirical models have not been met with universal acclaim. For example, "The only advantage of these models is that they're easy to calculate," says Philippe Huybrechts, an ice modeler at the Brussels Free University. "I think they're wrong" (Jones 2013). Process-based models also have been improved following the 2007 IPCC debacle and include more accurate estimates of how much heat the oceans receive as the atmosphere warms—and causes seawater to expand, a major factor in sea level rise. For the most part, the gaps in the 2007 models have been filled. As a result, estimates of sea level rise by 2100 have risen. The IPCC's models in 2014 called for three to 21 centimeters more than those in 2007 (a total rise of 28 to 97 centimeters). The wide range indicates continuing uncertainty—but even at that level some scientists believe the IPCC has short-changed the accelerating rate of ice melt over several decades. Semi-empirical estimates still call for higher rates of sea level rise, but the difference is narrowing. "I consider it something of a vindication," said Rahmstorf (Jones 2013). The IPCC in 2014 also anticipated three to 60 centimeters of sea level rise by 2050.

The major problems with the sea level rise forecasts now come from those that bedevil any attempt to anticipate the future: past performance does not guarantee future results. Scientists realize they do not know exactly how the Greenland and Antarctic ice sheets will behave as temperatures warm. Attempts to forecast such things sometimes contradict each other. The stakes at ground level are enormous. Melt all of the polar ice sheets and the sea level rises by some 65 meters (more than 200 feet). Mountain glaciers are a relatively minor contributor with a potential of half a meter of sea level rise.

"We were hugely criticized for being too conservative," said Jerry Meehl, a climate modeler in the U.S. National Center for Atmospheric Research in Boulder, Colorado, and one of the IPCC report's authors (Jones 2013). Even members of the IPCC admitted that they did not have the technical tools to accurately estimate changes in Greenland and Antarctica over several decades.

Uncertainty about the amount of sea level rise in coming years now is focusing on how ice sheets will react to the effects of a warmer climate and how they will interact with the oceans, explained Eric Rignot, professor of Earth-system science at the University of California–Irvine and a senior research scientist at NASA's Jet Propulsion Laboratory in Pasadena. "As a result of the acceleration of outlet glaciers over large regions, the ice sheets in Greenland and Antarctica are already contributing more and faster to sea level rise than anticipated," said Rignot. "If this trend continues, we are likely to witness sea level rise one meter or more by year 2100" ("Rising Seas" 2009).

"The ice loss in Greenland has accelerated over the last decade. The upper range of sea level rise by 2100 might be above one meter or more on a global average, with large regional differences depending where the source of ice loss occurs," said

Konrad Steffen, director of the Cooperative Institute for Research in Environmental Sciences at the University of Colorado, Boulder ("Rising Seas" 2009).

Sea Level Rise Ravages Coastlines

The net effect of sea level rise in any particular place must be adjusted for the gradual rise or fall of the land itself. Some areas that were covered with ice during the last glacial maximum are rising. An example is Stockholm, Sweden (Silver and DeFries 1990, 94). Parts of Canada and Scotland are also slowly rising following the melting of glaciers several thousand years ago, while much of the U.S. Atlantic coast has subsided about a foot during the 20th century. Along a relatively level, sandy shoreline, such a change may cause a beach to lose as much as 100 feet (Edgerton and the Natural Resources Defense Council 1991, 25). Many Atlantic beaches have been receding as much as three feet per year, and in some areas of the Gulf of Mexico coastline the rate averages five feet per year (Edgerton and the Natural Resources Defense Council 1991, 78; Silver and DeFries 1990, 92). By 2007, in the town of Surfside Beach, Texas, 65 miles south of Houston, the shoreline had retreated 200 feet in 25 years, leaving several homes that were once well landward of the ocean now raised on stilts but still awash in the tides (Hudson 2007, B1).

By 1990, coastal Louisiana was subsiding 0.4 inch per year, an unusually rapid rate, because of the removal of oil, gas, and groundwater (Silver and DeFries 1990, 96). Because it is subsiding so quickly, the Mississippi River delta in Louisiana is probably the area of the United States that is most vulnerable to sea level rises caused by global warming (Silver and DeFries 1990, 97). Land loss in Louisiana to subsidence and rising waters has been estimated at about 1 million acres during the 20th century; by 1990, roughly 50 square miles a year were being lost. The highest point in the city of New Orleans is only 13 feet above mean sea level, and the ground under the city continues to sink three feet per century.

Sea Level Rise Projections beyond the 21st Century

By 2016, several teams of scientists using paleoclimatic models were forecasting sea level rise beyond 2100, anticipating an acceleration that places many of the world's coastal cities on a carbon dioxide clock. Scientific debate no longer dwelled on whether current levels of greenhouse gas levels in the atmosphere would drown the coasts. Given the greenhouse gas levels already present, the question became, when and by how much?

A scientific team in February 2016 released a projection that sea levels would rise three to four feet by century's end. That was not startling. By that time, such projections were common (one was contained in a 2013 report by the Intergovernmental Panel on Climate Change), and "sunny-day flooding" already was afflicting coastal cities from Miami to Norfolk, Virginia. What was new in this report was a projection that sea level rise will likely be accelerating so quickly after 2100 that, as an account in *The New York Times* paraphrased it, "Experts say the situation would then grow far worse in the 22nd century and beyond, likely requiring the abandonment of many coastal cities (Gillis 2016, February 22). "I think we need a new way to think about most coastal flooding," said Benjamin H. Strauss, the primary author of one of two related studies. "It's not the tide. It's not the wind. It's us. That's true for most of the coastal floods we now experience" (Gillis 2016, February 22).

"I think we can definitely be confident that sea level rise is going to continue to accelerate if there's further warming, which inevitably there will be," said Stefan Rahmstorf, a professor of ocean physics at the Potsdam Institute for Climate Impact Research, in Germany, and coauthor of one of the papers published online in February 2016 by the *Proceedings of the National Academy of Sciences* (Kopp et al. 2016). This paper indicated that sea levels are extremely sensitive to small changes in temperatures. The frequency of tidal flooding already was accelerating by 2016. During the decade from 1955 to 1964 at Annapolis, Maryland, a tide gauge registered 32 days of flooding; from 2005 to 2014, the same gauge registered 394 days (Gillis 2016, February 22).

Geology professor David M. Harwood of the Department of Earth and Atmospheric Sciences at the University of Nebraska at Lincoln took part in the study and was able to elaborate in 2016:

> Given the present CO_2 levels in the atmosphere of 400 parts per million [ppm], once the ice sheets catch up to this rise, and adjust to an equilibrium state, there will be a loss of ice to equal 14 meters or 40 feet of sea level equivalent. So, we are already overdue for this sea level rise based on the 400 ppm value, based on geological evidence. . . . Basically, we got to 400 ppm so fast, the ice sheets have not had sufficient time to adjust to equilibrium, but in time they will. (Harwood 2016)

The *Proceedings of the United States National Academy of Sciences* (PNAS) study demonstrates this correlation not only once but several times in the past.

Because thermal inertia has not yet caught up with an atmospheric level of 400 ppm, said Harwood, "We are overdue for a sea level rise higher than seen in any previous interglacial" as recorded in ice cores from Antarctica covering at least the last 800,000 years. That is not the whole story. At 500 ppm, said Harwood (2016), "The results suggest up to 100 feet, or 30 meters, is possible within the next 2,000 to 3,000 years. In my view that is a lot, and that is fast."

From 20 to 30 Feet within 200 Years?

James Hansen and 18 coauthors published a study in the open-access journal *Atmospheric Chemistry and Physics* (from the European Geophysical Union), making a case that several meters in sea level rise could take place within a century, not the several hundred years projected by many scientists. This conclusion is based on a study of paleoclimate during the Eemian interglacial period 120,000 years ago, a situation analogous to today (except that temperature increases occurred less rapidly than now).

Hansen and colleagues assert that during the Eemian excess heat in the deep oceans rapidly eroded ice in Antarctica and Greenland at an accelerating pace. This melting was accelerated by a slowing of the Atlantic Ocean's meridional circulation that usually distributes heat through the world ocean. The slowing of ocean circulation caused stagnation of relatively warm water in some areas in and near the Arctic.

Hansen et al. (2016) pointed to maps of worldwide warming during the winter of 2015–2016 indicating that the only areas with below-average temperatures were over oceans adjacent to Greenland and Antarctica, where rapid melting of ice influenced the water's temperature level. "My interpretation is that this is the beginning," Hansen said of these cool patches. "And it's one or two decades sooner than in our model" (Gillis 2016, March 22). "I think almost everybody who's really familiar with both paleo and modern [global warming] is now very concerned that we are approaching, if we have not passed, the points at which we have locked in really big changes for young people and future generations," Hansen said (Mooney 2016).

Limiting global temperature rise to 2°C (3.6°F) over preindustrial levels as recommended by recent diplomatic efforts such as the 2015 Paris Accords, will not prevent climate-driven changes that will force evacuation of many coastal cities, Hansen and colleagues warned.

Hansen and colleagues (2016) hypothesized that:

Mass loss from the most vulnerable ice, sufficient to raise sea level several meters, is better approximated as exponential than by a more linear response. Doubling times of 10, 20, or 40 years yield multimeter sea level rise in about 50, 100, or 200 years. . . . These climate feedbacks aid interpretation of events late in the prior interglacial, when sea level rose to +6–9 meters with evidence of extreme storms while Earth was less than 1°C warmer than today. [Climate] modeling, paleoclimate evidence, and ongoing observations together imply that 2°C global warming above the preindustrial level could be dangerous, [including] growing sea level rise, reaching several meters over a timescale of 50–150 years. (Hansen et al. 2016, abstract)

"Some of the claims in this paper are indeed extraordinary," said Michael E. Mann, a climate scientist at Pennsylvania State University. "They conflict with the mainstream understanding of climate change to the point where the standard of proof is quite high." However, noting that Hansen and colleagues often have arrived at valid conclusions before many others have reached them, Mann said, "I think we ignore James Hansen at our peril" (Gillis 2016, March 22).

Within two weeks of Hansen and colleagues' publication, a study published in *Nature* (DeConto and Pollard 2016) projected a five- to six-foot rise in sea levels by the end of the 21st century, more than half of it from an accelerating collapse of the West Antarctic ice sheet. More melting would follow, according to this scenario, as much as 15 meters (approximately 50 feet) by 2500. By 2200, this study projects a rise of a foot or more per decade. The short-term projections in this study, which could cause shoreline flooding in many coastal cities worldwide, doubled the worst-case projections of the Intergovernmental Panel on Climate Change issued a few years previously.

In *The New York Times*, Justin Gillis wrote, "New York City is nearly 400 years old; in the worst-case scenario conjured by the research, its chances of surviving another 400 years in anything like its present form would appear to be remote. Miami, New Orleans, London, Venice, Shanghai, Hong Kong and Sydney, Australia, are all just as vulnerable as New York, or more so" (Gillis 2016, March 31).

Robert M. DeConto and David Pollard wrote,

> Polar temperatures over the last several million years have, at times, been slightly warmer than today, yet global mean sea level has been six to nine meters higher as recently as the Last Interglacial (130,000 to 115,000 years ago) and possibly higher during the Pliocene Epoch (about 3 million years ago). In both cases the Antarctic ice sheet has been implicated as the primary contributor, hinting at its future vulnerability. Here we use a model coupling ice sheet and climate dynamics—including previously underappreciated processes linking atmospheric warming with hydro-fracturing of buttressing ice shelves and structural collapse of marine-terminating ice cliffs—that is calibrated against Pliocene and Last Interglacial sea level estimates and applied to future greenhouse gas emission scenarios. Antarctica has the potential to contribute more than a meter of sea level rise by 2100 and more than 15 meters by 2500, if emissions continue unabated.

According to Hansen, so many people live on coastlines and within a few meters of sea level that a relatively small rise could endanger trillions of dollars worth of homes and other buildings. Additional warming already "in the pipeline" could take us halfway to paleoclimatic levels that raised the oceans five to six meters above current levels during the Eemian interglacial period some 120,000 to 130,000 years ago (Hansen 2004, 73). Past periods between ice ages (interglacials) have started with enough ice melt to raise sea levels roughly a meter every 20 years, "which was maintained for several centuries" (Hansen 2004, 73).

An important issue in global warming, wrote Hansen, is sea level change as related to "the question of how fast ice sheets can disintegrate" (Hansen 2004, 73). "In the real world," he wrote, "ice-sheet disintegration is driven by highly nonlinear processes and feedbacks." In nonscientific language, he is saying that seas can rise or fall quite rapidly in a short time without easily identifiable reasons.

Although glacial buildup is gradual, "once an ice sheet begins to collapse, its demise can be spectacularly rapid" (Hansen 2004, 74). The darkening of ice by black carbon aerosols (soot)—pollution associated with the burning of fossil fuels—also accelerates melting. Although the timing of melting is uncertain, wrote Hansen, "global warming beyond some limit will make a large sea level change inevitable for future generations" (Hansen 2004, 75). He estimated that such a limit could be crossed with about 1°C of additional worldwide warming. This amount is below even the most conservative estimates of the IPCC for the next 50 years.

Even with a 40-foot sea level rise, many of the world's costal urban areas would be severely inconvenienced. Beginning on the Gulf of Mexico, Houston and New

Orleans would be affected. Continuing along the Atlantic coast, and the effects would be strongly felt in Miami; Norfolk, Virginia; Washington,. D.C.; Baltimore, Maryland; New York City; and Boston. Cross the Atlantic and Pacific Oceans, and a wide range of cities from London to Shanghai, Mumbai, Calcutta, Sydney, Australia, Tokyo, and many others would have to deal with significant loss of coastline.

Further Reading

Alley, Richard B., et al. "Ice-Sheet and Sea-Level Changes." *Science* 310 (October 21, 2005): 456–460.

Blanchon, Paul, et al. "Rapid Sea-Level Rise and Reef Back-Stepping at the Close of the Last interglacial Highstand." *Nature* 458 (April 16, 2009): 881–884.

"Britain's Sea Levels 'Will Rise 5 Feet This Century.'" *Daily Mail* (London), December 17, 2007 (LEXIS).

Clark, Peter U., and Peter Huybers. "Interglacial and Future Sea Level." *Nature* 462 (December 17, 2009): 856–857.

DeConto, Robert M., and David Pollard. "Contribution of Antarctica to Past and Future Sea-Level Rise." *Nature* 531 (March 31, 2016): 591–597.

"Each Degree of Warming Will Raise Sea Levels 7.5 Feet." Environment News Service, July 15, 2013. http://ens-newswire.com/2013/07/15/each-degree-of-warming-will-raise -sea-levels-7-5-feet/.

Edgerton, Lynne T., and the Natural Resources Defense Council. *The Rising Tide: Global Warming and World Sea Levels.* Washington, DC: Island Press, 1991.

Eilperin, Juliet. "Clues to Rising Seas Are Hidden in Polar Ice." *Washington Post*, July 16, 2007, A6. http://www.washingtonpost.com/wp-dyn/content/article/2007/07/15/AR2007 071500882_pf.html.

Gillis, Justin. "Seas Are Rising at Fastest Rate in Last 28 Centuries." *The New York Times*, February 22, 2016. http://www.nytimes.com/2016/02/23/science/sea-level-rise-global -warming-climate-change.html.

Gillis, Justin. "Scientists Warn of Perilous Climate Shift within Decades, Not Centuries." *The New York Times*, March 22, 2016. http://www.nytimes.com/2016/03/23/science/global -warming-sea-level-carbon-dioxide-emissions.html.

Gillis, Justin. "Climate Model Predicts West Antarctic Ice Sheet Could Melt Rapidly." *The New York Times*, March 31, 2016. http://www.nytimes.com/2016/03/31/science/global -warming-antarctica-ice-sheet-sea-level-rise.html.

Hansen, James E. "Defusing the Global Warming Time Bomb." *Scientific American* 290(3) (March 2004): 68–77.

Hansen, James E. "Declaration of James E. Hansen." *Green Mountain Chrysler-Plymouth-Dodge-Jeep, et al., Plaintiffs v. Thomas W. Torti, Secretary of the Vermont Agency of Natural Resources, et al., Defendants.* Case Nos. 2:05-CV-302 and 2:05-CV-304, Consolidated. United States District Court for the District of Vermont. August 14, 2006. http://www .giss.nasa.gov/~dcain/recent_papers_proofs/vermont_14aug20061_textwfigs.pdf (no longer available).

Hansen, James E., et al. "Ice Melt, Sea Level Rise, and Superstorms: Evidence from Paleoclimate Data, Climate Modeling, and Modern Observations That 2°C Global Warming Could Be Dangerous." *Atmospheric Chemistry and Physics* 16(3) (March 22, 2016):761–3812. http://www.atmos-chem-phys.net/16/3761/2016/. doi: 10.5194/acp-16-3761-2016.

Harwood, David M. Personal correspondence, February 25, 2016.

Holly, Chris. "Sea-Level Rise Seen as Key Global Warming Threat." *The Energy Daily* 32:36 (February 25, 2004), n.p. (LEXIS).

Hudson, Kris. "Whose Beach Is This, Anyway?" *Wall Street Journal*, December 12, 2007, B1, B8.

Johannessen, Ola M., et al. "Recent Ice-Sheet Growth in the Interior of Greenland." *Science* 310 (November 11, 2005): 1013–1016.

Jones, Nicola. "Rising Tide: Researchers Struggle to Project How Fast, How High, and How Far the Oceans Will Rise." *Nature* 501 (September 19, 2013). http://www.nature.com/news/climate-science-rising-tide-1.13749.

Kopp, Robert E., et al. "Temperature-driven Global Sea-level Variability in the Common Era." *Proceedings of the National Academy of Sciences*, February 22, 2016. doi: 10.1073/pnas.1517056113.

Kopp, Robert E., et al. "Probabilistic Assessment of Sea Level During the Last Interglacial Stage." *Nature* 462 (December 17, 2009): 863–867.

Levermann, Anders, et al. "The Multimillennial Sea-Level Commitment of Global Warming." *Proceedings of the National Academy of Sciences* 110(34) (August 20, 2013): 13745–13750. http://www.pnas.org/content/110/34/13745.full.pdf.

Lynas, Mark. *Six Degrees: Our Future on a Hotter Planet*. London: Fourth Estate (HarperCollins), 2007.

Mathews-Amos, Amy, and Ewann A. Berntson. "Turning up the Heat: How Global Warming Threatens Life in the Sea." World Wildlife Fund and Marine Conservation Biology Institute, 1999. http://www.worldwildlife.org/news/pubs/wwf_ocean.htm.

Mooney, Chris. "We Had All Better Hope These Scientists Are Wrong about the Planet's Future." *Washington Post*, March 22, 2016. https://www.washingtonpost.com/news/energy-environment/wp/2016/03/22/we-had-all-better-hope-these-scientists-are-wrong-about-the-planets-future/?wpmm=1&wpisrc=nl_evening.

Overpeck, Jonathan T., et al. "Paleoclimatic Evidence for Future Ice-Sheet Instability and Rapid Sea-Level Rise." *Science* 311 (March 24, 2006): 1747–1750.

Pearce, Fred. "But Here's What They Didn't Tell Us." *New Scientist*, February 10–16, 2007, 6–9.

Rahmstorf, Stefan. "A Semi-Empirical Approach to Projecting Future Sea-Level Rise." *Science* 215 (January 19, 2007): 368–370.

"Rising Seas Could Swamp One in 10 People by 2100." Environment News Service, March 10, 2009. http://www.ens-newswire.com/ens/mar2009/2009-03-10-03.asp (no longer available).

Rohling, E. J., et al. "High Rates of Sea-Level Rise during the Last Interglacial Period." *Nature Geoscience* (December 16, 2007). http://www.nature.com/ngeo/journal/v1/n1/full/ngeo.2007.28.html. doi: 10.1038/ngeo.2007.28.

Rudolf, John Collins. "The Warming of Greenland." *The New York Times*, January 16, 2007. http://www.nytimes.com/2007/01/16/science/earth/16gree.html.

Schaeffer, Michiel, et al. "Long-Term Sea-Level Rise Implied by 1.5°C and 2°C Warming Levels." *Nature Climate Change* 2 (December 2012): 867–870.

Silver, Cheryl Simon, and Ruth S. DeFries. *One Earth, One Future: Our Changing Global Environment*. Washington, DC: National Academy Press, 1990.

Strauss, Benjamin, and Robert Kopp. "Rising Seas, Vanishing Coastlines." *The New York Times*, November 24, 2012. http://www.nytimes.com/2012/11/25/opinion/sunday/rising-seas-vanishing-coastlines.html.

See also: Biodiversity; Extinctions; Fisheries; Ocean Acidity, North America; Ocean Acidity, Worldwide; Oceans' Absorption of Heat; Sea Level Rise, Bangladesh; Sea Level Rise, Island Nations; Sea Level Rise, United States; Temperatures, Global

SEA LEVEL RISE, BANGLADESH

Bangladesh, one of the poorest countries on Earth, is likely to suffer disproportionately from global warming, largely because 90 percent of its land lies on floodplains. Regional cyclones historically have killed many people. Bangladesh has built an early warning system for cyclones, as well as a network of more than 2,500 raised concrete storm shelters that have reduced the number of deaths from storm surges after at least a half-million people died in Cyclone Bholoa in 1970. In April 1990, 130,000 people died in another cyclone. Less than one-fourth of Bangladesh's rural population has electricity; as a whole, the country emits less than 0.1 percent of the world's greenhouse gases, compared to 24 percent by the United States (Huq 2001). Bangladesh is planning to use solar energy for new energy infrastructure but lacks the money to build seawalls to fend off rising sea levels.

With 158 million people in an area the size of Louisiana in 2014, Bangladesh was facing rising sea levels, land subsidence, and erratic precipitation patterns all at once. Farmers bred salt-resistant strains of rice as the sea invaded coastal areas, turning former rice paddies into ponds in which they bred salable shrimp and crabs for export. Many people live on shifting sandbars in the Ganges River delta, prepared to move their homes on a few hours' notice as river channels shift. Many tend floating gardens. A special body of law protects people who must move around as the river shifts. "In previous times, this land was juicy, all rice fields," said Mohammad Hayat Ali, a 40-year-old farmer 30 miles from the sea, who tends land that is subject to tidal surges. "But now the weather has changed—summer is longer and hotter than it used to be, and the rains aren't coming as they should. The rivers are saltier than before, and any water we get from the ground is too salty to grow rice" (Belt 2011, 73).

Islands Disappear in the Ganges Delta

Shyamal Mandal lives at the edge of ruin on Ghoramara Island in the Ganges River delta; at two square miles, it has shrunk by half in fewer than 40 years. As described by Soimini Sengupta in *The New York Times*:

In front of his small mud house lies the wreckage of what was once his village on this fragile delta island near the Bay of Bengal. Half of it has sunk into the river. Only a handful of families still hang on so close to the water, and those that do are surrounded by reminders of inexorable destruction: an abandoned half-broken canoe, a coconut palm teetering on a cliff, the gouged-out remnants of a family's fish pond. All that stands between Mr. Mandal's home and the water is a rudimentary mud embankment, and there is no telling, he confessed, when it, too, may fall away. "What will happen next, we don't know," he said, summing up his only certainty. (Sengupta 2007)

The rising sea is only one reason why the island is disappearing. Rivers from the Himalayas emptying into the Bay of Bengal have swollen with glacial ice melt in recent years, changing the shape and size of islands in the Ganges delta. In 30 years, some 31 square miles of the islands have gone underwater, according to a study by Sugata Hazra, an oceanographer at Jadavpur University in Kolkata (formerly Calcutta), a city that shares the same delta. More than 600 families have lost their homes. Sheikh Suleman, now nearing 60, recalled at time when his harvest of coconuts was so plentiful that his wife would give them freely to their neighbors. Now, he said, she has to beg for a coconut. His eyes welled up with tears (Sengupta 2007).

By 2007, more than 600 families were displaced as fields were submerged. Two islands vanished entirely. Ironically, rising levels of atmospheric carbon dioxide are playing a major role in the drowning of these islands—and yet the people who live there use almost no fossil fuels. According to an account in *The New York Times*, several hundred families have moved to a displaced people's camp on Sagar, a nearby island (Sengupta 2007). The islands also are more exposed to intensifying cyclones from the Bay of Bengal in much the same way that the sinking of coastal wetlands of the Mississippi River exposed New Orleans to the storm surge of Hurricane Katrina during 2005. "Nature didn't create this place for humans to cut the forests and chase out the tigers and wildlife," said Tushar Kanjilal, founder of the Tagore Society for Rural Development. "We are killing the Sundarbans [a tidal mangrove forest and nature reserve]. Our government, the people themselves, we are all together killing it. After 50 years, will they exist?" (Sengupta 2007).

Further Reading

Sengupta, Somini. "Sea's Rise in India Buries Islands and a Way of Life." *The New York Times*, April 11, 2007. http://www.nytimes.com/2007/04/11/world/asia/11india .html.

Refugees in the Millions

"There are a lot of places in the world at risk from rising sea levels, but Bangladesh is at the top of everybody's list," said Rafael Reuveny, a professor in the School of Public and Environmental Affairs at Indiana University in Bloomington. "And the world is not ready to cope with the problems" (Harris 2014). Within the next 40 years, climate scientists' projections expect rising seas to inundate 15 percent to 20 percent of Bangladesh's land and displace as many as 18 million people, according to Atiq Rahman, one of the nation's leading climate scientists and executive director of the Bangladesh Center for Advanced Studies (Harris 2014).

Bangladesh, one of the most densely populated countries on Earth, lies on a web of 230 major rivers and streams feeding into the Ganges River delta, as well as a sinking, receding coastline on the Bay of Bengal. Because most of its rivers are polluted from upstream discharge (much of it from India), groundwater is most people's only source of drinking water. Removing water from the ground is causing land to subside as seas rise, saltwater intrudes, and floods increase. More and more people have abandoned low-lying villages, moving to Bangladesh's burgeoning urban areas and into slums in and around the capital, Dhaka, to escape floods, according to surveys by Rahman's Bangladesh Center for Advanced Studies (Harris 2014). Dhaka's slums are themselves built on sinking land. The entire area is prone to tropical cyclones that tend to push storm surges into the V-shaped Bay of Bengal. In addition, the Ganges River delta is eroding. "There are brick foundations torn in half, palm trees growing out of rivers, and rangy cattle grazing on island pastures the size of putting greens. Fields are dusted white with salt," wrote Harris (2014).

"Nature's Laboratory for Natural Disasters"

Many Bangladeshis live within perilous reach of receding shorelines—and destitution is always just one tropical storm away. Gardiner Harris, writing in *The New York Times* (2014), sketched the tenuous life of Jahanara Khatun, whose home was destroyed in Cyclone Aila, during which her home was destroyed and her husband killed. After the storm she became so destitute that she was forced to sell her son and daughter into bonded servitude.

> Ms. Khatun now lives in a bamboo shack that sits below sea level about 50 yards from a sagging berm. She spends her days collecting cow dung for fuel and struggling to grow vegetables in soil poisoned by saltwater. Climate scientists predict that this area will be inundated as sea levels rise and storm surges increase, and a cyclone or another disaster could easily wipe away her rebuilt life. But Ms. Khatun is trying to hold out at least for a while—one of millions living on borrowed time in this vast landscape of river islands, bamboo huts, heartbreaking choices and impossible hopes. (Harris 2014)

Ms. Khatun's husband, parents, and four children were swept away as "a nearby berm collapsed, and their mud and bamboo hut washed away in minutes. Unable to save her belongings, Ms. Khatun put her youngest child on her back and, with her husband, fought through surging waters to a high road. Her parents were swept away. "After about a kilometer, I managed to grab a tree," said Abddus Satter, Ms. Khatun's father. "And I was able to help my wife grab on as well. We stayed on that tree for hours" (Harris 2014).

John Pethick, retired professor of coastal science at Newcastle University in England, says that sea level are rising 10 times as quickly in Bangladesh compared to the global average, in part because attempts to solve the problem (such as dredging canals, pumping water, and raising levees) accelerate land subsidence. Pethick anticipates that apparent sea level in Bangladesh may rise 13 feet by the end of the 21st century, only a small fraction from melting ice and thermal expansion of seawater (Harris 2014). Approximately one-quarter of the county is less than seven feet above sea level. Other scientists disagree with Pethick's calculations. Robert E. Kopp, an associate director of the Rutgers University Energy Institute, uses data from Kolkata, also on the Ganges delta, that indicates the apparent sea level will rise five to six feet by 2100 (Harris 2014). Even so, a six-foot sea level rise may flood land now occupied by roughly 50 million people.

Saleemul Huq, chairman of the Bangladesh Center for Advanced Studies in Dhaka, Bangladesh and director of the Climate Change Program of the International Institute for Environment and Development in London, said that the world community has an obligation to pay serious attention to the views of people who stand to lose the most from climate change (Huq 2001). A sea level rise of half a meter (about 20 inches) could inundate 10 percent of Bangladesh's habitable land, which was home in 2004 to roughly 6 million people. A one-meter water level rise would put 20 percent of the country (and 15 million people) underwater (Radford 2004).

Bangladesh is "nature's laboratory for natural disasters," said Ainun Nishat, senior advisor at the International Union for Conservation of Nature in Dhaka. The country is prone to both drought and deluges that are expected to intensify with a warming climate. Laying in the delta of the Ganges River system, which is vulnerable to rising sea levels and salinity, Bangladesh by 2009 was becoming a beehive of global warming adaptation. Crops were being tested that could endure greater stresses of heat, variations in water supply, and salinity. Cost projections were being drawn up for systems of dikes, but these would likely be in the billions of dollars, where barely a few million dollars was available. The monsoon was becoming more erratic, with long, droughts interrupted by short, intense bursts of flooding rainfall. Limits were being forecast above which (usually 35°C) the staple rice crop may begin to go sterile. The search is on for varieties of rice that can endure worse floods, greater droughts, higher temperatures (especially at night), and saltier water (Inman 2009).

Around Dhaka, with a population of 14 million, many people are using intensive farming to feed themselves (McKibben 2010, 71–73, 170). The country was coping with more than rising sea levels. Dengue fever appeared in 2000 and has recurred each year since as the *Aedes aegypti* mosquito, which carries the disease, feeds more frequently as warmer temperatures decrease the size of its larvae, increasing incidence of the disease. The incubation period of the dengue type 2 virus lasts 12 days at 30°C but seven days at 32 to 35°C (McKibben 2010, 73). Thus, a few degrees of temperature rise can increase transmission of dengue by 300 percent ("Global Warming" 1998; McKibben 2010, 73). Dengue has no cure, and the mosquitoes feed during daylight hours, which makes bed netting useless.

Further Reading

Belt, Don. "The Coming Storm." *National Geographic*, May, 2011, 58–83.

"Global Warming Would Foster Spread of Dengue Fever into Some Temperate Regions." *Science Daily*, March 10, 1998.

Harris, Gardiner. "Borrowed Time on Disappearing Land." *The New York Times*, March 28, 2014. http://www.nytimes.com/2014/03/29/world/asia/facing-rising-seas-bangladesh -confronts-the-consequences-of-climate-change.html.

Huq, Saleemul. "Climate Change and Bangladesh." *Science* 294(November 23, 2001):1617.

Inman, Mason "Hot, Flat, Crowded—And Preparing for the Worst." *Science* 326 (October 30, 2009): 662–663.

McKibben, Bill. *Earth: Making a Life on a Tough New Planet*. New York: Times Books, 2010.
Radford, Tim. "2020: The Drowned World." *The Guardian* (U.K.), September 11, 2004, 10.

See also: Sea Level Rise; Sea Level Rise, Island Nations; Sea Level Rise, United States; Temperatures, Global

SEA LEVEL RISE, ISLAND NATIONS

On many small islands around the world, rising seas are everyday news, and global warming is on everyone's lips. Sea level rise provoked by global warming could imperil more than 3,000 small isolated islands grouped into 24 political entities in the Pacific Ocean and having a population of some 5 million people in 800 distinct cultures. Many coral atolls also contain permanent areas of freshwater (lagoons) that are vulnerable to salinization as sea levels rise.

An Urgent Issue for Kiribati

In no other place is global warming more urgent a practical issue than in equatorial Kiribati (pronounced "Keer-uh-boss"), a chain of coral atolls in the Pacific Ocean near the Marshall Islands. Kiribati is a republic, a member of the United Nations, and home to some 79,000 people. Its land at any point is no more than two meters above sea level. In addition, the living corals that have raised the islands above sea level are threatened with demise by warmer ocean waters. Ierimea Tabai, who became president of Kiribati after its independence from Britain in 1979, has said, "If the greenhouse effect raises sea levels by one meter, it will eventually do away with Kiribati. In 50 or 60 years, my country will not be here" (Webb 1998).

Kiribati's 277 square miles comprise 33 islands scattered across 5.2 million square kilometers (2 million square miles) of ocean, including three groups of islands: 17 Gilbert Islands, eight Line Islands, and eight Phoenix Islands. Kiribati includes Kiritimati (formerly Christmas Island), the world's largest coral atoll (150 square miles). The island of Kiribati, located roughly halfway between California and Australia, more than doubles its surface area at low tide.

The Alliance of Small Island States, including the Philippines, Jamaica, the Marshall Islands, the Bahamas, and Samoa, among others, has offered a strident voice at climate talks favoring swift, worldwide reductions in greenhouse gas emissions. Their self-interest is evident: with warming already "forced" but not yet fully worked into the world's temperature equilibrium, many small island states will lose substantial territory and economic bases within the next few decades. For every other country on Earth, competing economic interests drive climate negotiations. For the small island nations, the driving force is survival. "For us, it's a matter of death and life, whereas in terms of the citizens of the industrialized countries, [global warming] will affect their lifestyles basically, but not to the extent they will be disappearing," said Bikenibeu Paeniu, Tuvalu's prime minister (Webb 1998).

Columnist Nicholas Kristof of *The New York Times* described how Teunaia Abeta, a resident of Kiribati Island in the Pacific, watched in horror as a high tide came rolling in from the turquoise lagoon and did not stop. There was no typhoon, no rain, no wind, just an eerie rising tide that lapped higher and higher, swallowing up Abeta's thatched-roof home and scores of others in this Pacific Island nation. "This had never happened before," said Abeta, 73, who wore only his colorful *lava-lava*, a skirt-like garment, as he sat on the raised platform of his home fingering a home-rolled cigarette. "It was never like this when I was a boy." (Kristof 1997)

During 1997, the area was devastated by El Niño, which brought heavy rainfall, a half-meter rise in sea level, and extensive flooding. About 40 percent of the atolls' coral was killed by overheated water, and nearly all of Kiritimati Island's roughly 14 million birds died or deserted the island. During February 2009, President Anote Tong of Kiribati, in the equatorial Pacific Ocean, sought international aid to move the island nation's 100,000 people to a new homeland. He asked for an international fund that will buy land for the migration that will supplement their own funds. Many Kiribatians may migrate to New Zealand.

The Ocean Is No Longer Their Friend

Central lagoons of the small islands are both social centers and sources of food. According to Kristof, "Children play in the water from infancy. Many adults fish or sail for a living. The Kiribati men in their loincloths go out in outrigger canoes each day to catch tuna, and the women wade out on the coral reef to dive for shellfish and net smaller fish for dinner" (Kristof 1997). "People here think of the ocean as their source of livelihood, as their friend," said Ross Terubea, a Kiribati radio reporter. "It's hard to think that it would destroy us" (Kristof 1997).

By 1998, which was then an unprecedented record El Niño, rising sea levels were already swallowing some small Pacific islands and contaminating drinking water on others. The rising sea has endangered sacred sites and drowned some small islands near Kiribati and Tuvalu, including the islet of Tebua Tarawa, once a landmark for Tuvalu fishermen. Kiribati already has moved some roads inland on its main island as the rising Pacific Ocean eats into its shores.

Government of Maldives Islands
Meets Underwater

In October 2009, 13 of 16 members of the governing cabinet of the Maldive Islands held a meeting underwater wearing scuba gear to emphasize the globally threat to the nation's 350,000 people. President Mohammed Nasheed joined other officials at a table 20 feet below sea level near the island of Girifushi, with a coral reef in the background. "What we are trying to make people realize is that the Maldives is a frontline state. This is not merely an issue for the Maldives but for the world," Nasheed said ("Maldives Government" 2009).

The cabinet members then signed a statement urging all countries to reduce their greenhouse gas emissions.

The Maldives has almost 2,000 coral atolls, most of which are too small for human habitation or are home to clusters of tiny villages of coconut palms and small huts. In the capital of Male, however, which has been a trading crossroads 300 miles south of India for millennia (since about 300 BCE), 150,000 people are packed into one completely urbanized square mile. Population density has doubled in 30 years. Dogs are banned because no space exists for them, and pigeons are the only remaining wildlife.

Mohammed Nasheed, the first democratically elected president (in 2008) has announced plans for a sovereign wealth fund that may eventually pay for a new homeland. Nasheed replaced a long-lived dictatorship that had driven him into exile and banned his five satirical novels. "The Maldives," he says, "is the canary in the world's carbon coal mine" (McMahon 2009, 301).

Further Reading

"Maldives Government Dives for Climate Change." Associated Press in *The New York Times*, October 17, 2009. http://www.nytimes.com/aponline/2009/10/17/world/AP -AS-Maldives-Underwater-Cabinet.html (no longer available).
McMahon, Bucky. "Relocate! Relocate! Relocate!" *GQ* (*Gentleman's Quarterly*), December, 2009, 296–306.

Rising sea levels are seeping into soils on some islands as water tables rise. Soils in some areas are becoming too salty to grow most vegetables. In Tuvalu, according to a dispatch from Reuters, farmers "are beginning to grow their taro crops in tin containers filled with compost instead of traditional pits" (Webb 1998). Soil contamination is also an issue in the Bahamas, where locals fear that the limestone underlying the soil on many islands will absorb saline ocean water like a sponge. Given expectations that global temperatures will rise further in coming decades, the sea level rise of the 20th century is expected to be dwarfed by the rise during the 21st century. Scientists who attend to this issue typically estimate sea level rise possibilities in wide ranges because no one knows how much ice will melt given a certain rise in temperatures over the poles. A team led by J. J. Wells Hoffman, for example, in 1983 estimated that sea levels in the year 2100 would be between 58 and 368 centimeters higher than they were in the 1980s. In 1986, R. Thomas put the estimated range at 56 to 345 centimeters (Thomas 1986). The estimates vary by a factor of roughly seven, an indication of the uncertainty that plagues forecasts of how much the sea may rise under various assumptions about global warming.

Patrick Barkham of *The Guardian* (U.K.) reported from the South Pacific that islands widely imaged as paradises are falling victim to rising seas:

The dazzling white sand and dark green coconut palms of Tepuka Savilivili were much like those on dozens of other small islets within sight of Funafuti, the atoll capital of Tuvalu. But shortly after cyclones Gavin, Hina and Kelly had paid the tiny Pacific nation a visit, islanders looked across Funafuti's coral lagoon and noticed a gap on the horizon. Tepuka Savilivili had vanished. Fifty hectares of Tuvalu disappeared into the sea during the 1997 storms. The tiny country's precious 10 square miles of land were starting to disappear. (Barkham 2002)

Nothing living remains on the few square meters of sand that are still on the island, "only several odd flip-flops and a rusty tin." By 2002, Vasuaafua, another small island, had been reduced by the rising ocean to nine coconut palms on a narrow ribbon of sand. Several years earlier, the island had a sandy beach several hundred yards long (Barkham 2002). During the highest tides of 2001, the island's meteorological office near the airstrip on Funafuti was swamped by seawater (Barkham 2002). Before the mid-1980s, flooding tides usually arrived only in February. By 2002, however, floods became likely anytime between November and March.

Seas Rise as Atolls Sink

Many of these islands face two compounding problems: the sea is rising while the land itself is slowly sinking as 65-million-year-old coral atolls reach the end of their life spans. The atolls were formed as former volcanic peaks sank below the ocean's surface, leaving rings of coral. For five years, the government of Tuvalu has noticed many such troubling changes on its nine inhabited islands and concluded that, as one of the smallest and lowest-lying countries in the world, it is destined to become among the first nations to be sunk by a combination of sea level rise and slow erosion provoked by global warming. The evidence before their own eyes, including forecasts for a rise in sea level as much as 88 centimeters during the 21st century by international scientists, has convinced most of Tuvalu's 10,500 inhabitants that rising seas and more frequent violent storms are certain to make life unlivable on the islands if not for them, then for their children (Barkham 2002).

Residents of the islands have been seeking higher ground, often in other countries. The number of Tuvalu's residents living in New Zealand, for example, doubled from about 900 in 1996 to 2,000 in 2001, many of them fleeing the rising seas on their home islands. A sizable Tuvaluan community has grown up in West Auckland (Gregory 2003). The highest point on Tuvalu is only about three meters above sea level. "From the air," wrote Barkham, "its islands are thin slashes of green against the aquamarine water. From a few miles out at sea, the nation's numerous tiny uninhabited islets look smaller than a container ship and soon slip below the horizon" (Barkham 2002).

"As the vast expanse of the Pacific Ocean creeps up on to Tuvalu's doorstep, the evacuation and shutting down of a nation has begun," Barkham wrote. "With the curtains closed against the tropical glare, the prime minister, Koloa Talake, works in a flimsy Portakabin at the lagoon's edge on Funafuti. Talake, "who sits at his desk wearing flip-flops and bears a passing resemblance to Nelson Mandela," likens his task to the captain of a ship: "The skipper of the boat is always the last man to leave a sinking ship or goes down with the ship. If that happens to Tuvalu, the prime minister will be the last person to leave the island" (Barkham 2002).

Many Pacific island farmers report that their crops of swamp taro (*pulaka*), a staple food, are dying because of rising soil salinity (Barkham 2002). Another staple food, breadfruit (*artocarpus altilis*), also is threatened by saltwater inundation. The breadfruit are harvested from large evergreen trees with smooth bark and large thick leaves that reach a height of 20 meters (60 feet). The fruit, which is large and starchy, reaches edible maturity once or twice a year, depending on type of tree. Some varieties bear fruit all year. Breadfruit probably originated in the Moluccas, Philippines, and New Guinea (Field 2002). Diana Ragone, director of science at the National Tropical Botanical Garden in Hawaii, said that "On some atolls now the people are seeing total die-back of the breadfruit trees, one hundred percent" (Field 2002). Shallow-rooted breadfruit trees are especially vulnerable to increasing numbers of tropical storms, Ragone said. Two cyclones, Val and Ofa, wiped out breadfruit trees in parts of Samoa in 2002.

Living within a few meters of ocean level, island residents share a special fear of typhoons that roil the seas and drive storm surges through their villages. Before 1985, one or two serious tropical storms usually hit the islands each decade; during the 1990s, seven such storms ravaged them. More frequent El Niños (possibly intensified by global warming) now seem to be making storms more frequent.

Debilitating Drought in the Marshall Islands

More than 1,000 Marshall Islands on 29 coral atolls in the South Pacific by 2015 were eroding into a gradually rising ocean that regularly flooded many towns during high-tide cycles, inundating crops with salty water. Only the occasional bluff is more than six feet in elevation, even at low tide. Changing trade winds have raised sea levels a foot in 30 years (more during tidal cycles), with more coming. Damage was clearly evident in daily life. As Coral Davenport wrote in *The New York Times* (2015):

> Linber Anej waded out in low tide to haul concrete chunks and metal scraps to shore and rebuild the makeshift seawall in front of his home. The temporary barrier is no match for the rising seas that regularly flood the shacks and muddy streets with saltwater and raw sewage, but every day except Sunday, Mr. Anej joins a group of men and boys to haul the flotsam back into place. (Davenport 2015)

Davenport continued:

> In neighborhoods like Mr. Anej's, after the sewage-filled tides wash into homes, fever and dysentery soon follow. On other islands, the wash of saltwater has penetrated and salinated underground freshwater supply. On Majuro [the islands' capital], flooding tides damaged hundreds of homes in 2013. The

elementary school closed for nearly two weeks to shelter families. That same year, the airport temporarily closed after tides flooded the runway. Cemeteries near the ocean are being washed away. (Davenport 2015)

Phillip Muller, foreign minister of the Marshall Islands, said that islanders feel a special sense of helplessness because the climatic forces shaping their futures have often been beyond their control. In a sense, he said, this situation resembles the testing of 67 nuclear bombs on their islands by the United States between 1946 and 1958. "Now our residents are confronted by a different kind of atmospheric danger: the existential threat posed by climate change, he said (Muller 2013).

Rising seas are not the only problem. Recurring severe drought that causes a severe lack of drinking water is also creating trouble because residents have no freshwater from lakes, rivers, or streams. It rains, or they go thirsty. By 2013, many people were surviving on less than one liter of water per day. Staple crops that also depend on rainfall have been lost as well. As Muller wrote,

> My people are not only thirsty and hungry, they are also getting sick. The drying water wells are contaminated with bacteria and salt. Diarrhea, pink eye, flu, and other drought-related diseases are on the rise, particularly among children, and we are on the brink of a much wider outbreak. With no significant rain forecast . . . the situation is likely to get worse. (Muller 2013)

Mulller pointed out that with the atmospheric carbon dioxide level having passed 400 parts per million, sea levels are projected to rise three to seven feet by century's end. "For the world's lowest-lying countries, including my own, this is a death sentence, he wrote in the *Washington Post* (2013). "In the Pacific," Muller concluded, "We cannot afford to wait. Sadly, we are learning the terrible realities of living with climate change. My family built a seawall around our home, but it was destroyed by waves. The rising tides come closer every day" (Muller 2013).

The Philippines, Tropical Cyclones, and Climate-Change Diplomacy

The Philippines comprises more than 7,000 islands—from specks in the sea to Luzon, a mountainous mass larger than many nations and home to Manila, a metropolitan area with more people in 2015 (20 million) than New York City. This island is not about to be swamped by rising seas, although its shorelines are vulnerable as in the rest of the world. The Philippines has become a world leader in climate-change diplomacy, however, as some of the most intense tropical cyclones on record have raked its shores and killed thousands of people. The deadliest and most destructive storms have been supertyphoons Hagupit and Haiyan.

With scientists saying that the Philippines is among the most vulnerable countries in the world to climate change, its diplomats have been arguing that the richer and poorer nations should put aside their differences and prepare to reduce greenhouse gas emissions together before it is too late. At worldwide meetings of diplomats and scientists, some Philippine diplomats have gone on hunger strikes to make this point. Philippine climate diplomat Naderev Saño seized the stage in 2013 as Typhoon Haiyan ravaged parts of his island nation.

Mary Ann Lucille Sering, secretary of the Philippines climate change commission, said, "We have sustained losses every year since 2008 of 5 percent of GDP. We are experiencing tremendous damages and loss of life from typhoons" (Davenport 2014). The Philippines has taken a lead role in a negotiating bloc that has become known as the "most vulnerable countries." The country adopted this strategy in 2012 after Typhoon Hagupit ("Lasher" in English) tore through its islands.

According to Carol Davenport of *The New York Times* (2014), Saño

gave an impassioned, weeping speech that electrified the annual United Nations climate change summit meeting, demanding that the world wake up to the impact of fossil fuels on his homeland. Over the next year, he gained world attention as an advocate for his cause, staging a hunger strike, giving speeches to thousands and taking part in a 38-day climate awareness march from Manila to his flattened hometown Tacloban. (Davenport 2014)

According to another account, "Days after Typhoon Haiyan stuck the Philippines, Saño, who was leading the Philippines delegation at climate treaty talks in Warsaw, Poland, brought scientists and diplomats to tears as he advocated action on global warming, and went on a hunger strike until delegates made meaningful progress." The Environment News Service said that "he called on delegates at Warsaw National Stadium to make the connection between climate change and the increase in severe storms, rising sea levels, floods, and drought. He urged them to be the ones who inspire their countries to move the world out of the climate danger zone."

Saño called the world to account:

To anyone who continues to deny the reality that is climate change, I dare you to get off your ivory tower and away from the comfort of your armchair. I dare you to go to the islands of the Pacific, the islands of the Caribbean and the islands of the Indian ocean and see the impacts of rising sea levels; to the mountainous regions of the Himalayas and the Andes to see communities confronting glacial floods, to the Arctic where communities grapple with the fast dwindling polar ice caps, to the large deltas of the Mekong, the Ganges, the Amazon, and the Nile where lives and livelihoods are drowned, to the hills of Central America that confronts similar monstrous hurricanes, to the vast savannas of Africa where climate change has likewise become a matter of life and death as food and water becomes scarce. Not to forget the massive hurricanes in the Gulf of Mexico and the eastern seaboard of North America. And if that is not enough, you may want to pay a visit to the Philippines right now. ("Philippines Climate" 2013)

Saño asked diplomats to visit his country and survey the devastation of "this hellstorm called Super Typhoon Haiyan, which has been described by experts as the strongest typhoon that has ever made landfall in the course of recorded human history" ("Philippines Climate" 2013). He said that existing emissions limits are dangerously low, and that climate talks were stalling because of bickering along nationalistic lines. Saño received a standing ovation. China's delegation requested three minutes of silence among all attending the plenary hall in honor of everyone who had been killed by Typhoon Haiyan in the Philippines, southern China, and Vietnam. Olai Ngedikes, lead negotiator with a small island nations' alliance, said that Haiyan, "serves as a stark reminder of the cost of inaction on climate change and should serve to motivate our work in Warsaw" (Fountain and Gillis 2013).

At least 1,774 people died and more than 615,000 were displaced, according to official reports. Local officials estimated that Haiyan actually killed about 10,000 people in Leyte Province alone with its 13-foot storm surge and sustained winds of 175 to 200 miles per hour. It was, in essence, a 100-mile-wide EF-4 tornado with an added storm surge. The emotional impact of on the Warsaw conference proved transitory, however. The meeting as a whole did little to steer climate diplomacy in the direction that Saño had urged.

The Philippines' leadership emerged as scientific journals carried news of the role of warming oceans in providing the strength to now ever-more powerful tropical cyclones. For example, Dennis Normile wrote in *Science* (2013),

> Data collected by satellites and floating probes have chronicled a two-decade rise in the temperature and thickness of a layer of warm subsurface ocean water east of the Philippines. These warmer subsurface waters along Supertyphoon Haiyan's track fueled its fury. Meanwhile, strong easterly trade winds have pushed warmer water to the western North Pacific, raising sea level in the region. This phenomenon amplified Haiyan's storm surge and accounts for the largest share of the typhoon's death toll.

A single storm is not "proof" of a link to climate change, but as Kerry Emanuel, an atmospheric scientist at the Massachusetts Institute of Technology, noted, "As you warm the climate, you basically raise the speed limit on hurricanes" (Fountain and Gillis 2013). Tropical cyclones also are striking areas where they were once all but unknown such as Somalia on the horn of Africa, which experienced severe storms in both 2012 and 2013 that killed more than 100 people ("Rare Tropical" 2013).

Further Reading

Barkham, Patrick. "Going Down: Tuvalu, a Nation of Nine Islands" *The Guardian* (U.K.), February 16, 2002, 24.

Davenport, Carol. "Philippines Pushes Developing Countries to Cut Their Emissions." *The New York Times,* December 8, 2014. http://www.nytimes.com/2014/12/09/world/americas/philippines-pushes-developing-countries-to-cut-their-emissions-.html.

Davenport, Carol. "The Marshall Islands Are Disappearing." *The New York Times*, December 2, 2015. http://www.nytimes.com/interactive/2015/12/02/world/The-Marshall-Islands-Are-Disappearing.html.

Field, Michael. "Dying Pacific Breadfruit New Sign of Looming Disaster." *Agence France-Presse*, December 1, 2002 (in LEXS).

Fountain, Henry, and Justin Gillis. "Typhoon in Philippines Casts Long Shadow over U.N. Talks on Climate Treaty." *The New York Times*, November 11, 2013. http://www.nytimes

.com/2013/11/12/world/asia/typhoon-in-philippines-casts-long-shadow-over-un-talks
-on-climate-treaty.html.

Gregory, Angela. "Fear of Rising Seas Drives More Tuvaluans to New Zealand." *New Zealand Herald*, February 19, 2003, n.p. (LEXIS)

Kristof, Nicholas. "For Pacific Islanders, Global Warming Is No Idle Threat." *The New York Times*, March 2, 1997. http://sierraactivist.org/library/990629/islanders.html (no longer available).

Muller, Phillip. "Pacific Islands' Deadly Threat from Climate Change." *Washington Post*, May 30, 2013. http://www.washingtonpost.com/opinions/pacific-islands-face-a-deadly
-threat-from-climate-change/2013/05/30/86ff1956-c7a9-11e2-9245-773c0123c027
_print.html.

Normile, Dennis. "Clues to Super-Typhoon's Ferocity Found in the Western Pacific." *Science* 342 (November 29, 2013): 1027.

"Philippines Climate Negotiator Vows Hunger Strike." Environment News Service, November 12, 2013. http://ens-newswire.com/2013/11/12/philippines-climate-negotiator-vows
-hunger-strike/.

"Rare Tropical Cyclone Strikes Somalia." Environment News Service, November 12, 2013. http://earthobservatory.nasa.gov/IOTD/view.php?id=82377&src=eoa-iotd.

Thomas, R. "Future Sea-Level Rise and Its Early Detection by Satellite Remote Sensing." In Vol. 4, *Effects of Changes in Atmospheric Ozone and Global Climate*. New York: United Nations Environment Program/U.S. Environmental Protection Agency, 1986.

Webb, Jason. "Small Islands Say Global Warming Hurting Them Now." Reuters, 1998. http://
bonanza.lter.uaf.edu/~davev/nrm304/glbxnews.htm (no longer available).

See also: Sea Level Rise; Sea Level Rise, Bangladesh; Sea Level Rise, United States; Temperatures, Global

SEA LEVEL RISE, ITALY

Floods have plagued the northeastern Italian city of Venice for most of its history, but subsidence and slowly rising seas resulting from global warming worsened flooding during the late 20th and early 21st centuries. Venice, which sits atop several million wooden pillars pounded into marshy ground, has sunk by about 7.5 centimeters per century for the past 1,000 years. That rate is now accelerating.

Increased floods led to plans and initial construction of movable barriers across three separate entrances to Venice's lagoon. In June 2002, the water level in the city rose to 125 centimeters above sea level, a record for the month. At the beginning of the 20th century, St. Mark's Square, the center of the city, was flooded an average of nine times a year. During 2001 alone, it flooded almost 100 times. In 2003, the city flooded 111 times, more than any other year in its lengthy history. In another century, it will be flooded on a permanent basis ("Heavy Rains" 2002).

Since 1950, Venice has lost two-thirds of its population, and now only 60,000 people remain in the city. As a center of artistic interest, however, it annually hosts 12 million tourists who make their way over planks into buildings with foundations rotted by perennial flooding. At the Danieli, one of Venice's most luxurious hotels, tourists often arrive on wooden planks raised two feet above the marble floors amid a suffocating stench from the high water (Poggioli 2002).

"Acqua Alta"

Venice's residents and visitors have become accustomed to high-water drills for "acqua alta," or high water. A system of sirens much like the ones that convey tornado warnings in the U.S. Midwest sounds when the water surges. Restaurants have stocked Wellington boots and moved their dining rooms upstairs. Venetian gondoliers ask their passengers to shift fore and aft—and watch their heads—as they pass under bridges during episodes of high water (Rubin 2003). Some gondoliers have hacked off their boats' distinctive tailfins to pass below the bridges brought closer by rising waters. Amid considerable controversy but faced with rising waters, Venice officials proposed the construction of massive retractable dikes in an attempt to hold the water at bay. After 17 years of heated debate, the project known as MOSE (*Modulo Sperimentale Elettromeccanico*, or Experimental Electromechanical Module) was begun in 2003 with a cost then estimated at $1 billion. Ten years later, the cost had ballooned to $14 billion.

Some environmentalists asserted that the barriers would destroy the tidal movement required to keep local lagoon waters free of pollution and thereby damage marine life. Water quality near Venice is already precarious because pollution has leached into the lagoon from industry, homes, and motor traffic. The Italian Green Party favored shaping the lagoon's entrances to reduce the effects of tides along with raising pavements as much as a meter inside the city.

As of September 2016, the barriers being constructed at the three entrances to Venice's natural lagoon from the Adriatic Sea were nearly complete. Each barrier will house 79 "flippers" that can be adjusted like the flaps of an aircraft. Installed below the water line, they will be raised when the sea level rises by more than one meter, which at the turn of the millennium was occurring a dozen or so times a year (Watson 2001).

During normal tides, according to an account in *Scotland on Sunday*, "The hollow barriers will sit within especially constructed trenches in the bed of the channels connecting the lagoon to the open sea. When a dangerously high tide is forecast, compressed air will be forced into the flippers which will have the effect of squeezing seawater out. As they rise, more water will trickle out to be replaced by air" (Watson 2001).

MOSE May Run Full-Time

By the middle of the 21st century, the MOSE system may be running almost all the time, severing the city from the ocean and transforming its lagoon, according to one observer, "into a stagnant pond with devastating effects on marine life and health" (Poggioli 2002). Many Venetians said that Project MOSE would not help much because it will operate only when water rises at least 43 inches.

Environmentalists have long argued that the MOSE flood-control system was a construction boondoggle that would turn Venice into a toxic bathtub in which the city's canals would be laced with sludge from surrounding heavy industry, as well as the urban area's human waste. In particular, environmentalists focused attention on bacteria from animal and human waste in the waters surrounding the city (Petrillo

2003). "Venice has no sewage system; they just dump the stuff right out into the canals. It's not pretty," said Rick Gersberg, a microbiologist. "Normally, the tides come in and flush everything out. But when you cut off the tide, it just sits there" (Petrillo 2003). Venice's deputy mayor, Gianfranco Bettin, had called MOSE "expensive, hazardous, and probably useless" (Nosengo 2003, 608).

In 2012, James Atlas commented in *The New York Times*,

Is the Modulo Sperimentale Elettromeccanico—the project's official name—some engineer's fantasy? It was scheduled for completion this year but that has been put off until 2014. Even if, by some miracle, the gates materialize, they will be only a stay against the inevitable. Look at the unfortunate Easter Islanders, who left behind as evidence of their existence a mountainside of huge blank-faced busts, or the Polynesians of Pitcairn Island, who didn't leave behind much more than a few burial sites and a bunch of stone tools. Every civilization must go.

Further Reading

Atlas, James. "Is This the End?" *The New York Times*, November 24, 2012. http://www.nytimes.com/2012/11/25/opinion/sunday/is-this-the-end.html.

"Heavy Rains Threaten Flood-Prone Venice." *The Straits Times* (Singapore), June 8, 2002, n.p. (LEXIS).

Nosengo, Niccola. "Venice Floods: Save Our City!" *Nature* 424 (August 7, 2003): 608–609.

Petrillo, Lisa. "Turning the Tide in Venice." Copley News Service, April 28, 2003 (LEXIS).

Poggioli, Sylvia. "Venice Struggling with Increased Flooding." Morning Edition, National Public Radio, November 29, 2002 (LEXIS).

Rubin, Daniel. "Venice Sinks as Adriatic Rises." Knight-Ridder News Service, July 1, 2003, n.p. (LEXIS).

Watson, Jeremy. "Plan to Hold Back Tides of Venice Runs into Flood of Opposition from Greens." *Scotland on Sunday*, December 30, 2001, 18.

See also: Sea Level Rise; Sea Level Rise, Bangladesh; Sea Level Rise, United States; Temperatures, Global

SEA LEVEL RISE, UNITED STATES

Since 1880, the global mean sea level has risen an average of one foot. Actual figures, however, have varied widely, depending on other factors, such as whether the land itself is rising or subsiding. "It has climbed about a foot or more in some U.S. cities because of ocean currents and land subsidence—11 inches in New York and Boston, 12 in Charleston, 16 in Atlantic City, 18 in Norfolk and 25 in Galveston, Texas," Wendy Koch reported in *USA Today* using data from the National Oceanic and Atmospheric Administration (Koch 2013).

Hampton Roads, Virginia, near Norfolk, is exhibit A for East Coast sea level rise as the land in the area slowly sinks. Net sea level rise in that area has been around 2 feet over 75 years mainly because land has subsided. New Orleans is the only U.S. city that is more vulnerable. Paul Fraim, Norfolk's mayor since 1994, said that a category 2 hurricane could submerge Norfolk and neighboring urban areas, which lie on a nearly flat tidewater plain (Koch 2013, 2A). "Water moves downhill, but it's flat here. There's no place for rain to go," said Larry Atkinson, oceanography professor at Norfolk's Old Dominion University (Koch 2013, 2A). In November 2013, Norfolk's city government required that new and large renovated buildings be elevated three feet above the flood plain. The former requirement had been one foot.

In 2016, the real estate data firm Zillow said that a six-foot rise in sea levels would submerge 2 million homes worth $880 billion and displace several million people in the United States. Approximately half of these would be in Florida, including a third of Miami's housing stock. Some 190,000 would be in New Jersey and 97,000 in New York, with 80,000 in Louisiana. The seas may reach this level by the year 2100. Boston would lose 18 percent of its housing. Honolulu would lose a quarter, and sizable parts of Virginia's Tidewater urban area (Norfolk and environs) also would be submerged. In 2016, the U.S. Army Corps of Engineers projected a six-foot relative sea level rise by 2100 for parts of the Norfolk area.

Chesapeake Bay, Rising Sea Level

Relative sea levels are rising almost twice as fast in the Chesapeake Bay region as in most of the world, and waterside communities are spending millions to keep the water from eroding yards, marshes, and sandy beaches. Parts of Washington, D.C., near the bay are sinking; with sea level rise, the water is rising an eighth of an inch a year, double the global average. *Washington Post* reporter David A. Fahrenthold wrote in 2009, "To keep higher waves from washing away waterside property, homeowners and government agencies have spent millions to make the Chesapeake look like a high-sided swimming pool. About a quarter of Maryland's shoreline has been 'armored' with man-made seawalls or rock piles" (A1). "The beach just continues to erode right up to the wall," said James Titus, a lead author of a massive Environmental Protection Agency (EPA) study of sea level rise in the mid-Atlantic released in 2009. "The shore is moving. You put a wall in the way, and you have a wall and the water."

The problem is common all along the coast. For decades, wrote Fahrenthold, "Virginia Beach has dumped millions of tons of sand onto its beach to replenish what erodes. In Ocean City, Maryland, state, federal and local authorities spent $7 million in 2006 to deposit 100,000 dump trucks' worth of sand on its beach. And they say they believe it will need more sand next year" (Fahrenthold 2009, A1).

More than half of Chesapeake Bay's marshes are in danger of becoming open water, according to a University of Maryland study. Rising sea levels are flooding the marshes that filter pollutants in groundwater, guard against erosion, and provide habitat for ducks, geese, and other animals. Michael Kearney, an associate professor at University of Maryland, College Park, said the marshes could be gone in 20 to 30 years, and that time line could be shortened if nor'easters—northern latitude cyclones—become more severe. Kearney and colleague J. Court Stevenson (also of the University of Maryland) said that water levels in the bay generally have been rising since the year 1000 but that the rate of rise accelerated sharply (to more than 10 times the previous rate) during the 1990s. Sea level rise in the area is being made worse by subsidence of the land (McCord 2000, 1B).

Further Reading

Fahrenthold, David A. "Eco-Bills Come Due at Bay's Beaches Region Pays Dearly for Climate Change in Erosion, Abatement." *Washington Post*, March 19, 2009, A1. http://www.washingtonpost.com/wp-dyn/content/article/2009/03/18/AR200903 1804178.html

McCord, Joel. "Marshes in Decay Haunt the Bay." *Baltimore Sun*, December 6, 2000, 1B.

Wealth along Vulnerable Coasts

Making a seacoast less vulnerable to rising waters will require more than jacked-up foundations, however. It will also require a change in basic attitudes that still encourage a disconnect between real-estate values and vulnerability to ocean flooding. In 2012, New York City neighborhoods nearest the water ravaged by Hurricane Sandy had been growing in population and value. This pattern is common along the Atlantic Ocean and Gulf of Mexico coastlines. During the last few decades, the real-estate industry has created massive amounts of wealth along the coasts, with ranks of condos and million-dollar "McMansions." Tax assessors have joined in. How many storm surges will be required before we begin to adjust our notions of land value to account for rising climatic risks?

The situation has the attention of the U.S. Army Corps of Engineers, which made the Virginia Tidewater area the subject of a case study released late in 2013, which said that Naval Station Norfolk's infrastructure may be inundated by coastal storms by late in the 21st century (Kaufman 2010). The Department of Defense also said in 2013 that nationwide some 10 percent of its coastal facilities are at or near sea level and are "already vulnerable to flooding and inundation." A study by the Virginia legislature in 2013 found that "recurrent flooding" could raise sea levels more than elsewhere—mainly because of land subsidence—one to two feet by 2040 and three to nearly eight feet by 2100 (Koch 2013, 2A).

Sea Level Rise "Hot Spot"

Climate scientists led by Stefan Rahmstorf of Germany's Potsdam Institute said in November 2012 that worldwide sea level had risen 60 percent more than previous forecasts since 1993 (Rahmstorf et al. 2012). "Sea level rise is accelerating along the [United States] East Coast," said Asbury Sallenger, an oceanographer with the U.S. Geological Survey. The rise—taking into account sinking land as well as well as rising seas—has been most pronounced along a "hot spot" from Cape Hatteras, North Carolina, to Maine (Vergano 2012). In June 2012, Sallenger, an oceanographer at the U.S. Geological Survey, and colleagues publishing in *Nature Climate Change,* analyzed tidal-gauge records across North America between 1950 and 2009, finding that the rates of sea level rise from Cape Hatteras to Boston were increasing three to four times faster than rates of sea level rise globally—2 and 3.7 millimeters per year since 1980—compared to a worldwide increase of 0.6–1.0 millimeters per year during the same period. The apparent sea level rise in this area (including land subsidence) is expected to be close to 1 meter by 2100 and more if warming continues (Sallenger et al. 2012, 884).

They wrote:

Climate warming does not force sea level rise at the same rate everywhere. Rather, there are spatial variations of sea level rise superimposed on a global average rise. These variations are forced by dynamic processes, arising from circulation and variations in temperature and/or salinity, and by static equilibrium processes, arising from mass redistributions changing gravity and the Earth's rotation and shape. These sea level variations form unique spatial patterns, yet there are very few observations verifying predicted patterns or fingerprints[6]. Here, we present evidence of recently accelerated sea level rise in a unique 1,000-km-long hot spot on the highly populated North American Atlantic coast north of Cape Hatteras and show that it is consistent with a modeled fingerprint of dynamic sea level rise. Between 1950 to 1979 and 1980 to 2009, sea level rate increases in this northeast hot spot were about three to four times higher than the global average. Modeled dynamic plus steric sea level rise by 2100 at New York City ranges with Intergovernmental Panel on Climate Change scenario from 36 to 51 centimeters; lower emission scenarios project 24 to 36 centimeters. Extrapolations from data herein range from 20 to 29 centimeters of sea level rise superimposed on storm surge, wave run-up and set-up will increase the vulnerability of coastal cities to flooding, and beaches and wetlands to deterioration. (Sallenger et al. 2012, 884)

Stefan Rahmstorf and colleagues wrote in 2012:

We analyze global temperature and sea level data for the past few decades and compare them to projections published in the third and fourth assessment reports of the Intergovernmental Panel on Climate Change (IPCC). The results show that global temperature continues to increase in good agreement with the best estimates of the IPCC, especially if we account for the effects of short-term variability due to the El Niño/Southern Oscillation, volcanic activity and solar variability. The rate of sea level rise of the past few decades, on the other hand, is greater than projected by the IPCC models. This suggests that IPCC sea level projections for the future may also be biased low.

The Norfolk Area's Special Circumstances

As nearly all of the U.S. East Coast slowly sinks, the Norfolk, Virginia, area is subsiding at a faster rate than any other, partly because of its position at the edge of a former ice sheet, which depressed the areas it covered and provoking a rise in elevation at its edge. With the ice having melted, this discrepancy is now evening out. In addition, the Chesapeake Bay area was the site of a meteor strike 35 million years ago. Justin Gillis in *The New York Times* described "a collision so violent it may have killed nearly everything on the East Coast and sent tsunami waves crashing against the Blue Ridge Mountains. The meteor impact disturbed and weakened the sediments across a 50-mile zone. Norfolk is at the edge of that zone, and the ancient cataclysm may be one reason it is sinking especially fast" (Gillis 2014).

Sea Level Rise May Vary by Region

By 2009, scientists were beginning to calculate that relative sea level rise will not be equal. For example, acceleration of Greenland's melting ice may change ocean currents to send relatively warmer water toward the northeastern United States and maritime Canadian coasts. As the warming water expands, sea level could rise a foot or more in those areas in addition to the general sea level change, according to Scientists Aixue Hu and Gerald A. Meehl at the National Center for Atmospheric Research in Boulder, Colorado. Because the coast is flat in many areas, a few feet of sea level rise could send ocean water inland 100 feet or more. In recent years, Greenland melt rates have been increasing by 7 percent a year on average (Dean 2009; Hu et al. 2009).

One worldwide number for sea level rise obscures local differences, the largest of which is whether the land at a given location is rising or falling. Although much of the U.S. coast along the Atlantic Ocean and the Gulf of Mexico is subsiding (some areas more than others), Jeff Freymueller, a geophysicist at the University of Alaska Fairbanks, has measured a declining sea level in Alaska's Graves Harbor of as much as 3 centimeters per year as the land rebounds from the last ice age 10,000 years ago The worldwide annual average sea level rise averages 3.2 millimeters, but "[in] some places, sea level rise is ten times faster than the average," said Jerry Mitrovica, a geophysicist at Harvard University in Cambridge, Massachusetts (Jones 2013). Land near Hudson Bay in Canada is also rising about 1 centimeter per year because of glacial rebound. When river sediments are compacted and subsurface groundwater is removed, however, land sinks. Parts of China's Yellow River delta have been sinking as much as 25 centimeters per year, a major change when added up year after year.

Large volumes of ocean water can be shifted from location to location as air pressure, "winds, and currents can shove water in a given ocean to one side: since 1950, for example, a 1,000-kilometer stretch of the United States

Atlantic coast north of Cape Hatteras in North Carolina has seen the sea rise at 3–4 times the global average rate. In large part, this is because the Gulf Stream and the North Atlantic current, which normally push waters away from that coast, have been weakening, allowing water to slop back onto U.S. shores" (Jones 2013). Gravity plays a role in some locations. As Nicola Jones explained in *Nature* (2013),

> Finally, waters near big chunks of land and ice are literally pulled up onto shores by gravity. As ice sheets melt, the gravitational field weakens and alters the sea level. If Greenland melted enough to raise global seas by an average of 1 meter, for example, the gravitational effect would lower water levels near Greenland by 2.5 meters and raise them by as much as 1.3 meters far away.

Further Reading

Dean, Cornelia. "Sea's Rise May Prove the Greater in Northeast." *The New York Times*, May 28, 2009. http://www.nytimes.com/2009/05/28/science/earth/28warming .html.

Hu, A., et al. "Transient Response of the MOC and Climate to Potential Melting of the Greenland Ice Sheet in the 21st Century." *Geophysical Research Letters* 36 (May 29, 2009). doi:10.1029/2009GL037998.

Jones, Nicola. "Rising Tide: Researchers Struggle to Project How Fast, How High and How Far the Oceans Will Rise." *Nature* 501 (September 19, 2013). http://www .nature.com/news/climate-science-rising-tide-1.13749.

Norfolk, which lies west of the mouth of Chesapeake Bay, is bordered on three sides by water and built on former marshlands that have been settling and compacting. In addition, the land is sinking and sea levels slowly rising, producing a net sea level rise of 14.5 inches from 1930 to 2010 at the Sewells Point naval station. By 2013, coastal flooding had become so frequent in the Norfolk area that some houses became unsalable. Norfolk is now spending money to improve drainage and raise streets, but the amount of money required to fully address the problem (more than $1 billion) is beyond the city's reach (Gillis 2014). "In the last couple or three years, there's really been a change," said William A. Stiles Jr., head of Wetlands Watch, a Norfolk environmental group. "What you get now is people saying, 'I'm tired of driving through saltwater on my way to work, and I need some solutions'" (Gillis 2013).

Insurance providers have been watching storm surges as well. In October 2012, Congress began to reduce subsidies for the federal flood insurance program, which has become heavily indebted as more than a million homeowners faced sharply higher premiums that were more in line with actual risk. "There will be a slow exodus" from the coasts as property values sink, said oceanographer John Englander, author of *High Tide on Main Street* (Koch 2013, 2A).

Two days after Sandy's storm surge struck during October 2012, New York City Mayor Michael Bloomberg, a Republican, endorsed Barack Obama and said he was doing it in the hopes that Obama would do something about global warming because of its role in a growing roster of natural disasters. Bloomberg's *Business-Week* news service headlined, "It's Global Warming, Stupid!" Two weeks later, another severe storm tracked along the same path, intensifying the devastation.

Sea Level Rise: Maine to Florida

In 2009, the New York City Panel on Climate Change issued a prophetic report. "In the coming decades, our coastal city will most likely face more rapidly rising sea levels and warmer temperatures, as well as potentially more droughts and floods, which will all have impacts on New York City's critical infrastructure," said William Solecki, a geographer at Hunter College and a member of the panel (Atlas 2012).

The movement of land today may be governed by past events. Much of the U.S. mid-Atlantic cast is sinking today because it was on the edge of a massive ice sheet 20,000 years ago. As Justin Gillis explained in *The New York Times* (2014):

As a massive ice sheet, more than a mile thick, grew over what are now Canada and the northern reaches of the United States, the weight of it depressed the crust of the Earth. Areas away from the ice sheet bulged upward in response, as though somebody had stepped on one edge of a balloon, causing the other side to pop up. Now that the ice sheet has melted, the ground that was directly beneath it is rising, and the peripheral bulge is falling.

Land is sinking from southern Maine to northern Florida, coupled with actual sea level rise to make the apparent rising of the sea much larger than what is being provoked by melting ice and the thermal expansion of warming water alone. The rate of subsidence, Gillis wrote, "is fastest in the Chesapeake Bay region [where] whole island communities that contained hundreds of residents in the 19th century have already disappeared. Holland Island, where the population peaked at nearly 400 people around 1910, had stores, a school, a baseball team and scores of homes. But as the water rose and the island eroded, the community had to be abandoned." "Eventually," wrote Gillis, "Just a single, sturdy Victorian house, built in 1888, stood on a remaining spit of land, seeming at high tide to rise from the waters of the bay itself" (Gillis 2014).

Some soils compress and sink over time. The New Jersey coast is sinking in this manner. In addition, removal of subsurface water for human use can cause land to sink. At times, this subsidence expresses itself suddenly and violently when a surface collapses in sinkholes that swallow cars and even houses. After a modest sea level rise of eight inches by 2050 was forecast, a team from Rutgers University projected in 2013 that the apparent rise will be 14 inches at the Battery (on the southern tip of Manhattan Island) and 15 inches on the New Jersey shore. Even if the global sea level rises only eight more inches by 2050, a moderate forecast, the

Rutgers group foresees relative increases of 14 inches at bedrock locations such as the Battery and 15 inches along the New Jersey coastal plain, where the sediments are compressing. By 2100, the group's analysis indicates, there will be an average rise in the global ocean of 28 inches, with 36 inches at the Battery and 39 inches on the New Jersey coastal plain (Miller et al. 2013).

Bloomberg (who finished his last term as New York City's mayor on December 31, 2013) commissioned exhaustive research on climate change, including the use of expanded wetlands to absorb rising tides. Even so, New York, a city of peninsulas and islands with 520 miles of shorelines, runs a risk of inundation as seas rise. As Mireya Navarro wrote in *The New York Times* (2012), "critics say New York is moving too slowly to address the potential for flooding that could paralyze transportation, cripple the low-lying financial district and temporarily drive hundreds of thousands of people from their homes." Areas such as the South Bronx and Sunset Park in Brooklyn—with their long industrial waterfronts that contain chemical- and oil-storage sites, manufacturing plants, and garbage-transfer stations that often handle hazardous materials—need to be protected from storm surges such as Hurricane Sandy's or the city will face neighborhoods inundated with toxic, salty water. "A lot of attention is devoted to Lower Manhattan, but you forget that you have real industries on the waterfront" elsewhere in the city, said Eddie Bautista, executive director of the New York City Environmental Justice Alliance, which represents low-income residents. "We're behind in consciousness-building and disaster planning" (Navarro 2012).

In the United States, New York City is second only to New Orleans in the number of people (200,000) living within four feet of high tide. Some residents have advocated a series of seawalls and ocean barriers to keep storm surges at bay, but the complexity of the city's topography, with its many islands and peninsulas, makes such a thing nearly impossible. A study issued in 2004 by the Storm Surge Research Group at the State University of New York at Stony Brook "recommended installing movable barriers at the upper end of the East River, near the Throgs Neck Bridge; under the Verrazano-Narrows Bridge; and at the mouth of the Arthur Kill, between Staten Island and New Jersey" (Navarro 2012). Even if such barriers can be built, they would harm the ecosystem by interfering with outflow of polluted water and could ultimately fail as storm surges become more severe as oceans rise and the land subsides. The initial cost was estimated at $10 billion.

Florida's Sea-Swept Future

With the exception of Orlando's fantasy castles, universities, and parts of the panhandle, most of Florida's real-estate assets (and tax base), are on or near its coasts, home to ranks of waterfront condos, businesses, and other homes that are perilously vulnerable to sea level rise. Will real-estate values (and the tax base) crumble before rising seas and subsiding shorelines wipes them out?

Florida has by far the highest risk of any U.S. state for flooding in a warming world and has almost half of the country's at-risk population. The population lives mainly on or near the coast, most of it on porous limestone that makes seawalls all but useless because seawater easily flows under or around any obstruction. (Coastal sections of Louisiana, Virginia, Maryland, New York, and New Jersey also are vulnerable.) Three-quarters of Florida's 18 million people live in coastal counties that generate 80 percent of the state's economy. A five-foot sea level rise would flood 1 million homes in the state. In the metropolitan Miami area, the shoreline will move inland 500 to 2,000 feet for every foot of sea level rise (Parker 2015, 114–116).

"Sunny-Day Flooding"

As in Norfolk, Virginia, Miamians have seen "sunny-day flooding" become common with lunar high tides. "During high tide one recent afternoon," reported Coral Davenport in *The New York Times* (2014)

> Eliseo Toussaint looked out the window of his Alton Road laundromat and watched bottle-green saltwater seep from the gutters, fill the street and block the entrance to his front door. "This never used to happen,' Mr. Toussaint said. "I've owned this place eight years, and now it's all the time." Down the block at an electronics store it is even worse. Jankel Aleman, a salesman, keeps plastic bags and rubber bands handy to wrap around his feet when he trudges from his car to the store through ever-rising waters.

"It's remarkable. We get calls from people asking: 'It didn't rain, so why is my street underwater?'" said Broward County Commissioner Kristin Jacobs, noting the region's decades-old water drains are now routinely overflowing. "I have a photo of a man swimming—doing the backstroke—in his cul-de-sac," she says, adding that 30 percent of her county—just north of Miami—is five feet or fewer above sea level (Koch 2014). According to the World Bank, in terms of monetary costs, Miami is the city most at risk for flood damage from sea level rise in the world ("10 Coastal Cities" 2013).

Even with densely populated land within a few feet of sea level, constructing seawalls is nearly useless because the area is underlain by porous limestone that allows "the rising seas to soak into the city's foundation, bubble up through pipes and drains, encroach on freshwater supplies and saturate infrastructure. County governments estimate that the damages could rise to billions or even trillions of dollars" (Davenport 2014). Miami ranks first worldwide among cities by number of residents (4.8 million in 2014) who risk coastal flooding by 2070, according to a report issued in 2012 by the Organization for Economic Cooperation and Development.

A King tide provoked by a "supermoon" already has raised ocean waters enough around Miami to strand an octopus in a parking garage ("Findings," 2017, 96). How long will it be before dangerous day-to-day reality causes property values to sink along Miami's coastline? Miami's city government in 2014 was considering spending $400 million to improve its drainage system amid bickering factions. By often remaining silent, Republicans have asserted that the problem is an illusion or not the fault of humans, whereas Democrats claim that the city's fix is too little, too late. Public hearings on the issue have packed auditoriums in Miami Beach.

The Prospective Toll of Rising Seas

"In the most dire predictions," according to a report in *The New York Times*, "South Florida's delicate barrier islands, coastal communities and captivating subtropical beaches will be lost to the rising waters in as few as 100 years" (Madigan 2013). One estimate of property assets within three feet of high tide along Florida's 1,197 miles of coastline is $156 billion; 2,555 miles of road, 35 public schools, one power plant, and several sites listed by the Environmental Protection Agency as hazardous waste dumps and sewage plants are in danger (Madigan 2013). The same article noted that most real-estate agents in the state are still rather oblivious to global warming's effect on sea level rise because a waterfront location still adds quite a bit to the value of houses and condos.

"Further inland," noted the *Times*, "the Everglades, the river of grass that gives the region its freshwater, could one day be useless, some scientists fear, contaminated by the inexorable advance of the salt-filled ocean. The Florida Keys, the pearl-like strand of islands that stretches into the Gulf of Mexico, could be mostly submerged alongside their exotic crown jewel, Key West" (Madigan 2013). "I don't think people realize how vulnerable Florida is," said Harold R. Wanless, chairman of the geological sciences department at the University of Miami. "We're going to get four or five or six feet of water, or more, by the end of the century. You have to wake up to the reality of what's coming" (Madigan 2013).

Wanless predicts that because so much seawater is seeping through Miami's limestone base insurance companies will be forced to quit selling policies on luxury condos that line Miami Beach's Biscayne Bay. This will happen after banks realize that pumps will not solve the problem and refrain from writing mortgages (Kolbert 2015, 43). By 2015, the city of Hallandale Beach north of Miami already had closed most of its drinking-water wells because of saltwater inundation. In some parts of the Miami area, storm drains built to convey rainwater from streets to the ocean are backing up with saltwater that is flowing into neighborhoods. "The infrastructure we have is built for a world that doesn't exist anymore," said Nicole Hernandez Hammer, a researcher with the Union of Concerned Scientists (Kolbert 2015, 45).

"The issue," wrote *New York Times* reporter Nick Madigan (2013), "appears to be similarly opaque to segments of the community—business, real estate, tourism—that have a vested interest in protecting South Florida's bustling economy." "The business community for the most part is not engaged," said Wayne Pathman, a Miami land-use lawyer and Chamber of Commerce board member. "They're not affected yet." Ultimately, said Pathman, the most salient indicator of the crisis "will be the insurance industry's refusal to handle risk in coastal areas here and around the country that are deemed too exposed to rising seas." Most people who live in Florida "haven't grasped the possibilities," said Pathman (Madigan 2013).

Regarding those possibilities, rather tepid suggestions have been proposed that the government do its best to advise forward-thinking coastal residents that they would be better off moving inland. Anyone who makes such a suggestion has not thought much about why most Floridians live on or near the coast—views of the ocean and ready access to beaches. Inland locations mean heat and humidity in summer, swarms of mosquitoes, swamps, and various other forms of natural entertainment such as perpetually hungry alligators. One might as well advise Floridians to live without air conditioning in an increasingly hot and humid world. Without it, the big sand spit we call Florida would probably revert to its population density before Freon was invented for air conditioning in the 1920s. By that time, real estate agents may have noticed that something has gone seriously wrong with the climate.

Storm-Surge Perils

Sea level rise is not simply linear. Over time, the ocean level rises slowly as land also slowly sinks, but these rises are punctuated by storm surges. The real nightmare scenario at Daytona Beach would be a landfalling major hurricane such as a category 5 coming ashore off the Atlantic a few miles to the south with a 160 mile-an-hour east wind pushing a 15- to 20-foot storm surge across the barrier islands. This is not imaginary. Typhoon Haiyan did precisely that in November 2013 in the Philippines. Closer to home, Hurricane Andrew did it south of Miami in 1992. Both storms were so strong that the land-based analog would be a 100-mile-wide tornado pushing a wall of water. "As you warm the climate, you basically raise the speed limit on hurricanes," said Kerry A. Emanuel, an atmospheric scientist at the Massachusetts Institute of Technology (Fountain and Gillis 2013).

To counter pell-mell coastal development, attention has been directed to the Coastal Barrier Resources Act, which was signed in 1982 by President Ronald Reagan, who called it a "triumph for natural resource conservation and federal fiscal responsibility" (Gillis 2013). The act's intent was to protect a large swath of the U.S. coast. This law has discouraged construction on 1.3 million acres of U.S. coastlines. In our time, a proposal has been floated to expand this law to cover *all* coastlines and thus get Uncle Sam out of the business of covering flood risks that private insurers will not touch as climate-related risks rise. Such proposals aim to take coastal development off welfare and let the market readjust property values to fit *real* risk. This would be tough love for the real-estate industry, which thrives on development-related subsidies.

Sea Level Rise: Local and Regional Effects

"Measurements around the world show that sea level has risen almost 20 centimeters (7.87 inches) since 1880," explained Professor Stefan Rahmstorf of the Potsdam Institute for Climate Impact Research. The rate of sea level rise is closely linked to temperature—sea level rises faster the warmer the atmosphere gets, Rahmstorf explained. "If sea level keeps rising at a constant pace, we will end up in the middle of that 18–59 centimeter IPCC range by 2100," says Rahmstorf. "But based on past experience I expect that sea level rise will accelerate as the planet gets hotter" ("Rising Seas" 2009). Heat not only melts ice, but also because the molecular structure of liquid water expands as it is heated. The same amount of liquid water will occupy more volume as it warms. "Sea level rise is like an invisible tsunami, building force while we do almost nothing," said Benjamin H. Strauss, an author, with other scientists, of two new papers outlining the research. "We have a closing window of time to prevent the worst by preparing for higher seas" (Gillis 2012).

In April 2016, Sean Becketti, chief economist for Freddie Mac, the mortgage agency backed by the federal government, said that it is only a matter of time before sea level rise and storm surges become so common along the U.S. East Coast that people will vacate coastal properties in large numbers as insurance rates rise to unbearable levels and property values decline. By 2016, prices were already trending downward in Atlantic City, New Jersey; Norfolk, Virginia; and St. Petersburg, Florida, according to local real estate agents. "Some residents will cash out early and suffer minimal losses," he wrote. "Others will not be so lucky" (Urbina 2016).

Sandy Beaches Receding

Barrier islands and sand spits of land along the Atlantic and Gulf Coasts are among the areas most vulnerable areas to rises in sea level. Most coastal barrier islands are long, narrow strands of sand with ocean on one side and bay on the other. Typically, the oceanfront shore of any given island usually ranges from two to four meters above high tide; the bay side is typically less than a meter above high water. Thus, even a one-meter rise in sea level would threaten many of these places (with their ranks of hotels, businesses, and homes) with saltwater inundation. Erosion, moreover, threatens the high parts of these islands and is generally viewed as a more immediate problem than the inundation of barrier islands' bay sides.

Coastal erosion is an additional problem in the Houston–Galveston area of Texas. At Sargent Beach, the shoreline eroded about 1,000 feet between 1956 and the early 1990s. Galveston built a seawall to protect against oceanic flooding during the early 1900s after a historically disastrous hurricane and storm surge devastated the city in 1900. When the seawall was constructed, it faced a beach that averaged about 300 feet wide. Ensuing years have seen the beach erode; by about 1940, more rocks were required to protect the original seawall. Most of these rocks sank during the following few years, requiring even more rocks for reinforcement (North et al. 1995, 172).

At least 70 percent of sandy beaches around the world have been receding; in the United States, roughly 86 percent of East Coast barrier beaches (excluding evolving spit areas) have experienced erosion during the last century (Zhang et al. 2004, 41). "We're losing the battle," said Stanley Riggs, a geologist at East Carolina University in Greenville, North Carolina (Boyd 2001). North Carolina's Outer Banks have been eroding rapidly. "Highway 12 is falling into the ocean. What was once the third row of houses [on the beach] is now the first row," Riggs told a National Academy of Sciences conference on coastal disasters (Boyd 2001). A French and American space satellite named "Jason" was launched in 2001 to monitor the upward creep of the seas. Orbiting 830 miles above Earth, Jason's radar altimeter will be able to calculate the sea level within an accuracy of one inch, according to Ghasser Asrar, NASA's associate administrator for Earth science (Boyd 2001).

"It is virtually certain," the report said, that coastal headlands, spits, and barrier islands will erode faster than they have in the past (Dean 2009). A sea level rise of two feet in a century (as forecast by the Intergovernmental Panel on Climate Change) "is likely [to mean] some barrier islands in this region will cross a threshold" and begin to break up, the report said. The islands along the Outer Banks of North Carolina are most vulnerable (Dean 2009). Such a rise also will swamp many coastal wetlands.

During the 20th century, for example, mean sea levels rose 12.3 inches in New York City; 8.3 inches in Baltimore; 9.9 inches in Philadelphia; 7.3 inches in Key West, Florida.; 22.6 inches in Galveston, Texas; and 6 inches in San Francisco (Boyd 2001). The rate of sea level rise has been accelerating over time. In the port of Baltimore at the head of Chesapeake Bay, for example, the water level crept up at only about one-tenth of an inch per year for much of the 20th century. After 1989, however, the level rose by half an inch per year, according to Court Stevenson, a researcher at the University of Maryland's Center for Environmental Science (Boyd 2001). Sea levels have risen 12 to 20 inches on the Maine coast and as much as two feet along Nova Scotia in 250 years, according to an international team of researchers. Global warming is the main factor, said Roland Gehrels of England's University of Plymouth. He said the rate of sea level rise accelerated during the 20th century "as industrialization swept the globe" ("Global Warming Blamed" 2001).

Like the Mississippi River south of New Orleans, the Nile River north of Cairo in Egypt is also eroding because of subsidence and rising water levels. The Nile River delta, which is approximately the size of Delaware, is home to more than 50 million people, one of the most densely populated areas in the world and one of the most intensely cultivated. Until construction of the Aswan High Dam in 1960, 100 million tons of sediment coursed down the river each year building and refreshing the delta. By 1970, this flow had stopped, with the sediment becoming trapped in Lake Nasser behind the dam. Instead of replenishing delta farmlands with soil, farmers now use chemical fertilizers. The land also is subsiding at a rate that in 2010 placed 30 percent of it less than one meter above sea level. The level of the Mediterranean Sea also is expected to rise because of global warming. A one-meter rise in sea level with continued subsidence of the land could place half the delta below sea level in a century (Bohannon 2010).

Sea Levels Rise around the World

On Cape Sable on the far southwestern edge of Florida, boaters, sportsmen, and scientists have watched as a rising sea level has swallowed a freshwater marsh. A field of saw grass is now a saltwater mangrove swamp. The endangered Cape Sable seaside sparrow has fled northward. With old photographs and tidal gauge records, University of Miami professor Harold Wanless, chairman of the university's geology department, found that since the 1930s sinking land and rising seas have seen the relative water level rise nine inches, he said. "Freshwater marshes on Cape Sable are now evolving into more or less open marine waters," he said. "We're not talking about global warming as something that will happen in the future. Its happening right now. All the king's horses and all the king's men won't be able to put Cape Sable together again" (Harden and Eilperin 2006).

A Vietnamese government report in August 2009 said that a three-foot sea level rise could submerge a third of the Mekong River delta, an area that houses 17 million people and grows almost half the country's rice. Lesser sea level rise could drown a fifth of the delta, according to Tran Thuc of the Vietnam National Institute for Hydrometeorology and Environmental Sciences and lead author of the report. Increasingly intense typhoons and storm surges could complicate sea level rise still further, he said (Mydans 2009).

Alarm over rising sea levels and subsidence in Shanghai, China's largest city (population 16 million), has prompted officials to consider building a dam across its main river, the Huangpu. "Its main function is to prevent the downtown areas from being inundated with floods," Shen Guoping, an urban planning official, told the *China Daily* ("Shanghai Mulls" 2004). Rising water levels of the Huangpu, provoked by rising sea levels as well as sinking land, has resulted in construction of floodwalls hundreds of kilometers in length. At the same time, subsidence caused by removal of groundwater and rapid construction of skyscrapers has averaged more than 10 millimeters a year.

By 2020, a one-half to one-meter rise in sea levels could submerge three of India's biggest cities—Mumbai, Kolkata, and Chennai—according to Rajiv Nigam, a scientist with India's Geological Oceanography Division. Nigam said that a one-meter rise in sea level could cause 5 trillion rupees ($108 billion U.S.) worth of damage to property in India's Goa state alone. "If this is the quantum of damage in a small state like Goa that has only two districts, imagine the extent of property loss in metros like Bombay," Nigam added at a workshop in the National College in Dirudhy, Tamil Nadu ("Warming Could Submerge" 2003).

By the year 2000, rising sea levels were nibbling up to 150 meters a year from the low-lying, densely populated Nile River delta. At Rosetta, Egypt, a seawall two stories high has slowed the march of the sea, which is compounded by land subsidence in the delta, but "seawalls cannot stop the rising [salinity of] the palm groves and fields adjoining the shore" (Bunting 2000). A one-meter rise in sea level could drown most of the Nile delta, which was 12 percent of Egypt's arable land and home to 7 million people in 2004.

As sea levels rise, cities along the U.S. East and Gulf Coasts (and other coastal locales around the world) may find saline water seeping into their drinking-water supplies. Cities most at risk in the United States may be New York City and Philadelphia, where a sea level rise of a third of a meter could require a 12-percent rise in reservoir capacity to prevent saltwater intrusion into system intakes on the Delaware River (Cline 1992, 127). Other cities around the world that lie near seacoasts may face similar problems: Amsterdam, Rotterdam, Liverpool, Istanbul, Venice, Barcelona, Gothenburg, St. Petersburg, among others.

Along the U.S. Eastern Seaboard, oceanfront property, some of the most valuable real estate in the country, may be vulnerable to sea level rises. "In Massachusetts," wrote author and activist Bill McKibben, "Between [3,000 and 10,000] acres of ocean-front land worth between $3 billion and $10 billion might disappear by

2025, and that figure does not include land lost to growing ponds and bogs [created] as the rising sea lifts the water table" (McKibben 1989, 112–113). In Chesapeake Bay, many small islands, which have been important rookeries for several species of birds, are eroding. The Cape Hatteras lighthouse, the tallest such structure in the United States was 1,600 feet from shore when it was built in 1803, but120 feet from open water when it was moved in 1999 ("Cape Hatteras" 1999).

Texas may face special hazards from sea level rise along the Gulf Coast. A report outlining possible effects of the greenhouse effect on Texas sketched severe problems with land subsidence and sea level rise along parts of the state's coast: "The Houston–Galveston urban region, with its high groundwater table, subsidence, and long history of severe flooding along its coastline and bayous, will be particularly vulnerable, and could experience permanent loss of urban land in sensitive areas, necessitating extensive relocation programs" (North et al. 1995, 170).

In addition, a rise in sea level may imperil drinking water supplies in the coastal regions of Texas, the state's fastest-growing region in terms of human population and development. The Houston–Galveston area is additionally vulnerable because the land is relatively flat for several miles inland and has been experiencing subsidence (sinking) as groundwater has been removed for human consumption. The cities of Beaumont, Port Arthur, Orange, Freeport, and Corpus Christi subsided as much as a foot between 1906 and 1974 (North et al. 1995, 171), and some small areas sank as much as 10 feet during that period. Subsidence rates could increase as more groundwater is withdrawn.

The Netherlands Responds to Rising Sea Levels

The Dutch fear that rising storm surges could inundate much of the Netherlands, large areas of which have been reclaimed from the sea over the centuries. Fears have been expressed that the country's western provinces may flood. The Hague, for example, may become uninhabitable as low-lying suburbs of Amsterdam return to marshland or open water.

The Dutch already have been forced to anticipate surrendering 200,000 hectares of farmland to river floodplains. A major construction program of floating homes has started. Pieter van Geel, the Dutch environment secretary, said, "Half of our country is below sea level and so beyond a certain level, it is not possible to build dikes anymore. If we have a sea level rise of two meters, we have no control, no possibility of solving that. It's unthinkable. I fear the problem is going to catch us within 25 years, and that is a very short period," he said (Evans-Pritchard 2004).

In mid-2008, a Dutch commission recommended spending $144 billion to reinforce the country's sea defenses through 2100 as a precaution against sea level rise. The measures include widening dunes that face the North Sea and raising the height of dikes along the coastline and rivers.

The Dutch are also meeting the threat of sea level rise with amphibious houses. Anne van der Molen's two-bedroom, two-story house in Maasbommel cost about $420,000 and "rest[s] on land but built to rise with the water level. It sits on a hollow concrete foundation and is attached to six iron posts sunk into the lake bottom. Should the river swell, as it often does in the rain, the house will float up as much as 18 feet, held in place by two horizontal mooring posts that connect it to the neighboring house, and then float back down as the water subsides." "Dutch people have always had to fight against the water," she said. "This is another way of thinking about it. This is a way to enjoy the water, to work with it instead of against it" (Lyall 2007).

Van der Molen's home is one of 46 built in a single development that is designed to anticipate relentlessly rising sea levels and a heightened chance of flooding rains because of climate change, said Steven de Boer, a concept developer at Dura Vermeer, the company that developed the project. In 1995, local rivers flooded and forced 250,000 people to evacuate. Local dikes have since been raised to counter a similar flood, but because the Netherlands is essentially a large river delta, caution is the byword for the future.

"All the universities are united in one big program with the government; we have a team of some 500 people working on climate-proofing the Netherlands," said Pier Vellinga, a professor of climate change at the University of Amsterdam. "Whatever happens—Greenland melting or tropical storms surging on the Atlantic—we are here to stay. That is becoming our national slogan" (Lyall 2007).

Further Reading

Atlas, James. "Is This the End?" *The New York Times*, November 24, 2012. http://www.nytimes.com/2012/11/25/opinion/sunday/is-this-the-end.html.

Bohannon, John. "The Nile Delta's Sinking Future." *Science* 327 (March 19, 2010): 1444–1447.

Boyd, Robert S. "Rising Tides Raises Questions; Satellites Will Provide Exact Measurements." *Pittsburgh Post-Gazette*, December 9, 2001, A3.

Bunting, Madeleine. "Confronting the Perils of Global Warming in a Vanishing Landscape: As Vital Talks Begin at the Hague, Millions Are Already Suffering the Consequences of Climate Change." *The Guardian* (U.K.), November 14, 2000, 1.

"Cape Hatteras, N.C. Lighthouse Lights Up Sky from New Perch." Omaha *World-Herald*, November 14, 1999, p. A16.

Cline, William R. *The Economics of Global Warming*. Washington, DC: Institute for International Economics, 1992.

Davenport, Coral. "Miami Finds Itself Ankle-Deep in Climate Change Debate," *The New York Times*, May 8, 2014. http://www.nytimes.com/2014/05/08/us/florida-finds-itself-in-the-eye-of-the-storm-on-climate-change.html.

Dean, Cornelia. "Study Warns of Threat to Coasts from Rising Sea Levels." *The New York Times*, January 17, 2009. http://www.nytimes.com/2009/01/17/science/earth/17sea.html.

Evans-Pritchard, Ambrose. "Dutch Have Only Years before Rising Seas Reclaim Land: Dikes No Match against Global Warming Effects." *Daily Telegraph* (London), October 30, 2004, 1.

"Findings." *Harper's*, February, 2017, 96.

Fountain, Henry, and Justin Gillis. "Typhoon in Philippines Casts Long Shadow over U.N. Talks on Climate Treaty." *The New York Times*, November 11, 2013. http://www.nytimes.com/2013/11/12/world/asia/typhoon-in-philippines-casts-long-shadow-over-un-talks-on-climate-treaty.html.

Gillis, Justin. "Rising Sea Levels Seen as Threat to Coastal U.S." *The New York Times*, March 13, 2012. http://www.nytimes.com/2012/03/14/science/earth/study-rising-sea-levels-a-risk -to-coastal-states.html.

Gillis, Justin. "Rebuilding the Shores, Increasing the Risks." *The New York Times,* April 8, 2013. http://www.nytimes.com/2013/04/09/science/earth/rebuilding-our-shores-incr easing-the-risks.html.

Gillis, Justin. "The Flood Next Time." *The New York Times*, January 13, 2014. http://www .nytimes.com/2014/01/14/science/earth/grappling-with-sea-level-rise-sooner-not-later .html.

"Global Warming Blamed for Rising Sea Levels." *Omaha World-Herald*, November 25, 2001, 20A.

Harden, Blaine, and Juliet Eilperin. "On the Move to Outrun Climate Change; Self-Preservation Forcing Wild Species, Businesses, Planning Officials to Act." *Washington Post*, November 25, 2006, A3.

Kaufman, Leslie. "Front-Line City in Virginia Tackles Rise in Sea." *The New York Times*, November 25, 2010. http://www.nytimes.com/2010/11/26/science/earth/26norfolk.html.

Koch, Wendy. "Rising Sea Levels Torment Norfolk, Va., and Coastal U.S." *USA Today*, December 18, 2013, 1A, 2A. http://www.usatoday.com/story/news/nation/2013/12/17/sea-level -rise-swamps-norfolk-us-coasts/3893825/.

Koch, Wendy. "Miami is One of USA's Top Hot Spots for Climate Change." *USA Today*, May 7, 2014. http://www.usatoday.com/story/news/nation/2014/05/07/miami-south-florida -climate hot-spots/8803849/.

Kolbert, Elizabeth. "The Siege of Miami: As Temperatures Rise, So Will Sea Levels." *The New Yorker,* December 21 and 28, 2015, 42–50.

Lyall, Sarah. "At Risk from Floods, but Looking Ahead with Floating Houses." *The New York Times*, April 3, 2007. http://query.nytimes.com/gst/fullpage.html?res=9B0CE6DB1 F30F930A35757C0A9619C8B63.

Madigan, Nick. "South Florida Faces Ominous Prospects from Rising Waters." *The New York Times*, November 11, 2013. http://www.nytimes.com/2013/11/11/us/south-florida-faces -ominous-prospects-from-rising-waters.html.

McKibben, Bill. *The End of Nature.* New York: Random House, 1989.

Miller, Kenneth G., et al. "A Geological Perspective on Sea-level Rise and Its Impacts along the U.S. Mid-Atlantic Coast." *Earth's Future* 1(1) (December 5, 2013). http://onlinelibrary .wiley.com/doi/10.1002/2013EF000135/abstract.

Mydans, Seth. "Vietnam Finds Itself Vulnerable If Sea Rises." *The New York Times*, September 24, 2009. http://www.nytimes.com/2009/09/24/world/asia/24delta.html.

Navarro, Mireya. "New York Is Lagging as Seas and Risks Rise, Critics Warn." *The New York Times*, September 10, 2012. http://www.nytimes.com/2012/09/11/nyregion/new-york -faces-rising-seas-and-slow-city-action.htm (no longer available).

North, Gerald R., Jurgen Schmandt, and Judith Clarkson. *The Impact of Global Warming on Texas.* Austin: University of Texas Press, 1995.

Parker, Laura. "Treading Water." *National Geographic*, February 2015, 106–128.

Rahmstorf, Stefan, Grant Foster, and Anny Cazenave. "Comparing Climate Projections to Observations up to 2011." *Environmental Research Letters* 7(4) (2012). http://iopscience .iop.org/1748-9326/7/4/044035/article.

"Rising Seas Could Swamp One in 10 People by 2100." Environment News Service, March 10, 2009. http://www.ens-newswire.com/ens/mar2009/2009-03-10-03.asp.

Sallenger, Asbury, Jr., Kara S. Doran, and Peter A. Howd. "Hotspot of Accelerated Sea-Level Rise on the Atlantic Coast of North America." *Nature Climate Change* 2(2012): 884–888.

"Shanghai Mulls Building Dam to Ward off Rising Sea Levels." Agence France Presse, February 9, 2004. http://www.spacedaily.com/2004/040209065658.wbolr45c.html.

"10 Coastal Cities at Greatest Flood Risk as Sea Levels Rise." Environment News Service, September 3, 2013. http://ens-newswire.com/2013/09/03/10-coastal-cities-at-greatest-flood-risk-as-sea-levels-rise/.

Urbina, Ian. "Perils of Climate Change Could Swamp Coastal Real Estate." *The New York Times*, November 24, 2016. http://www.nytimes.com/2016/11/24/science/global-warming-coastal-real-estate.html.

Vergano, Dan. "Sandy Revives Debate over Rising Sea Level." *USA Today*, November 28, 2012, 3A.

"Warming Could Submerge Three of India's Largest Cities: Scientist." Agence France Presse, December 6, 2003. http://www.spacedaily.com/2003/031206151721.ha5156ha.html.

Zhang, Keqi, Bruce C. Douglas, and Stephen P. Leatherman. "Global Warming and Coastal Erosion." *Climatic Change* 64(1 and 2) (May 2004): 41–58.

See also: Adaptation, Animals, Lizards, and Warming Habitats; Sea Level Rise; Sea Level Rise, Bangladesh; Sea Level Rise, Island Nations; Sea Level Rise, Italy; Temperatures, Global

THERMOHALINE CIRCULATION

Scientists studying past climate epochs have found times when the oceans' circulation lost its vital character and perhaps even shut down. Analyses of ice cores, deep-sea sediment cores, and other geologic evidence have clearly demonstrated that the global conveyor has abruptly slowed or halted many times in Earth's past. That has caused the North Atlantic region to cool significantly and brought long-term drought conditions to other areas of the Northern Hemisphere—and over time spans as short as years to decades ("Study Reports" 2003). According to climate scientist Thomas F. Stocker,

> Evidence from paleo-climatic archives suggests that the ocean atmosphere system has undergone dramatic and abrupt changes with widespread consequences in the past. Climatic changes are most pronounced in the North Atlantic region where annual mean temperatures can change by 10 [°C] and more within a few decades. Climate models are capable of simulating some features of abrupt climate change. These same models also indicate that changes of this type may be triggered by global warming. (Stocker et al. 2001, 277)

Meltwater Releases

At the end of the last ice age (between 8,200 and 12,800 years ago), ice-core records from Greenland indicate abrupt temperature declines at about the same time that massive amounts of cold freshwater were released from a huge body of glacial meltwater that glaciologists today call Lake Agassiz. The meltwater reached from today's Canadian prairies eastward to Quebec and southward to Minnesota. Reaching the Atlantic Ocean through the St. Lawrence Valley and Hudson Bay, "such a large amount of low-density freshwater would have reduced the density of the

North Atlantic surface water considerably, preventing it from sinking and thus slowing down (or perhaps even shutting off completely) the Gulf Stream," wrote glaciologist Doug Macdougall in *Frozen Earth: The Once and Future Story of Ice Ages* (2004, 110).

Studies by William Patterson of the University of Saskatchewan and colleagues released in 2009 indicated that changes in North Atlantic circulation can force the Northern Hemisphere into a mini "ice age" within a few months, contradicting earlier estimates of a decade or more. They use the Younger Dryas cold period of some 12,800 years ago as a reference. Then the Northern Hemisphere experienced a spike of colder weather that lasted 1,300 years after "a sudden influx of freshwater, when the glacial Lake Agassiz in North America burst its banks and poured into the North Atlantic and Arctic Oceans. This vast pulse, a greater volume than all of North America's Great Lakes combined, diluted the North Atlantic conveyor belt and brought it to a halt" ("Big Freeze" 2009). Another such disruption occurred about 19,000 years ago (Clark et al. 2004, 1141).

Jochen Erbacher studied an anoxic event "in the restricted basins of the western Tethys and North Atlantic" during the mid-Cretaceous Period 112 million years ago that appears to have been caused by "increased thermohaline stratification" (Erbacher et al. 2001, 325). The Western Tethys was a sea formed as continents were separating on the site of the present-day North Atlantic Ocean. "Ocean anoxic events were periods of high carbon burial that led to drawdown of atmospheric carbon dioxide, lowering of bottom-water oxygen concentrations and, in many cases, significant biological extinctions," commented Erbacher and colleagues (2001, 325). "We suggest that that the partial tectonic isolation of the various basins in the Tethys, and Atlantic, a low sea level, and the initiation of warm global climates may be important factors in setting up oceanic stagnation" during this event (Erbacher et al. 2001, 327).

An outstanding climate anomaly 8200 years before the present (BP) in the North Atlantic is commonly postulated to be the result of weakened overturning circulation triggered by a freshwater outburst. New stable isotopic and sedimentological records from a northwest Atlantic sediment core reveal that the most prominent Holocene anomaly in bottom-water chemistry and flow speed in the deep limb of the Atlantic overturning circulation begins at ~8.38 thousand years BP, coeval with the catastrophic drainage of Lake Agassiz. The influence of Lower North Atlantic Deep Water was strongly reduced at our site for ~100 years after the outburst, confirming the ocean's sensitivity to freshwater forcing. The similarities between the timing and duration of the pronounced deep circulation changes and regional climate anomalies support a causal link. (Kleiven et al. 2008, 60).

Ocean "Weather" and "Climate"

Ocean circulation is not unlike surface weather. It varies widely over both short and long periods. Ocean circulation experiences a kind of "weather," as well as a longer-term "climate." Scientists have studied past epochs and found that the deep

Atlantic Ocean during the height of the last ice age appears to have been quite different from today. One group reviewed observations that implied

> that Atlantic meridional overturning circulation during the Last Glacial Maximum was neither extremely sluggish nor an enhanced version of present-day circulation. The distribution of the decay products of uranium in sediments is consistent with a residence time for deep waters in the Atlantic only slightly greater than today. However, evidence from multiple water-mass tracers supports a different distribution of deepwater properties, including density, which is dynamically linked to circulation. (Lynch-Stieglitz et al. 2007, 66)

Thermohaline Circulation and East Coast Sea Level Rise

In combination with shifts in the Gulf Stream, the thermohaline circulation also influences sea levels along the U.S. East Coast. Relative sea levels are rising faster here than in most areas. As world sea levels slowly rise because of melting ice worldwide, much of the U.S. East Coast is also subsiding. "Within a two-year period, the sea level was found to rise by 128 mm [millimeters], a 1 in 850 year event," Srokocsz and Bryden wrote. They continued: "The event caused persistent and widespread coastal flooding and beach erosion almost on a level with that due to a hurricane" (2015, 1300; Mooney 2016).

During the early 1970s, the overturning circulation began to weaken as sea ice melting in the Arctic flowed southward. It strengthened somewhat during the 1990s but then weakened again after 2000, with melting from Greenland a major influence. By 2014, the Atlantic meridional overturning circulation (AMOC) had weakened 15 percent to 20 percent from historical averages.

Chris Mooney of *The Washington Post* explained the effect on East Coast sea level rise:

> Sea level is "anomalously low" off the U.S. East Coast due to the motion of the Gulf Stream. This is for at least two reasons. First, explains Rahmstorf's co-author Michael Mann of Penn State University, there's the matter of temperature contrast: Waters to the right or east of the Gulf Stream, in the direction of Europe, are warmer than those on its left or west. Warm water expands and takes up more area than denser cold water, so sea level is also higher to the right side of the current, and lower off our coast. "So if you weaken the 'Gulf Stream' and weaken that temperature contrast . . . sea level off the U.S. east coast will actually rise!" explained Mann.

Another factor involves the "geostrophic balance of forces" in the ocean. This gets complicated, but the bottom line result is that "sea surface slope perpendicular to any current flow, like the Gulf Stream, has a higher sea level on its right hand side, and the lower sea level on the left hand slide," said Rahmstorf. (This would only be true in the Northern Hemisphere; in the Southern Hemisphere it would be the opposite.) Indeed, researchers recently found a sudden, four-inch sea level rise of the U.S. East Coast in 2009 and 2010 that they attributed to a slowdown of the AMOC. Rahmstorf said that "for a big breakdown of the circulation, [sea level rise] could amount to one meter, in addition to the global sea level rise that we're expecting from global warming." (Mooney 2015)

Further Reading

Mooney, Chris. "Global Warming Is Now Slowing Down the Circulation of the Oceans—with Potentially Dire Consequences." *Washington Post*, March 24, 2015. http://www.washingtonpost.com/news/energy-environment/wp/2015/03/23 /global-warming-is-now-slowing-down-the-circulation-of-the-oceans-with-poten tially-dire-consequences/.
Srokosz, M. A., and H. L. Bryden. "Observing the Atlantic Meridional Overturning Circulation Yields a Decade of Inevitable Surprises." *Science* 348 (June 19, 2015): 1330. http://science.sciencemag.org/content/348/6241/1255575. doi: 10.1126/ science.1255575.

M. Latiff and colleagues asked in the *Journal of Climate* (2006, 4631–4637), "Is the Thermohaline Circulation Changing?" Their answer is that it has changed considerably during the last century, mainly as a result of natural multidecadal climate variability; it may also change in coming decades because of melting Arctic ice freshening the Nordic Sea. However, these researchers do believe that "such a weakening will not exceed the range of multidecadal variability [which they take to be 40 percent] for several decades" (Latif 2006, 4635).

The most prominent cold period during the Holocene 8,200 years ago spanned several hundred years and was caused by an influx of meltwater into the North Atlantic. Evidence from a North Atlantic deep-sea sediment core reveals that the largest climatic perturbation in our current interglacial is marked by two distinct cooling events in the subpolar North Atlantic at 8,490 and 8,290 years ago. An associated reduction in deep flow speed provides evidence of a significant change to a major down-welling limb of the Atlantic meridional overturning circulation. The existence of a distinct surface freshening signal during these events strongly suggests that the sequenced surface and deep ocean changes were forced by pulsed meltwater outbursts from a multistep final drainage of the proglacial lakes associated with the decaying Laurentide ice sheet margin.

A study of North Atlantic ocean circulation published in late 2005 (Quadfasel 2005) reported a 30-percent reduction in the meridional overturning (thermohaline) circulation at 26.5° north latitude based on readings taken in 1957, 1981, 1992, and 1998. Harry Bryden at the Southampton Oceanography Center in the United Kingdom strung buoys in a line across the Atlantic from the Canary Islands to the Bahamas "and found that the flow of water north from the Gulf Stream into the North Atlantic had faltered by 30 percent since the 1990s." Less ocean water was going north on the surface and less was coming south along the bottom (Pearce 2007, 147). Bryden said at the time that he was not sure whether the change he described is temporary or part of a long-term trend. This analysis is, however, "the first observational evidence that such a decrease of the oceanic overturning circulation is well underway" (Quadfasel 2005, 565).

A Cyclical Variation?

By 2007, scientific consensus was leaning away from Bryden's analysis as changes that had been taken to be long-term erosion were now being regarded as part of cyclical variation (Schiermeier 2007). Stuart Cunningham of the National Oceanography Center at Southampton in the United Kingdom and colleagues found that the thermohaline circulation varied by 25 percent in one year (Cunningham et al. 2007).

The idea that changes in thermohaline circulation might cool Europe substantially as most of the world warms has been losing traction among scientists, even as popular entertainment promotes it. Movies such as *The Day After Tomorrow* (2004) presented the idea in cinematic fashion. Reality intruded, however, as temperatures rose in Europe and climate models forecast more of the same. Richard Kerr wrote in *Science* that the ocean circulation system is "prone to natural slowdowns and speedups. Furthermore, researchers are finding that even if global warming were slowing the conveyor and reducing the supply of warmth to high latitudes, it would be decades before the change would be noticeable above the noise" (Kerr 2006, November 17). These findings came from the Rapid Climate Change (RAPID) program after researchers moored 19 lightweight cables with instruments along 26.5° north latitude from western Africa to the Bahamas to measure ocean flows.

"The concern had previously been that we were close to a threshold where the Atlantic circulation system would stop," said Susan Solomon, a senior scientist at the National Oceanic and Atmospheric Administration. "We now believe we are much farther from that threshold, thanks to improved modeling and ocean measurements. The Gulf Stream and the North Atlantic Current are more stable than previously thought" (Gibbs 2007).

In summarizing 23 climate models in 2007, the U.N. Intergovernmental Panel on Climate Change said that substantial cooling of Europe and that a marked disruption of thermohaline circulation is unlikely during the 21st century. The IPCC assessment did say that the circulation probably would weaken by some 25 percent through 2100 given increased melting of the Greenland ice cap and more rainfall in the Arctic that will cause more freshwater to flow southward. However, any resulting cooling of Europe will probably be overwhelmed by a general worldwide warming trend. "The bottom line is that the atmosphere is warming up so much that a slowdown of the North Atlantic Current will never be able to cool Europe," said Helge Drange, a professor at the Nansen Environmental and Remote Sensing Center in Bergen, Norway (Gibbs 2007).

Thermohaline circulation is only one contributor to Europe's mild climate. Prevailing winds also play a part. In addition, the Greenland ice cap would have to melt rapidly to stop the ocean conveyor, according to current theories. "The ocean circulation is a robust feature, and you really need to hit it hard to make it stop," said Eystein Jansen, a paleoclimatologist who directs the Bjerknes Center for Climate Research, also in Bergen. "The Greenland ice sheet would not only have to melt, but to dynamically disintegrate on a huge scale across the entire sheet" (Gibbs 2007). Any collapse would probably require centuries.

Present-Day Evidence of Breakdown

Inhibition of global ocean circulation by a warming atmosphere is not a mere theory or a paleoclimatic curiosity. Evidence is accumulating that ocean circulation already is breaking down, although researchers are unsure whether this is evidence of a natural cycle, a provocation of warming seawaters, or both (Häkkinen and Rhines 2004, 559). Analysis has been restricted by scanty data prior to 1978. "These observations of rapid climatic changes over one decade [the 1990s] may merit some concern," according to informed observers (Häkkinen and Rhines 2004, 559).

By 2014, scientists had been monitoring the Atlantic meridional overturning circulation (AMOC) in a sustained, organized manner for a decade, and patterns were beginning to emerge. As predicted, the circulation's transport of heat between latitudes was declining as temperatures warmed, but it was doing so with a surprising degree of year-to-year variability, sometimes exceeding 30 percent over a year or two. In 2009–2010, for example, transport of heat from the tropics to northerly latitudes declined 30 percent over a year and then bounced back.

Overall, the AMOC's strength has declined in recent years. In reviewing the literature, however, M. A. Srokosz and H. L. Bryden (2015), cautioned that a 10-year record is insufficient to draw long-term conclusions. However, they wrote, "This result is robust with respect to the inclusion or exclusion of the 2009–2010 AMOC event. . . ." During the 2009–2010 decline, heat transport to higher latitudes faltered because the heat was retained in the tropics. This may have been associated with an active Atlantic hurricane season.

As Srokosz and Bryden also observed, "The 30 percent decline in the AMOC during 2009–2010 was totally unexpected and exceeded the range of interannual variability found in climate models used for the IPCC assessments. . . . Paleoclimatic evidence suggests that AMOC can undergo rapid changes that are difficult to reproduce with climate models" (Srokocsz and Bryden 2015).

"Given the surprises and insights into the Atlantic circulation that observations have produced to date," they concluded in their 2015 review of more than 100 studies related to changes in the AMOC, "it is not too much to expect that with the new observations there will be future inevitable surprises" (Srokocsz and Bryden 2015).

Evidence suggests that the North Atlantic has cooled while the rest of the world has been warming—a probable signature of thermohaline disruption. During the last half of the 20th century, research reports indicate a "dramatic" increase in freshwater released into the North Atlantic by melting ice. This "freshening" is well under way (Speth 2004, 61). According to scientists at the Woods Hole Oceanographic Institution, this is "the largest and most dramatic oceanic change ever measured in the era of modern instruments" ("Abrupt Climate," 2003). By 2002, the amount of freshwater entering the Arctic Ocean was 7 percent more than during the 1930s (Speth 2004, 61).

Overturning circulation in the subpolar North Atlantic Ocean is not a story with a single plot line, however. It can be erratic. Its flow was shallow or nonexistent after 2000, a possible consequence of a warmer climate. During the winter of 2007–2008, however, deep convection in the subpolar gyre in both the Labrador and Irminger Seas returned. Scientists analyzing a variety of in situ, satellite, and reanalysis data in *Nature Geoscience* showed that "contrary to expectations the transition to a convective state took place abruptly, without going through a phase of preconditioning" involving "changes in hemispheric air temperature, storm tracks, the flux of freshwater to the Labrador Sea and the distribution of pack ice all contributed to an enhanced flux of heat from the sea to the air, making the surface water sufficiently cold and dense to initiate deep convection. Given this complexity, we conclude that it will be difficult to predict when deep mixing may occur again" (Våge et al. 2009).

Saltier Water

According to oceanographer Ruth Curry, sea surface waters in tropical regions have become dramatically saltier over the past 50 years while surface waters at higher latitudes, especially in Arctic regions, have become much fresher. These changes in salinity accelerated during the 1990s as global temperatures warmed. "This is the signature of increasing evaporation and precipitation" because of warming, Curry said, "and a sign of melting ice at the poles. These are consequences of global warming, either natural, human-caused or, more likely, both" (Cooke 2003).

Writing in *Science*, Richard A. Kerr said, "To Curry and her colleagues, it's looking as if something has accelerated the world's cycle of evaporation and precipitation by 5 percent to 10 percent, and that something may well be global warming" (Kerr 2004). These results indicate that freshwater has been lost from the low latitudes and added at high latitudes at a pace exceeding the ocean circulation's ability to compensate, the authors said ("Study Reports" 2003).

Curry led a team that examined salinity "on a long transect [(50° south to 60° north), or between Iceland in the north and the tip of South America] through the western basins of the Atlantic Ocean between the 1950s and the 1990s" (Curry et al. 2003, 826). They found "systematic freshening at both poleward ends contrasted with large increases of salinity pervading the upper water columns at lower latitudes." The authors asserted that their data extends "a growing body of evidence indicating that shifts in the oceanic distribution of fresh and saline waters are occurring worldwide in ways that suggest links to global warming and possible changes in the hydrologic cycle of the Earth" (Curry et al. 2003, 826).

Evaporation and Precipitation

Curry's study suggests that relatively rapid oceanic changes and recent climate changes, including warming global temperatures, may be altering the fundamental planetary system that regulates evaporation and precipitation and cycles freshwater around the globe. An acceleration of Earth's global hydrological cycle could affect global precipitation patterns that govern the distribution, severity, and frequency of droughts, floods, and storms. The same pattern could exacerbate global warming by rapidly adding more water vapor—itself a potent, heat-trapping greenhouse gas—to the atmosphere. It could also continue to freshen northern North Atlantic Ocean waters to a point that could disrupt ocean circulation and trigger further climate changes ("Study Reports" 2003).

Japanese and Canadian scientists reported in *Nature* (Fukasawa et al. 2004, 825) that the deepest waters of the North Pacific Ocean have warmed significantly across the entire width of the ocean basin. Masao Fukasawa of the Japan Marine Science and Technology Center in Yokosuka, Japan, and five other researchers measured temperature changes by going to sea on three research vessels and measuring deep-water temperatures across the North Pacific. Then they compared those temperature measurements to measurements made by researchers in 1985.

The Fukasawa team's findings indicated a warming in the deep North Pacific over a "shorter time scale and larger spatial scale than [has] ever been believed," Fukasawa said. "As [far] as I know, our result is the first which shows such a large-scale temperature change in the global thermohaline circulation," that is, in global heat and salt circulation via ocean currents (Davidson 2004). It is much too soon to blame global warming for the deep-sea warming, Reid and other experts cautioned. "To go from this (observation) to say, 'This is global warming,' is just a guess," Reid said. In any case, "this (Fukasawa result) is something that should be watched very carefully" (Davidson 2004).

M. J. McPhaden and D. Zhang, writing in *Nature*, also made a case that thermohaline circulation is slowing:

> Some theories ascribe a central role [in global climate change] to the wind-driven meridional overturning circulation between the tropical and subtropical oceans. Here we show, from observations over the last 50 years, that this overturning circulation has been slowing down since the 1970s, causing a decrease in upwelling of about 25 percent in an equatorial strip between 9 degrees north and 9 degrees south. This reduction in equatorial upwelling of relatively cool water . . . is associated with a rise in equatorial sea surface temperatures of about 0.8 degrees C. (McPhaden and Zhang 2002, 603)

Speed of Changes in the Past

Paleoclimatic records indicate that once climate change is "tripped," change may be quite rapid. For example, various proxy records from such sources as ice cores indicate temperature changes of up to 16°C in Greenland in a decade (Stocker et al.

2001, 289). Could such a rapid change, triggered by global warming, shut down the oceans' circulation system within a matter of decades?

Terrence Joyce, chairman of the physical oceanography department at Woods Hole Oceanographic Institution in Massachusetts, has been trying to raise awareness about this possibility. He is "not predicting an imminent climate change—only that once it started (and it is getting more likely) it could occur within 10 years," he says (Cowen 2002). Woods Hole director Robert Gagosian feels an urgency to settle the question. He sees enough disturbing information in the North Atlantic data that oceanographers from Woods Hole and other institutions have gathered to elicit "strong evidence that we may be approaching a dangerous threshold" (Cowen 2002).

Stocker and colleagues have written, "Although the magnitude of the change is highly uncertain, the models agree that the thermohaline circulation in the Atlantic will reduce due to the gain of buoyancy associated with the warming and a stronger hydrological cycle" (Stocker et al. 2001, 290). R. B. Thorpe et al. consider several model variations that indicate various degrees of global warming may decrease thermohaline circulation in the Atlantic between 20 percent and 60 percent because of temperature change (greater warming at high latitudes) and 40 percent to salinity change (Thorpe et al. 2001, 3102).

Writing in *Nature*, B. Dickson and colleagues noted,

From observations it has not been possible to detect whether the ocean's over-turning circulation is changing, but recent evidence suggests that the transport over the sills [in the ocean] may be slackening. Here we show, through the analysis of hydrographic records, that the system of overflow and entrainment that ventilates the deep Atlantic has steadily changed over the past four decades. We find that these changes have already led to sustained freshening of the deep ocean. (Dickson et al. 2002, 832)

Dickson and colleagues' ideas have received support from other studies. An international team of hydrologists and oceanographers reported in *Science* that the flow of freshwater from Arctic rivers into the Arctic Ocean has increased significantly over recent decades. If the trend continues, some scientists believe it could impact the global climate, perhaps leading to cooling in Northern Europe ("Study Reveals" 2002).

Discharge Data from the Six Largest Eurasian Rivers

Bruce J. Peterson of the Marine Biological Laboratory's Ecosystems Center led a research team of scientists from the United States, Russia, and Germany whose members analyzed discharge data from the six largest Eurasian rivers that drain into the Arctic Ocean. These rivers, all with headwaters in Russia, account for more than 40 percent of total freshwater inflows from rivers into the Arctic Ocean. Peterson and colleagues found that combined annual discharge from the Russian rivers increased by 7 percent from 1936 to 1999. They contend that this measured increase in runoff is an observed confirmation of what climatologists have been saying for years—that freshwater flow to the Arctic Ocean and North Atlantic will increase with global warming ("Study Reveals" 2002).

"If the observed positive relationship between global temperature and river discharge continues into the future, Arctic river discharge may increase to levels that impact Atlantic Ocean circulation and climate within the 21st century," said Peterson. Annual discharge of the rivers into the Arctic increased about 128 cubic kilometers per year during the 63-year study period. The study's authors warned that increasing river discharge, coupled with ice melt from Greenland, could have important effects on thermohaline circulation during the 21st century ("Study Reveals" 2002; Peterson 2002, 2171–2172).

During mid-2001, other scientists also disclosed evidence that the thermohaline circulation is indeed breaking down. Researchers in Denmark's Faroe Islands, halfway between Iceland and the northern tip of Britain, found evidence of a 20-percent drop since 1950 in the volume of deep cold water flowing south from the Arctic region through one of several channels into the North Atlantic. Most of the decrease has occurred during the past 30 years, and the rate of decline has accelerated in the past five years. This study, reported in *Nature*, combined water-flow readings from the ocean floor after 1995 with 50 years of temperature and salinity records from weather ships in the area.

Rising Volume of Freshwater in the Norwegian Sea

More evidence of thermohaline breakdown has come from a rising volume of freshwater in the Norwegian Sea from melting sea ice (Connor 2001). "The seas are warmer and there is more freshwater not just from melting sea ice but from the Siberian rivers. The water has to be salty and dense or it just won't sink," researcher Bogi Hansen said (Connor 2001). "Estimating the volume flux conservatively, we find a decrease by at least 20 percent relative to 1950. If this reduction in deep flow from the Nordic seas is not compensated by increased flow from other sources," wrote Hansen et al. (2001, 927). "It implies a weakened global thermohaline circulation and reduced inflow of Atlantic water to the Nordic seas."

"This is yet another very important brick in a very solid wall of evidence about the reality of global climate change," said Andrew Weaver, who holds the Canada Research Chair in Atmospheric Studies at the University of Victoria (Calamai 2001).

The findings most likely indicate an equal drop in the amounts in warm surface water flowing north through the same passage, said Hansen. "The motor driving the ocean conveyor belt appears to be slowing down," Hansen said (Calamai 2001; Hansen et al. 2001, 927). According to a report by Steve Connor in *The Independent* (London), "The research centered on measurements taken of water movements in a deep-sea channel that separates the Faroes from northern Scotland. The current in the Faroe Bank channel pushes about 2 million cubic meters of water every second into the Atlantic, an amount equivalent to about twice the total flow of all the rivers of the world combined" (Conner 2001). "If this reduction we have seen in the Faroe Bank channel is also seen in the Denmark Strait then we can be sure that the North Atlantic flow has been reduced," Hansen said (Conner 2001).

Thermohaline Circulation: Debating Points

The idea that failure of the thermohaline circulation could cause colder temperatures in landmasses around the North Atlantic Ocean is highly debatable and politically charged. The debate was given extra force when one of the idea's major proponents backed away from it. Wallace S. Broecker of Columbia University, heretofore an advocate of a belief that the "business as usual" fossil fuel track [would] run the risk, late in the 21st century, of triggering an abrupt reorganization of the Earth's thermohaline circulation" (Broecker 2001, 83) has maintained that doubling atmospheric levels of CO_2 could "cripple the ocean's conveyor circulation" (Broecker 2001, 83). In 1997, Broecker suggested that if the Gulf Stream was blocked. winter temperatures in the British Isles could fall by an average of 11°C, plunging Liverpool or Berwick to the same temperatures as Spitsbergen inside the Arctic Circle. Any dramatic drop in temperature could have devastating implications for agriculture and Europe's ability to feed itself (Radford 2001).

In a book published by the American Association of Petroleum Geologists, Broecker basically repudiated his earlier position:

> The recent discovery by Gerard Bond that the 1,500-year cycle that paced these glacial disruptions continued in a muted form during times of interglaciation casts a new light on this situation. . . . It leads me to suspect that the large and rapid atmospheric changes of glacial time were driven by a sea ice amplifier. If so, then, because little sea ice will remain at the time of a greenhouse-induced thermohaline reorganization, perhaps the threat will be far smaller than I had previously envisioned. (Broecker 2001)

Broecker said, "I apologize for my previous sins" of overemphasizing the Gulf Stream's role (Kerr 2002).

As is so often the case regarding global warming's putative effects on Earth's ecosystem, the premises under examination here are open to dispute on several fronts. Is the Gulf Stream *really* the main climate driver warming Europe's winters? In October 2002, a team of scientists writing in the *Quarterly Journal of the Royal Meteorological Society* asserted that this popular assumption is incorrect. Rather, they maintained, Europe is warmed "by atmospheric circulation tweaked by the Rocky Mountains [and] . . . summer's warmth lingering in the North Atlantic" (Kerr 2002).

Richard Seager of Columbia University's Lamont–Doherty Earth Observatory in Palisades, New York, and David Battisti of the University of Washington headed this study, which sought to determine the relative influences of various influences on European climate. They noted that winds carry five times as much heat out of the tropics to the mid-latitudes than oceanic currents. They also estimated that roughly "80 percent of the heat that cross-Atlantic winds picked up was summer heat briefly stored in the ocean rather than heat carried by the Gulf Stream" (Kerr

2002). Seager and colleagues relegate the Gulf Stream to the role of a minor player in Europe's wintertime climate. They asserted, however, that the Gulf Stream *does* play a major role in warming Scandinavia and keeping the far northern Atlantic free of ice (Seager 2002, 2563).

Thermohaline Circulation and Plankton Depletion

The decline of plankton also may be related to the changing nature of the thermohaline circulation (the Atlantic meridional overturning circulation, or AMOC). Writing in *Nature*, Andreas Schmittner reported on results of models simulating disruption of the AMOC that led to a collapse of the North Atlantic plankton stocks to less than half their initial biomass "owing to rapid shoaling of winter mixed layers and their associated separation from the deep ocean nutrient reservoir." Schmittner wrote, "These model results are consistent with the available high-resolution paleorecord, and suggest that global ocean productivity is sensitive to changes in the Atlantic meridional overturning circulation" (Schmittner 2005, 628).

Adding more support to the idea that phytoplankton populations are declining around the world because of changing ocean circulation, British scientists examining the Atlantic Ocean south of Iceland found that populations of zooplankton, which feed many larger ocean species, have declined by as much as 90 percent in four decades. This may portend population reductions for larger species from cod and haddock to whales and dolphins. "This is deeply worrying," said marine biologist Dr. Phil Williamson of East Anglia University. "We don't know why zooplankton numbers have plummeted, though global warming looks [like] the best candidate. What is certain is that removing the bottom link from the ocean food chain could have profound and unpleasant results" (McKie 2001; Knutti et al. 2004, 851–854).

Updating a major survey of zooplankton levels in the North Atlantic completed during 1963, a team of British scientists in November 2001 set out in the marine research vessel *Discovery* to measure changes in these levels. Using automated equipment, the scientists sampled concentrations of *Calanus finmarhicus*, the principal type of Atlantic zooplankton. Having carried out 800 samplings in an area 1,000 miles south of Iceland, the scientists found 5,000 to 10,000 zooplankton per square meter instead of the 50,000 average found in the earlier survey (McKie 2001). The scientists believe that gradually increasing sea temperatures are playing a major role in the zooplankton's decline.

Whether the decline of plankton is directly related to increased ocean temperatures is not yet certain, according to Watson W. Gregg, a NASA biologist at the Goddard Space Flight Center in Greenbelt, Maryland, because other factors also affect the productivity of phytoplankton such as availability of iron. According to Gregg, the greatest loss of phytoplankton has occurred where ocean temperatures have risen most significantly between the early 1980s and the late 1990s.

During summer in the North Atlantic, sea surface temperatures rose about 1.3°F during that period, Gregg said, while surface temperatures in the North Pacific rose some 0.7°F (Perlman 2003). "This research shows that ocean primary productivity is declining, and it may be the result of climate changes such as increased temperatures and decreased iron deposition into parts of the oceans," Gregg said. "This has major implications for the global carbon cycle" (Perlman 2003).

Is Thermohaline Circulation Related to U.S. East Coast Sea Level Rise?

AMOC slowing may be a factor not only in changing the path of the Gulf Stream but also in sea level rise along the U.S. mid-Atlantic Coast. According to one group of scientists in the *Journal of Geophysical Research* (Ezer et al. 2013), "Coastal sea level variations were found to be strongly influenced by variations in the [Gulf Stream] on time scales ranging from a few months to decades. It appears that the [Gulf Stream] has shifted from a six- to eight-year oscillation cycle to a continuous weakening trend since about 2004 and that this trend may be responsible for recent acceleration in local sea level rise."

"There have been several papers showing [sea level rise] acceleration," said lead author Tal Ezer, of Old Dominion University's Center for Coastal Physical Oceanography. "This . . . paper confirms the hypothesis for why it's happening" (Lemonick 2013). The U.S. East Coast is at high risk for sea level rise not only for this reason but also because the land is slowly sinking in most locations. Storm surges from hurricanes add another risk.

The Gulf Stream tends to run a meter or two higher (offshore) than surrounding water, because it is relatively warm and buoyant—or it did before weakening in recent decades. "It keeps coastal sea level a meter or a meter and a half lower than the rest of the ocean," Ezer said. In recent years, however, satellite imagery has shown that the midpoint of the Gulf Stream is not as high as it used to be and the edges are not quite as low—again, evidence that the stream itself is starting to slow down (Lemonick 2013). "Theory says this is just what should be happening. Ordinarily, the Gulf Stream brings warm surface water from the tropics up along the U.S. coast, and then across to the eastern North Atlantic, where it cools and sinks to the bottom of the sea," wrote Michael D. Lemonick for Weather.com (2013).

The relatively cold bottom water then flows south to the tropics, where it gradually warms, rises to the surface, and begins flowing north again. This constant flow, which meanders through all of the world's oceans, is sometimes called the *global ocean conveyor belt*; its branch in the North Atlantic is the Atlantic meridional overturning circulation.

Lemonick explained:

In a warming world, two things happen to throw a monkey wrench into the conveyor belt. First, melting ice, mostly from Greenland, dilutes the surface waters where the Gulf Stream reaches its northernmost extent. Since freshwater is less dense than salty water, the water has a more difficult time sinking to begin its journey southward. Second, the surface water is warmer than it used to be, and since warm water is less dense than cold water, this just adds to the problem. Put the two together and you start to jam up the works, with the result that the whole conveyor belt slows down. And the water along the Atlantic coast of the U.S. begins to rise at an accelerating rate. (Lemonick 2013)

Further Reading

"Abrupt Climate Change: Should We Be Worried?" Woods Hole Oceanographic Institution, February 10, 2003. https://www.whoi.edu/page.do?pid=83339&tid=3622&cid=9986.

"Big Freeze Plunged Europe into Ice Age in Months." NASA Earth Observatory, November 29, 2009. http://earthobservatory.nasa.gov/Newsroom/view.php?id=41518&src=eoa -manews (no longer available).

Broecker, Wallace S. "Are We Headed for a Thermohaline Catastrophe?" Pp. 83–95 in Lee C. Gerhard, William E. Harrison, and Bernold M. Hanson, eds., *Geological Perspectives of Global Climate Change*. Tulsa, OK: American Association of Petroleum Geologists, 2001.

Calamai, Peter. "Atlantic Water Changing: Scientists." *Toronto Star*, June 21, 2001, A18.

Clark, Peter U., et al. "Rapid Rise of Sea Level 19,000 Years Ago and Its Global Implications." *Science* 304 (May 21, 2004): 1141–1144.

Connor, Steve. "Britain Could Become as Cold as Moscow." *The Independent* (London), June 21, 2001, 14.

Cooke, Robert. "Waters Reflect Weather Trend; Study Finds Warming Effects." *Newsday*, December 18, 2003, A2.

Cowen, Robert C. "Into the Cold? Slowing Ocean Circulation Could Presage Dramatic— and Chilly—Climate Change." *Christian Science Monitor*, September 26, 2002, 14.

Cunningham, Stuart A., et al. "Temporal Variability of the Atlantic Meridional Overturning Circulation at 26.5°N." *Science* 317 (2007): 935–937. doi: 10.1126/science.1141304.

Curry, Ruth, Bob Dickson, and Igor Yashayaev. "A Change in the Freshwater Balance of the Atlantic Ocean over the Past Four Decades." *Nature* 426 (December 18, 2003): 826–829.

Davidson, Keay. "Going to Depths for Evidence of Global Warming; Heating Trend in North Pacific Baffles Researchers." *San Francisco Chronicle*, March 1, 2004, A4.

Dickson, B., et al. "Rapid Freshening of the Deep North Atlantic Ocean over the Past Four Decades." *Nature* 416 (April 25, 2002): 832–836.

Erbacher, Jochen, et al. "Increased Thermohaline Stratification as a Possible Cause for an Ocean Anoxic Event in the Cretaceous Period." *Nature* 409 (January 18, 2001): 325–327.

Ezer, Tal, et al. "Gulf Stream's Induced Sea Level Rise and Variability along the U.S. Mid-Atlantic Coast." *Journal of Geophysical Research* 118(2) (February 2013): 685–697. doi: 10.1002.

Fukasawa, Masao, et al. "Bottom Water Warming in the North Pacific Ocean." *Nature* 427 (February 26, 2004): 825–827.

Gibbs, Walter. "Scientists Back Off Theory of a Colder Europe in a Warming World." *The New York Times,* May 15, 2007. http://www.nytimes.com/2007/05/15/science/earth /15cold.html.

Häkkinen, Sirpa, and Peter B. Rhines. "Decline of Subpolar North Atlantic Circulation During the 1990s." *Science* 304 (April 23, 2004): 555–559.

Hansen, Bogi, William R. Turrell, and Svein Østerhus. "Decreasing Overflow from the Nordic Seas into the Atlantic Ocean through the Faroe Bank Channel since 1950." *Nature* 411 (June 21, 2001): 927–930.

Kerr, Richard A. "European Climate: Mild Winters Mostly Hot Air, Not Gulf Stream." *Science* 297 (September 27, 2002): 2202.

Kerr, Richard A. "Climate Change: Sea Change in the Atlantic." *Science* 303 (January 2, 2004): 35.

Kerr, Richard A. "Atlantic Mud Shows How Melting Ice Triggered an Ancient Chill." *Science* 312 (June 30, 2006): 1860.

Kerr, Richard A. "False Alarm: Atlantic Conveyor Belt Hasn't Slowed Down After All." *Science* 314 (November 17, 2006): 1064.

Kleiven, Helga (Kikki) Flesche, et al. "Reduced North Atlantic Deep Water Coeval with the Glacial Lake Agassiz Freshwater Outburst." *Science* 3129 (January 4, 2008): 60–64.

Knutti, R., et al. "Strong Hemispheric Coupling of Glacial Climate through Freshwater Discharge and Ocean Circulation." *Nature* 430 (August 19, 2004): 851–856.

Latif, M., et al. "Is the Thermohaline Circulation Changing?" *Journal of Climate* 19(18) (September 15, 2006): 4631–4637.

Lemonick, Michael D. "East Coast Faces Rising Seas from Slowing Gulf Stream." Weather .com. February 13, 2013. http://www.climatecentral.org/news/east-coast-faces-rising-seas -from-slowing-gulf-stream-15587.

Lynch-Stieglitz, Jean, et al. "Atlantic Meridional Overturning Circulation during the Last Glacial Maximum." *Science* 316 (April 6, 2007): 66–69.

Macdougall, Doug. *Frozen Earth: The Once and Future Story of Ice Ages*. Berkeley: University of California Press, 2004.

McKie, Robin. "Dying Seas Threaten Several Species; Global Warming Could Be Tearing Apart the Delicate Marine Food Chain, Spelling Doom for Everything from Zooplankton to Dolphins." *The Observer* (London), December 2, 2001, 14.

McPhaden, M. J., and D. Zhang. "Slowdown of the Meridional Overturning Circulation in the Upper Pacific Ocean." *Nature* 415 (February 6, 2002): 603–607.

Pearce, Fred. *With Speech and Violence: Why Scientists Fear Tipping Points in Climate Change*. Boston: Beacon Press, 2007.

Perlman, David. "Decline in Oceans' Phytoplankton Alarms Scientists; Experts Pondering Whether Reduction of Marine Plant Life Is Linked to Warming of the Seas." *San Francisco Chronicle*, October 6, 2003, A6.

Peterson, Bruce J., et al. "Increasing River Discharge to the Arctic Ocean." *Science* 298 (December 13, 2002): 2171–2173.

Quadfasel, Detlef. "Oceanography: The Atlantic Heat Conveyor Slows." *Nature* 438 (December 1, 2005): 565–566.

Radford, Tim. "As the World Gets Hotter, Will Britain Get Colder? Plunging Temperatures Feared after Scientists Find Gulf Stream Changes." *The Guardian* (U.K.), June 21, 2001, 3.

Schiermeier, Quirin. "Ocean Circulation Noisy, Not Stalling." *Nature* 448 (August 23, 2007): 844–845.

Schmittner, Andreas. "Decline of the Marine Ecosystem Caused by a Reduction in the Atlantic Overturning Circulation." *Nature* 434 (March 31, 2005): 628–633.

Seager, R., et al. "Is the Gulf Stream Responsible for Europe's Mild Winters?" *Quarterly Journal of the Royal Meteorological Society* 128 (2002): 2563–2586.

Speth, James Gustave. *Red Sky at Morning: America and the Crisis of the Global Environment*. New Haven, CT: Yale University Press, 2004.

Srokosz, M. A., and H. L. Bryden. "Observing the Atlantic Meridional Overturning Circulation Yields a Decade of Inevitable Surprises." *Science* 348 (June 19, 2015): 1330. http:// science.sciencemag.org/content/348/6241/1255575. doi: 10.1126/science.1255575.

Stocker, Thomas F., Reto Knutti, and Gian-Kasper Plattner. "The Future of the Thermohaline Circulation—A Perspective." Pp. 277–293 in Dan Seidov, Bernd J. Haupt, and Mark

Maslin, eds., *The Oceans and Rapid Climate Change: Past, Present, and Future.* Washington, DC: American Geophysical Union, 2001.

"Study Reports Large-Scale Salinity Changes in Oceans; Saltier Tropical Oceans, Fresher Ocean Waters Near Poles Are Further Signs of Global Warming's Impacts on Planet." *Ascribe Newsletter*, December 17, 2003 (LEXIS).

"Study Reveals Increased River Discharge to Arctic Ocean; Finding Could Mean Big Changes to Global Climate." Ascribe Newsletter, December 12, 2002 (LEXIS).

Thorpe, R. B., et al. "Mechanisms Determining the Atlantic Thermohaline Circulation Response to Greenhouse Gas Forcing in a Non-Flux-Adjusted Coupled Climate Model." *Journal of Climate* 14 (July 15, 2001): 3102–3116.

Våge, Kjetil, et al. "Surprising Return of Deep Convection to the Subpolar North Atlantic Ocean in Winter 2007–2008." *Nature Geoscience* 2(1) (January 2009). http://www.nature.com/ngeo/journal/v2/n1/abs/ngeo382.html.

See also: Migrations; Ocean Circulation; Temperatures, Global; Temperatures, Greenhouse Gas Levels and

Index

Bold indicates volume numbers.